NEXT GENERATION SEQUENCING AND MICROBIAL ECOLOGY

NEXT GENERATION SEQUENCING AND MICROBIAL ECOLOGY

Edited by:

Dr Prashant Pant, Dr Pratibha Pant and Dr Prerna Pandey

Delve
PUBLISHING

www.delvepublishing.com

Next Generation Sequencing and Microbial Ecology

Dr Prashant Pant, Dr Pratibha Pant and Dr Prerna Pandey

Delve Publishing

2010 Winston Park Drive,

2nd Floor

Oakville, ON L6H 5R7

Canada

www.delvepublishing.com

Tel: 001-289-291-7705
001-905-616-2116

Fax: 001-289-291-7601

Email: orders@arclereducation.com

ISBN: 978-1-77361-194-5 (Hardcover)

DECLARATION

Some content or chapters in this book are open access copyright free published research work, which is published under Creative Commons License and are indicated with the citation. We are thankful to the publishers and authors of the content and chapters as without them this book wouldn't have been possible.

ABOUT THE EDITORS

Dr. Prashant Pant, is an Assistant Professor of Botany in Dyal Singh College, University of Delhi, New Delhi, India. He has been teaching for more than 5 years now and his main areas have been Bioinformatics, Molecular Biology and Genomics at the college. His PhD was on understanding the Microbial Succession using a Metagenomic approach in a Barren mined-out site undergoing ecological restoration. For this, he deployed many molecular and computational Biology tool and techniques. He has several publications (articles and book chapters) in International Journals and International Publications. Currently, he is also Chief Editor of Bioinformatics Review and a co-founder/Executive Chief Editor of a Science Journal in Hindi Language "Swadeshi Vigyan" for the native language speaking communities. Both the Journals are a service to the scientific community as they are fully open access.

Dr. Pratibha Pant, is an Assistant Professor of Botany in Daulat Ram College, University of Delhi, New Delhi, India and has a teaching experience of more than 6 years. She is a plant virologist and biotechnologist by training and has done her PhD in Characterization of Begomoviral genomes from Papaya from Department of Botany, University of Delhi. During her PhD, she utilized a combination of molecular biology and Computational Biology tools to analyze begomoviral genomes and to her credit, a couple of novel begomoviruses have been isolated, sequenced and reported in public databases.

Dr. Prerna Pandey, a biotechnologist with several years of wet lab research experience, worked at International Center for Genetic Engineering and Biotechnology, New Delhi. Her field of research in PhD included isolation and molecular characterization of geminiviruses, genome sequencing, gene annotation, and gene silencing using the RNA interference technology. She has also worked at Transasia Biomedicals and Advance Enzyme Technologies as a scientist. Prerna has published papers in peer-reviewed journals, and has submitted a number of annotated Geminiviral genome sequences in the GenBank, including two novel ones. She has also completed her editing and proof-reading courses from Society for the promotion of Editors and Proof-readers, UK and now works as a freelance scientific editor and writer. When Prerna is not busy with her assignments, she loves to explore historical places.

TABLE OF CONTENTS

LIST OF CONTRIBUTORS

Prerna Pandey (Editor)

Dr. Prerna Pandey, a biotechnologist with several years of wet lab research experience, worked at International Center for Genetic Engineering and Biotechnology, New Delhi. Her field of research in PhD included isolation and molecular characterization of geminiviruses, genome sequencing, gene annotation, and gene silencing using the RNA interference technology. She has also worked at Transasia Biomedicals and Advance Enzyme Technologies as a scientist. Prerna has published papers in peer-reviewed journals, and has submitted a number of annotated Geminiviral genome sequences in the GenBank, including two novel ones. She has also completed her editing and proof-reading courses from Society for the promotion of Editors and Proof-readers, UK and now works as a freelance scientific editor and writer. When Prerna is not busy with her assignments, she loves to explore historical places.

Eleonora Egidi (Author)

Dr. Eleonora Egidi is a mycologist interested in different aspects of fungal research, including fungal ecology, biology and phylogeny. Eleonora has an international background and has been involved in collaborative projects for the identification of new fungal barcoding regions, study of fungal communities from extreme environments, and use of fungi in phytoremediation. Her current scientific interests focus on exploring the biodiversity of Australian fungal flora and their role as drives of plant diversity in natural ecosystems, as well as understanding how these systems respond to environmental changes.

Prashant Pant (Editor)

Dr. Prashant Pant, is an Assistant Professor of Botany in Dyal Singh College, University of Delhi, New Delhi, India. He has been teaching for more than 5 years now and his main areas have been Bioinformatics, Molecular Biology and Genomics at the college. His PhD was on understanding the Microbial Succession using a Metagenomic approach in a Barren mined-out site undergoing ecological restoration. For this, he deployed many molecular and computational Biology tool and techniques. He has several publications (articles and book chapters) in International Journals and International Publications. Currently, he is also Chief Editor of Bioinformatics Review and a co-founder/Executive Chief Editor of a Science Journal in Hindi Language "Swadeshi Vigyan" for the native language speaking communities. Both the Journals are a service to the scientific community as they are fully open access.

Pratibha Pant (Editor)

Dr. Pratibha Pant, is an Assistant Professor of Botany in Daulat Ram College, University of Delhi, New Delhi, India and has a teaching experience of more than 6 years. She is a plant virologist and biotechnologist by training and has done her PhD in Characterization of Begomoviral genomes from Papaya from Department of Botany, University of Delhi. During her PhD, she utilized a combination of molecular biology and Computational Biology tools to analyze begomoviral genomes and to her credit, a couple of novel begomoviruses have been isolated, sequenced and reported in public databases.

Jennifer Wood (Author)

Jennifer Wood is post-doctoral fellow in Dr. Ashley Franks' Applied and Environmental Microbiology Laboratory at La Trobe University, Australia. She has recently completed her Doctoral research in Microbial Ecology at the same university. Her broad research interest is in understanding the ecology of soil-microbial communities and their interactions with macro-communities. She is interested in developing trait-based approaches to interpret and predict how environmental and anthropogenic changes impact microbial community ecology and function.

Surajit De Mandal

Department of Biotechnology, Mizoram University, Aizawl-796004, Mizoram, India

Amrita Kumari Panda

Department of Zoology, Kumaun University, Nainital-263002, Uttarakhand, India

Satpal Singh Bisht

Department of Zoology, Kumaun University, Nainital-263002, Uttarakhand, India

Nachimuthu Senthil Kumar

Department of Biotechnology, Mizoram University, Aizawl-796004, Mizoram, India

Daniel C. Richter
ZBIT- Center for Bioinformatics Tu¨bingen, University of Tu¨bingen, Tu¨bingen, Germany

Felix Ott
ZBIT- Center for Bioinformatics Tu¨bingen, University of Tu¨bingen, Tu¨bingen, Germany

Alexander F. Auch
ZBIT- Center for Bioinformatics Tu¨bingen, University of Tu¨bingen, Tu¨bingen, Germany

Ramona Schmid
Boehringer Ingelheim Pharma GmbH & Co. KG, Biberach, Germany

Daniel H. Huson
ZBIT- Center for Bioinformatics Tu¨bingen, University of Tu¨bingen, Tu¨bingen, Germany

Stilianos Louca
Institute of Applied Mathematics, University of British Columbia, Vancouver, Canada

Michael Doebeli
Department of Zoology, University of British Columbia, Vancouver, Canada

Jorge Barriuso,
Centro Nacional de Biotecnología (CSIC), c/Darwin 3, 28049 Madrid, Spain

Jose R Valverde
Centro Nacional de Biotecnología (CSIC), c/Darwin 3, 28049 Madrid, Spain

Rafael P Mellado
Centro Nacional de Biotecnología (CSIC), c/Darwin 3, 28049 Madrid, Spain

Catherine H. Ravenscroft
Department of Biology, Syracuse University, 107 College Place, Syracuse, NY 13244, USA

Raj Whitlock
Institute of Integrative Biology, University of Liverpool, The Biosciences Building, Crown Street, Liverpool L69 7ZB, UK
Department of Animal and Plant Sciences, University of Sheffield, Western Bank, Sheffield S10 2TN, UK

Jason D. Fridley
Department of Biology, Syracuse University, 107 College Place, Syracuse, NY 13244, USA

Luis E. Castañeda
Instituto de Ciencias Ambientales y Evolutivas, Facultad de Ciencias, Universidad Austral de Chile, Valdivia, Chile

Olga Barbosa
Instituto de Ciencias Ambientales y Evolutivas, Facultad de Ciencias, Universidad Austral de Chile, Valdivia, Chile
Instituto de Ecologı´a y Biodiversidad (IEB-Chile), Santiago, Chile

Marcin Gołębiewski
Chair of Plant Physiology and Biotechnology, Nicolaus Copernicus University, Lwowska 1, 87-100 Toruń, Poland
Centre of Modern Interdisciplinary Technologies, Nicolaus Copernicus University, Wilenska 4, 87-100 Toruń, Poland
Institute of Oceanology PAS, Powstańców Warszawy 55, 81-712 Sopot, Poland

Edyta Deja-Sikora
Chair of Plant Physiology and Biotechnology, Nicolaus Copernicus University, Lwowska 1, 87-100 Toruń, Poland

Marcin Cichosz
Department of Chemical Pro-ecological Processes, Nicolaus Copernicus University, Gagarina 7, 87-100 Toruń, Poland

Andrzej Tretyn
Chair of Plant Physiology and Biotechnology, Nicolaus Copernicus University, Lwowska 1, 87-100 Toruń, Poland
Centre of Modern Interdisciplinary Technologies, Nicolaus Copernicus University, Wilenska 4, 87-100 Toruń, Poland

Borys Wróbel
Institute of Oceanology PAS, Powstańców Warszawy 55, 81-712 Sopot, Poland
Evolutionary Systems Laboratory, Adam Mickiewicz University, Umultowska 89, Room S1, 61-614 Poznań, Poland

Agnieszka Kalwasińska
Department of Environmental Microbiology and Biotechnology, Faculty of Biology and Environmental Protection, Nicolaus Copernicus University, Torun´, Poland

Tamás Felföldi
Department of Microbiology, Eo¨tvo¨s Lora´nd University, Budapest, Hungary

Attila Szabó
Department of Microbiology, Eo¨tvo¨s Lora´nd University, Budapest, Hungary

Edyta Deja-Sikora
Department of Environmental Microbiology and Biotechnology, Faculty of Biology and Environmental Protection, Nicolaus Copernicus University, Torun´, Poland

Przemysław Kosobucki
Chair of Environmental Chemistry and Bioanalytics, Faculty of Chemistry, Nicolaus Copernicus University, Torun´, Poland

Maciej Walczak
Department of Environmental Microbiology and Biotechnology, Faculty of Biology and Environmental Protection, Nicolaus Copernicus University, Torun´, Poland

Hon Lun Wong
School of Biotechnology and Biomolecular Sciences, The University of New South Wales, Sydney 2052, Australia
Australian Centre for Astrobiology, University of New South Wales, Sydney 2052, Australia

Aria Ahmed-Cox
School of Biotechnology and Biomolecular Sciences, The University of New South Wales, Sydney 2052, Australia

Brendan Paul Burns
School of Biotechnology and Biomolecular Sciences, The University of New South Wales, Sydney 2052, Australia
Australian Centre for Astrobiology, University of New South Wales, Sydney 2052, Australia

PREFACE

The book titled "**Next Generation Sequencing and Microbial Ecology**" is a very timely and succinct compilation of the current developments in nucleic acid sequencing methods and their applications especially in helping understand microbial interaction and their form, community structure and function in various biotic communities. The book introduces the readers to different methods of sequencing DNA - from conventional to next generation sequencing, and discussing invariably the limitation and advantages of each of these methods, with respect to microbial ecology studies. The book highlights several gene families which have been favourably taken up in the recent past for the purpose of dissecting the communities and their possible roles. In this book, we also discuss some very novel aspects of microbial ecology, like "the metagenomics approach", which scientist have been able to work out mainly by culture independent approaches. We will highlight how this is helping us understand the structural and functional aspect i.e., "who is out there" and "what they are doing" in a number of different environments and the ecological significance of these aspects.

We hope that the book will find use among students, researchers and teachers as a reference for choosing an appropriate sequencing approach, with respect to the gene family and analytical methods for studying microbes and their role in a community/ecosystem. Suggestions for improving the book are cordially invited.

Looking forward,

Editors and Authors

CHAPTER 1

CURRENT STATE OF NUCLEIC ACID SEQUENCING: LIMITATIONS AND ADVANTAGES

Prerna Pandey[1], Eleonora Egidi[1], Prashant Pant[2], Pratibha Pant[3]

[1]Dr Prerna Pandey, a biotechnologist with several years of wet lab research experience, worked at International Center for Genetic Engineering and Biotechnology, New Delhi. Her field of research in PhD included isolation and molecular characterization of geminiviruses, genome sequencing, gene annotation, and gene silencing using the RNA interference technology. She has also worked at Transasia Biomedicals and Advance Enzyme Technologies as a scientist. Prerna has published papers in peer-reviewed journals, and has submitted a number of annotated Geminiviral genome sequences in the GenBank, including two novel ones. She has also completed her editing and proof-reading courses from Society for the promotion of Editors and Proof-readers, UK and now works as a freelance scientific editor and writer. When Prerna is not busy with her assignments, she loves to explore historical places.

[1]Dr. Eleonora Egidi is a mycologist interested in different aspects of fungal research, including fungal ecology, biology and phylogeny. Eleonora has an international background and has been involved in collaborative projects for the identification of new fungal barcoding regions, study of fungal communities from extreme environments, and use of fungi in phytoremediation. Her current scientific interests focus on exploring the biodiversity of Australian fungal flora and their role as drives of plant diversity in natural ecosystems, as well as understanding how these systems respond to environmental changes.

[2]Dr. Prashant Pant, is an Assistant Professor of Botany in Dyal Singh College, University of Delhi, New Delhi, India. He has been teaching for more than 5 years now and his main areas have been Bioinformatics, Molecular Biology and Genomics at the college. His PhD was on understanding the Microbial Succession using a Metagenomic approach in a Barren mined-out site undergoing ecological restoration. For this, he deployed many molecular and computational Biology tool and techniques. He has several publications (articles and book chapters) in International Journals and International Publications. Currently, he is also Chief Editor of Bioinformatics Review and a co-founder/Executive Chief Editor of a Science Journal in Hindi Language "Swadeshi Vigyan" for the native language speaking communities. Both the Journals are a service to the scientific community as they are fully open access.

[3]Dr. Pratibha Pant, is an Assistant Professor of Botany in Daulat Ram College, University of Delhi, New Delhi, India and has a teaching experience of more than 6 years. She is a plant virologist and biotechnologist by training and has done her PhD in Characterization of Begomoviral genomes from Papaya from Department of Botany, University of Delhi. During her PhD, she utilized a combination of molecular biology and Computational Biology tools to analyze begomoviral genomes and to her credit, a couple of novel begomoviruses have been isolated, sequenced and reported in public databases

INTRODUCTION

Microorganisms are very diverse and include all bacteria, archaea, fungi and most protozoa. These organisms play several crucial functions in many environments, ranging from their contribution to photosynthesis to aiding in the digestion of complex substances, as well as being responsible for the cycling of many fundamental elements for plants, such as carbon and nitrogen. In addition, the role of groups of microbes as a community in the cleaning up or digestion of oils and chemicals has been reported. Microbial communities are also known to possess endurance to drastic conditions such as extreme temperatures and pH, making them able to withstand and survive to the most extreme environments on Earth (Lasken and McLean, 2014).

SEQUENCING: A GLIMPSE

Prior to the advent of sequencing technologies, isolation and identification of microbes based on morphological and physiological differences, were paramount to investigate microbial diversity. The development of DNA-based molecular approaches targeting the ribosomal RNA gene as universal marker for microbes (16S region for bacteria, the Internal Transcribed Spacer -ITS- region for fungi) has revolutionized the field of microbial ecology. Indeed, the study of microbial diversity based on genetic rather than phenotypic differences, has greatly improved our understanding of the true extent of the microbial world (Walters and Knight, 2014) It is only thanks to this molecular approach that it was, for example, possible to discriminate Archaea as a new and separate group among the microbes. Novel applications of the traditional sequencing technologies, such as the Sanger dideoxynucleotide termination method, greatly improved our ability to identify several microbial groups from different environments using these biomarkers, which are now considered the "gold standard' for microbial molecular identification. The Sanger method of sequencing and its following implementations, such as the use of fluorescence *in situ* detection and several fingerprinting techniques, occupied a major chunk of experiments in microbial ecology for many decades. However, the efficiency of this technique was mainly limited to profiling microbial communities or investigating single individuals within the community, due to several technical hurdles, including the requirement of cloning each of the sequences and its associated cost, which hampered the possibility to explore complex microbial assemblages. Another limitation was a grouping of sequences using bioinformatics tools like BLAST to yield phylogenetic structures, that offer challenges in deciphering some underlying relationships. However, the last decade saw the development of high-throughput sequencing technologies (next generation sequencing, NGS), which opened the possibility to gain a deeper understanding of microbial community instead of single members, as well as to extend the investigation of microbes also to the not-cultivable fraction of the community.

NEXT GENERATION SEQUENCING TECHNOLOGIES (NGS)

As previously mentioned, Sanger sequencing technology had a great impact on the early stage of microbial community studies. Since then, the sequencing yield and sequence length have changed a lot (Table 1). Currently, Sanger sequencing can retrieve up to 96 sequences per run with an average length of 650 bp. In contrast, low-cost platforms known as Next Generation Sequencing technologies (NGS) are capable of parallel sequencing millions of DNA molecules with different yields and sequence lengths (Niedringhaus et al., 2011; Logares et al., 2012; Sanchez-Flores and Abreu-Goodger, 2014). NGS allows for the sequencing of DNA in a parallel manner using a flow cell in a large scale. Every single molecule is sequenced in either a stepwise or in real time manner (Pareek et al., 2011).

There are several platforms available:

Pyrosequencing

The technique was first developed by Pal Nyren and his student Mostafa Ronhaghi at the Royal Institute of Technology in 1996, available through Roche 454 Life Technologies (Nikolaki and Tsiamis, 2013). This method represented the first alternative to the conventional Sanger method for *de novo* DNA sequencing. This method employs the detection of pyrophosphate that is released during primer-directed DNA polymerase catalyzed nucleotide incorporation using luminometry (Fakruddin et al., 2012).

This technology uses a one-by-one nucleotide addition cycle, where the pyrophosphate (PPi) released from the DNA polymerization reaction is translated as a luminous signal. The light emission from a plate with millions of microwells containing a given DNA fragment is then detected and translated to nucleotides (Margulies et al., 2005). The pyrosequencing technique employs beads starts off with a single template molecule, which is subject to amplification with emulsion PCR (emPCR). A multitude of beads are loaded on picolitre plates with a bead in a well. The reagents include ATP sulfurylase, luciferase enzymes, adenosine 5′ phosphosulfate and luciferin substrates are then flowed on the surface of the plate.

FLX pyrosequencer has a reported 100 flows for a nucleotide in a span of 8 hours, with 250 nucleotides as the average read length. The use of filters removes mixed and low-quality sequences with additional quality controls that discard sequences lacking initiating TCGA sequence (Reviewed by Nikolaki and Tsiamis, 2013).

There are several advantages reported with the technique that include:

- The method facilitates the programming of the dispensing of nucleotides that permit the study of aid in the study of alterations such as deletions or insertions.
- The method can be automated with ease for the study of a large number of samples apart from its study in real time (Gharizadeh et al., 2003).
- Pyrosequencing can generate signals immediately downstream of the primer when compared to Sanger that reads several bases away from the primer.
- The length of the sequences can be long > 200bp.
- Apart from easy incorporation of PCR protocols in the system, the technique can also offer easy sample preparation and simple frequency data.
- The cost incurred in the reagents used is lesser for small segments of DNA.

However, certain challenges that are reported include:

- The high costs, a reported high error rate of 0.0098 and a bias induced with the use of fusion primers (Fakruddin et al., 2012)
- The current proprietor of the technology: Roche reported the closing down of the technology to end the support to the platform by 2016 (Karow, 2013).

Illumina

It is the second commercial NGS platform that was developed by Solexa and later transferred under the aegis of Illumina. It employs a principle based on synthesis: the reversible-termination sequencing by synthesis with fluorescently labeled nucleotides (Escobar-Zepeda et al., 2015).

A solid glass surface is used to adhere molecules, followed by bridge PCR to amplify DNA into small clusters of identical molecules. Using 3′-O-Fluorophore-labeled nucleotides in a method similar to Sanger, reversible terminators are applied (Nikolaki and Tsiamis, 2013). Following the incorporation of the fluorescently labeled nucleotides, a laser excites the chemical and the resulting fluorescent signal is recorded. The nucleotide is then washed off to permit the incorporation of the next nucleotide to facilitate sequencing from one or both sides.

To date, this method allows to obtain up to 600 bp per run (300 base pairs per paired-end read) (Bennett, 2004), ranking it first in the the second-generation sequencing technologies (Escobar-Zepeda et al., 2015).

The method has several advantages that include:

- A low error rate of 1% and hence improved accuracy.
- A higher output is reported
- Like the 454 technology, it is impervious to homopolymer reads (Tang et al., 2009)

SOLID

SOLiD is another technology that uses ligation to sequence the genetic information. The reported read lengths for the technology are 25 and 35 bp to yield 2 to 4 Gb of data (Cloonan et al., 2008 and Tang et al., 2009).

Ion Torrent

This technology uses a similar approach as 454 but circumvents the use of cameras, lasers or fluorescent dyes. It uses the detection of hydrogen ions in place of pyrophosphatase and uses microchip technology (Nikolaki and Tsiamis, 2013).

The Ion Torrent PGM is considered as the smallest potentiometer, permitting detection in the change in hydrogen potential generated each time a proton is released following addition of a nucleotide during the sequencing reaction occurring in millions of microwells (Rothberg et al., 2011). This technology uses highly dense microwell arrays, with each well serving as a site of DNA polymerization. The wells contain the fragment to be sequenced and a polymerase, beneath which is an ion-sensitive layer with highly dense array of field effect transistors. Following the cycling of the four nucleotides on the wells, the incorporation of a nucleotide leads to pH shift-read and is converted to voltage by the field effect transistors (Nikolaki and Tsiamis, 2013).

Advantages of Ion Torrent:

- The method has a reported yield of about 500 million reads with a mode length of 400 bp (Escobar-Zepeda et al., 2015).
- An additional advantage is a decrease in the cost: a reported 10% dip in the cost as compared to pyrosequencing (Whiteley et al., 2012).
- Ion proton, a newly designed model, employs bigger chips to report a value of 10 Gb per run, suggesting its use in whole genomes and exomes sequencing.
- Ion Torrent delivers 80–100 Mb/hour: the fastest throughput coupled with the smallest time run of 3 hours (Nikolaki and George Tsiamis, 2013).

The challenges in Ion Torrent include:

- The improper washing of wells between the cycles can lead to error accumulation at a rate of 1.78%.
- The method is limited to short reads and not long repeated regions (Nikolaki and Tsiamis, 2013).

SMRT

The single molecule real time (SMRT) platform has been developed by Pacific Biosciences, and has been reported as the most popular third generation sequencing method. It uses microfabricated nanostructures: tiny holes in a metallic film covering a chip called zero-mode waveguides. The use of apertures that have diameters less than wavelength of incident light causes the light to decay allowing the it to light up the bottom of the wells (Heather and Chain, 2016)

The system uses γ-phosphate fluorescently-labeled nucleotides with the zero-mode waveguides that does not allow penetration of light through the waveguide, but permit entry of energy to excite the fluorophores attached to the nucleotides. Following incorporation of each nucleotide, fluorescence detection is performed in real time (Nikolaki and Tsiamis, 2013).

The system offers advantages such as:

- Sequencing of single molecules in short time.
- There are reports to long reads of up to 10 kb for *de novo* genome assemblies.
- The detection of modified bases is permitted as kinetic data is also generated (Heather and Chain, 2016)

A 2013 study attempted to examine the error rate associated with the method as compared to other systems that reported that the system had high accuracy. It was additionally observed that the system offered advantages such as such as improvements of genomes, data of methylation patterns obtained, and deduction of functional activity of their methyltransferase genes (Roberts et al., 2013).

A TOOLKIT FOR SEQUENCING ANALYSIS

The throughput of current sequencing techniques is being pitched as several folds more than the traditional sequencing (Metzker, 2010). In microbial ecology, the use of sequencing is extended to amplicon studies of 16S rRNA, metatranscriptomics and metagenomics, aiding in the understanding of microbial communities that cannot be cultured. To facilitate these studies, several bioinformatics tools have been reported such as libraries that use Python or R as building blocks, and automated pipelines (e.g. such as QIIME, or Mothur, or Megan and MG-RAST). Such pipelines allow to perform the following steps: assignment of raw sequences to samples in accordance to bar codes followed by filtering for quality. Operational taxonomic units (OTUs) are constructed from clustering of sequences that are then compared against a database to study the identity of such sequences (Walters and Knight R, 2014).

A tool called Bio-Community was reported in 2014. This tool is reported as a Perl object-oriented toolkit that is comprised of a group of entities from the same geographic area. These entities correspond to individual genomes, genes, taxa or operational taxonomic units from amplicon and meta-omic surveys.

The authors reported that the tool can aid in the analysis of sequences rapidly with the use of many file formats apart from holistic microbial studies (Angly et al., 2014).

APPLYING NGS IN MICROBIAL ECOLOGY: CASE STUDIES

Microbial ecology studies, especially sequencing of environmental samples, have used mass sequencing to reveal communitys structures and drivers of diversity in many environments. The results of high-volume sequencing of environmental microorganisms often allow to capture sequences that occur only rarely and are often overlooked by traditional cultivation methods, allowing for the revealing of the least abundant members of the community. This "rare biosphere" sometimes are phylogenetically more diverse than the more abundant organisms. We report here some case studies that are exemplificative examples on how NGS technologies can aid to reveal some of these patterns.

Freshwater Ecology

A 2013 study implemented a 454-pyrosequencing approach targeting the 16S rDNA gene (V3-V4 regions) to investigate bacterioplankton communities. The study aimed at decoding the changes in the plankton in Antarctic lakes near the coast in response to environmental changes. It was revealed that new bacterial classes were established with shifts in the community patterns of the predecessors of the organisms in response to variations in the salt levels (Logares et al., 2013).

Similarly, another 2011 study explored the use of NGS in the form of pyrosequencing to quantify populations of fungi in freshwater lakes, revealing that the organization of the community mirrored the physical and chemical properties of the environment, suggesting the occurrence of a strong microbe-environment relationship (Monchy et al., 2011)

Soil Ecology

There are more microbes in a teaspoon of soil than there are people on the earth. Soils contain about billions of bacteria, fungi, protozoa, nematodes, earthworms, and arthropods, which play crucial roles in many ecosystem functions, such nutrient cycling and pathogens regulation. Deep sequencing demonstrated to be pivotal to reveal the true extent of microbial diversity in soil. For example, a pioneering 2009 study on a forest soil revealed an unprecedented abundance of fungi, with Agaricomycetes being the most copious class in this system (Buée et al., 2009). A more recent work focusing on investigating the fungal biogeographical patterns of diversity at a global level, revealed that species' latitudinal ranges increase toward the poles (Rapoport's rule) and diversity increases toward the equator. An increasing body of research is now focusing on disentangling the relationship of microbes, soil and plants, opening for potential application of microbial community's vegetation monitoring, conservation and restoration, as well as improving agricultural outcomes.

Marine Ecology

The study of microbial communities is essential to reveal distribution and occurrence of new microbes especially in those environments that are more difficult to reach and explore, such as marine systems. For example, a 2012 study implemented the 18S rDNA pyrosequencing to investigate diversity patterns and activity of unculturable marine heterotrophic flagellates. These newly discovered and characterized organisms were then named MArine STramenopiles (Logares et al., 2012). Deep sequencing was also used to reveal that, following the Deepwater Horizon oil spill in the Gulf of Mexico in 2010, a bloom of uncultivable microbes within the *Oceanospirillales* occurred (Mason et al., 2012). Subsequent sequencing studies showed the presence of enzymes that could degrade n-alkane and cycloalkane (Lasken and McLean, 2014).

Among the sponge communities, a strong association with microbes belonging to the phylum Candidatus Poribacteria has been observed. Sequencing yielded genomes of Poribacteria with special metabolic potential to serve as scavengers and recyclers of a particular carbon compounds that are unique to sponges (Kamke et al., 2013).

Aerobiology

An instance of the use of next generation sequencing is the study of the diversity of microbes in the air in New York City subway platforms. Although the study did not report microbes that were threatening to public health, the majority of the identified fungi were organisms that are associated with wood rot. In contrast, the bacteria reported in the study were mainly *Staphylococcus epidermidis*: the most copious organism present on the human skin, as well as *S. hominis, S. cohnii, S. caprae,* and *S. haemolyticus* (Robertson et al., 2013). This study revealed an unprecedented and unexpected diversity of microbes contributing to the subway aerosol microbiota, with consequences for the air monitoring and surveillance strategies.

Human Ecology

The possibility to have access to information on complex microbial community structure and composition in a relatively cheap and rapid way, has led to several insights in understanding the relationship between microbial communities and immunity of humans. There are over 10 trillion microbes that inhabit the human body, collectively called the microbiome. Most of them live in the gut and intestines, where they help us to digest food, synthesize vitamins, and fight off infection. A 2010 study demonstrated that the method of child-birth (normal or Caesarian) largely influenced the microbial community in infants and, consequently the immune system. Indeed, the method of delivery was found to affect the microbial colonization patterns, with repercussions for the immune health (Dominguez-Bello MG et al., 2010). Another study investigating the variability of the microbes associated with humans found that the class of microbes that occupy gut- or skin of humans were substantially different across individuals with reference to the location on the body or with time (Caporaso et al., 2011).

Other projects, including American Gut Project and Earth Microbiome Project, have looked at various factors that associate the microbial communities to their roles in the human metabolism and health, allowing us to gain new insight on the paramount role of microbes in determining human fitness (Walters and Knight, 2014).

METAGENOMICS: GENOMES SEQUENCING FOR ECOLOGY

The studies and approaches reported so far mainly targeted the microbial ribosomal RNA gene, which contains hypervariable (V1 to V8) regions that allow the identification of bacteria, fungi and archaea at the species level with a relatively good level resolution and certainty. Despite being crucial to reveal the diversity and abundance of microbial communities in many ecosystems and environments, this approach only allows to retrieve a very small portion of the genome (about 600-700 bp), and heavily relies on the presence of described taxa in public databases for the identification of the obtained sequences, limiting the possibility to explore and characterize the metabolic potential of novel organisms. However, recent advances in the NGS applications, such as metagenomics, are now allowing us to investigate the genetic abilities of complex microbial assemblages. Indeed, the metagenomic method permits the retrieval of microbial genomes in a single environmental sample, thus facilitating the study of the entire microbial gene-pool associated with a particular ecosystem.

Thanks to the possibility to access information on entire genome sequences (millions of bp) simultaneously, this method hold promises to revolutionize our approach to microbial ecology. A giant study in 2013 looked at the sequencing of uncultivated bacteria and archaea that were referred to as microbial dark matter, at the US Department of Energy Joint Genome Institute. The use of metagenomics was applied on several environmental samples such as freshwater, brackish, marine etc. Following MDA using 16S rRNA gene, genomes were partially assembled. The very first genomic dataset was assembled for several Archaea, such as *Nanoarchaeota*, as well as many bacterial phyla such as WS1, WS3, SAR406, Oct-Spa1-106 etc, giving new insights into the ecological and functional role of these organisms in the ecosystem (Rinke et al., 2013).

Metagenomics has also been widely applied to investigate the role of microbes in humans (*The Human Microbiome Project, HMP).* The sequencing of uncultivated organisms from human stool and oral swabs was performed and funded by the US National Institutes of Health. The samples were subjected to MDA and sequencing, allowing to retrieve around 250 genome drafts deposited in HMP Reference Genomes. Such studies aid in the understanding of the ecology of microbes present in a human body that have possible contributions to health or disease (Lasken and McLean, 2014).

Nevertheless, given the richness in diversity across microbial ecologies, the metagenomic approach requires a technological suite for its maximum use. In this case, the combination of NGS along with multiple displacement amplification (MDA) and bioinformatics serves as a tool in the analysis of microbial communities (Lasken and McLean, 2014)

LATEST TECHNOLOGIES

In order to facilitate the application of this field in understanding microbial ecology, several advances in technology have been necessary. We report some here:

- ***Mini Metagenomics***: It uses an approach common to both single cells and large populations of microbes. Cells from environments are sorted by FACS to limit them and then subjected to MDA and sequencing.
- ***Immunomagnetic Separation***: The technique avoids the culture of organisms to be used along with MDA. Magnetic beads bound to antibody bind the pathogen or cell of choice that can be sequenced to characterize new species.
- ***Shotgun Metagenomics***: In order to facilitate the understanding of the metabolism of a metagenomic study, shotgun libraries can be prepared from the total DNA, allowing for the discovery of new genes and their annotation. Parallel-meta is a multithreading software that extracts ribosomal marker genes from such data from metagenomics. Other programs include MOCAT and AMPHORA that use multiple phylogenetic markers to find use in taxonomic annotation.

Challenge and solution in the shotgun metagenomics:

- Pseudogenes: are reported to account for 35% in prokaryotic genomes that interfere with annotation of genes. This can be lessened by the use of resources such as BactPepDB and Pseudogene.org.
- ORFans: refer to protein sequences that do not have ascribed annotations or functions. Metagenomics may yield data that has these ORFans as the incidence is more as the diversity increases. The use of of possible non-synonymous and synonymous substitutions (ka/ks) apart from prediction tools

such as QUARK or I-TASSER are warranted.

- The bias arising out of amplification can be reduced by the use of laboratory methods such as growth of the cells in agarose or the culture of cells in microfluidic chambers or even the blocking of cell division. Bioinformatics can also find application in this reduction of bias: a software called SPAdes has been reported that eliminates low coverage regions. Another tool, IDBA-UD aids in rectifying uneven reads from sequencing data (Reviewed by Escobar-Zepeda et al., 2015)
- The technique of shotgun metagenomics does enable the observation of all the genes of the total DNA extracted. However, samples that are contaminated with host tissues could lead to complications. For instance, lung biopsies contain a major of human DNA and lower microbial content. Such challenges can be overcome by the use of studying RNA transcripts called as Meta-transcriptomics. (Walters and Knight, 2014)

Single Cell Genomics (SCG)

The technique enables the isolation of cells from several sources to be amplified and then sequenced. Following the preservation of the samples, the cells are sorted using micromanipulation or fluorescence-activated cell sorting (FACS) though the latter has an advantage of lesser contamination. The sorting is followed by lysis of cells and amplification of the genome using Multiple Displacement Amplification (MDA).

The application of such a system has led to studies in phyla such as Poribacteria, Picobiliphytes and divergent groups of aquatic Proteobacteria, Flavobacteria, and Archaea. (Nikolaki and Tsiamis, 2013)

A 2013 study examined the "microbial dark matter" that includes several uncultivated bacteria and archaea. The sequencing and analysis of 201 genomes enabled the development of relationships. For instance, 2 new super phyla were reported: Terrabacteria, composed of atypical single membranes, and the Patescribacteria, characterized by a decrease in metabolic activity (Rinke et al., 2013). Similarly, the genomes of uncultured organisms in complicated ecosystems such as microbes (mainly bacterioplankton and Flavobacteria) involved in cleaning up of

Deepwater Horizon oil spill was studied using SCG (Raghunathan et al., 2005; Woyke et al., 2010).

Metagenomics and SCG: A combination of these techniques can validate the results of each other. The reads of a single cell output could generate 20% more metagenomic data from uncultured organisms in the giant study of 2013 (Rinke et al., 2013). (Reviewed by Lasken and McLean, 2014)

SUMMARY

With reports of microbes occupying a plethora of environments, the field of microbial ecology that deals with the various aspects of members of a population and their inter-relationships carries an increasing and exciting new body of research. Several of the latest sequencing platforms were discussed in this chapter, such as Pyrosequencing, Illumina, SoLID, Ion torrent and SMRT, which have already largely contributed to our understanding of microbial ecology. Additional approaches, such as metagenomics, with its various applications including shotgun metagenomics, hold the potential to further increase our knowledge on the functions associated with the microbial fraction, opening new perspectives on the importance of microbes in every ecosystem, including humans.

REFERENCES

1. Angly FE, Fields CJ, Tyson GW. (2014). The Bio-Community Perl toolkit for microbial ecology. *BioABACUS*. 30(13): 1926–1927. doi:10.1093/bioinformatics/btu130.
2. Bennett S. (2004). Solexa Ltd. *Pharmacogenomics*. 5: 433–438. doi: 10.1517/14622416.5.4.433.
3. Buée M, Reich M, Murat C, et al. (2009). 454 Pyrosequencing analyses of forest soils reveal an unexpectedly high fungal diversity. *New Phytol*. 184(2): 449–456.
4. Caporaso JG, Lauber CL, Costello EK, Berg-Lyons D, Gonzalez A, Stombaugh J, et al. (2011). Moving pictures of the human microbiome. *Genome Biol*. 12:R50.
5. Cloonan N, Forrest ARR, Kolle G, et al. (2008). Stem cell

transcriptome profiling via massive-scale mRNA sequencing. *Nat Methods*. 5(7): 613–619.

6. Dominguez-Bello MG, Costella EK, Contreras M, Magris M, Hidalgo G, Fierer N, et al. (2010). Delivery mode shapes the acquisition and structure of the initial microbiota across multiple body habitats in newborns. *Proc Natl Acad Sci USA*. 107:11971–11975.
7. Escobar-Zepeda A, de León AV-P, Sanchez-Flores A. (17 December 2015). The Road to Metagenomics: From Microbiology to DNA Sequencing Technologies and Bioinformatics. *Front Genet*. https://doi.org/10.3389/fgene.2015.00348.
8. Fakruddin M, Chowdhury A, Hossain MN, Shahnewaj bin Mannan K, Mazumdar RM. (apr-jun 2012). Pyrosequencing- principles and applications. *Int J Life Sci Pharm Res*. 2(2): 65-76.
9. Gharizadeh B, Ohlin A, Molling P, Backman A, Amini B, Olcen P, Nyren P. (2003). Multiple group-specific sequencing primers for reliable and rapid DNA sequencing. *Mol Cell Probes*. 17: 203-210.
10. Heather JM, Chain B. (2016). The sequence of sequencers: The history of sequencing DNA. *Genomics*. 107(1):1-8. doi:10.1016/j.ygeno.2015.11.003.
11. Kamke J, et al. (2013). Single-cell genomics reveals complex carbohydrate degradation patterns in poribacterial symbionts of marine sponges. *ISME J*. 7:2287–2300.
12. Karow J. (2013). *Following Roche's Decision to Shut Down 454, Customers Make Plans to Move to Other Platforms*. GenomeWeb. (https://www.genomeweb.com/sequencing/following-roches-decision-shut-down-454-customers-make-plans-move-other-platform)
13. Lasken RS, McLean JS. (2014). *Nature Rev Genetics*. 15(9):577-584. doi:10.1038/nrg3785.
14. Logares R, Audic S, Santini S, Pernice MC, de Vargas C, R. Massana. (2012). Diversity patterns and activity of uncultured marine heterotrophic flagellates unveiled with pyrosequencing. *ISME J*. 6(10): 1823–1833.
15. Logares R, Lindström ES, Langenheder S, et al. (2013). Biogeography of bacterial communities exposed to progressive

long-term environmental change. *The ISME J.* 7(5): 937–948.

16. Margulies M, Egholm M, Altman WE, Attiya S, Bader JS, Bemben LA, et al. (2005). Genome sequencing in open microfabricated high density picoliter reactors. *Nature.* 437: 376–380. doi: 10.1038/nature03959.
17. Mason OU, et al. (2012). Metagenome, metatranscriptome and single-cell sequencing reveal microbial response to Deepwater Horizon oil spill. *ISME J.* 6:1715–1727.
18. Metzker ML. (2010). Sequencing technologies—the next generation. *Nat Rev Genet.* 11, 31–46.
19. Monchy S, Sanciu G, Jobard M, et al. (2011). Exploring and quantifying fungal diversity in freshwater lake ecosystems using rDNA cloning/sequencing and SSU tag pyrosequencing. *Environ Microbiol.* 13(6): 1433–1453.
20. Niedringhaus TP, Milanova D, Kerby MB, Snyder MP, Barron AE. (2011). Landscape of next-generation sequencing technologies. *Anal Chem.* 83(12): 4327–4341.
21. Nikolaki S, Tsiamis G. (2013). Microbial Diversity in the Era of Omic Technologies. *BioMed Res Int.* 2013, Article ID 958719. doi:10.1155/2013/958719.
22. Pareek CS, Smoczynski R, Tretyn A. (2011). Sequencing technologies and genome sequencing. *J Appl Genet.* 52(4): 413–435.
23. Raghunathan A, Ferguson HR Jr, Bornarth CJ, Song W, Driscoll M, Lasken RS. (2005). Genomic DNA amplification from a single bacterium. *Appl Environ Microbiol.* 71(6):3342-3347.
24. Rinke C, Schwientek P, Sczyrba A, et al. (2013). Insights into the phylogeny and coding potential of microbial dark matter. *Nature.* 499(7459): 431–437.
25. Rinke C, Schwientek P, Sczyrba A, Ivanova NN, Anderson IJ, Cheng JF, et al. (2013). Insights into the phylogeny and coding potential of microbial dark matter. *Nature.* 499(7459):431-437.
26. Roberts RJ, Carneiro MO, Schatz MC. (2013). The advantages of SMRT sequencing. *Genome Biol.* 14(7):405. doi:10.1186/gb-2013-14-7-405.

27. Robertson CE, Baumgartner LK, Harris JK, et al. (2013). Culture-independent analysis of aerosol microbiology in a metropolitan subway system. *Appl Environ Microbiol.* 79(11): 3485–3493.

28. Rothberg JM, Hinz W, Rearick TM, Schultz J, Mileski W, Davey M, et al. (2011). An integrated semiconductor device enabling non-optical genome sequencing. *Nature.* 475: 348–352. doi: 10.1038/nature10242.

29. Tang F, Barbacioru C, Wang Y, et al. mRNA-Seq whole-transcriptome analysis of a single cell. *Nat Methods.* 6(5): 377–382.

30. Walters WA, Knight R. (2014). Technology and Techniques for Microbial Ecology via DNA Sequencing. *Ann Am Thorac Soc.* 11(Suppl 1):S16-S20. doi:10.1513/AnnalsATS.201306-160MG.

31. Whiteley AS, Jenkins S, Waite I, Kresoje N, Payne H, Mullan B, et al. (2012). Microbial 16S rRNA Ion Tag and community metagenome sequencing using the Ion Torrent (PGM) Platform. *J Microbiol Methods.* 91: 80–88. doi: 10.1016/j.mimet.2012.07.008.

32. Woyke T, Tighe D, Mavromatis K, et al. (2010). One bacterial cell, one complete genome. *PLoS One.* 5(4), Article ID e10314, 2010.

CHAPTER 2

MICROBES IN ENVIRONMENT AND INTRODUCTION TO MICROBIAL ECOLOGY

Eleonora Egidi[1], Jennifer Wood[2], Prashant Pant[3], Prerna Pandey[4], Pratibha Pant[5]

[1]Dr. Eleonora Egidi is a mycologist interested in different aspects of fungal research, including fungal ecology, biology and phylogeny. Eleonora has an international background and has been involved in collaborative projects for the identification of new fungal barcoding regions, study of fungal communities from extreme environments, and use of fungi in phytoremediation. Her current scientific interests focus on exploring the biodiversity of Australian fungal flora and their role as drives of plant diversity in natural ecosystems, as well as understanding how these systems respond to environmental changes.

[2]Jennifer Wood is post-doctoral fellow in Dr. Ashley Franks' Applied and Environmental Microbiology Laboratory at La Trobe University, Australia. She has recently completed her Doctoral research in Microbial Ecology at the same university. Her broad research interest is in understanding the ecology of soil-microbial communities and their interactions with macro-communities. She is interested in developing trait-based approaches to interpret and predict

how environmental and anthropogenic changes impact microbial community ecology and function.

[3]Dr. Prashant Pant, is an Assistant Professor of Botany in Dyal Singh College, University of Delhi, New Delhi, India. He has been teaching for more than 5 years now and his main areas have been Bioinformatics, Molecular Biology and Genomics at the college. His PhD was on understanding the Microbial Succession using a Metagenomic approach in a Barren mined-out site undergoing ecological restoration. For this, he deployed many molecular and computational Biology tool and techniques. He has several publications (articles and book chapters) in International Journals and International Publications. Currently, he is also Chief Editor of Bioinformatics Review and a co-founder/Executive Chief Editor of a Science Journal in Hindi Language "Swadeshi Vigyan" for the native language speaking communities. Both the Journals are a service to the scientific community as they are fully open access.

[4]Dr Prerna Pandey, a biotechnologist with several years of wet lab research experience, worked at International Center for Genetic Engineering and Biotechnology, New Delhi. Her field of research in PhD included isolation and molecular characterization of geminiviruses, genome sequencing, gene annotation, and gene silencing using the RNA interference technology. She has also worked at Transasia Biomedicals and Advance Enzyme Technologies as a scientist. Prerna has published papers in peer-reviewed journals, and has submitted a number of annotated Geminiviral genome sequences in the GenBank, including two novel ones. She has also completed her editing and proof-reading courses from Society for the promotion of Editors and Proof-readers, UK and now works as a freelance scientific editor and writer. When Prerna is not busy with her assignments, she loves to explore historical places.

[5]Dr. Pratibha Pant, is an Assistant Professor of Botany in Daulat Ram College, University of Delhi, New Delhi, India and has a teaching experience of more than 6 years. She is a plant virologist and biotechnologist by training and has done her PhD in Characterization of Begomoviral genomes from Papaya from Department of Botany, University of Delhi. During her PhD, she utilized a combination of molecular biology and Computational Biology tools to analyze begomoviral genomes and to her credit, a couple of novel begomoviruses have been isolated, sequenced and reported in public databases

INTRODUCTION

Being microorganisms omnipresent, they invariably affect the environment in which they grow. The effects can be harmful or beneficial. In this chapter, we define the role of microbes in many ecosystem services, such as nutrients cycling, decomposition, nitrogen fixation and primary production, as well as their biotechnological applications. Moreover, the concept of microbial ecology is introduced.

Recycling the primary elements that make up all living systems, especially carbon (C), oxygen (O) and nitrogen (N) is one of the most significant effect of the microorganisms on Earth. These elements are the basic constituents of many organisms' nutrients, and they make up for most of the living material, together with sulfur (S), phosphorus (P), potassium (K), iron (Fe), sodium (Na), calcium (Ca) and magnesium (Mg). C, H, O, N, S, and P are the primary constituents of organic material. Microbes play crucial roles in the cycling of many of these nutrients in the environment.

Decomposition

Decomposition is the breakdown of complex organic materials to easily usable forms of carbon. When this process is mediated by an organism, it is known as biodegradation. Microbial enzymes are able to break down the organic (solid) materials, which are eventually converted to a few organic acids and carbon dioxide (Lovely, 2003). Through microbial fermentation and respiration, the organic molecules are eventually broken down to CO_2 which is returned to the atmosphere. The activities of microbes in the carbon cycle found an important application in the waste management of compost, landfills or sewage treatment facilities. Indeed, almost all naturally-occurring organic compounds can be degraded by some microbes. However, some synthetic compounds such as Teflon, Styrofoam, plastics, insecticides and pesticides are broken down slowly or not at all (Joshi, 2007), posing issues for their accumulation in the environment.

Nitrogen Fixation

Nitrogen fixation is a process by which nitrogen in the Earth›s atmosphere is converted into ammonia (NH_3) or other molecules available to all forms of life. This process is essential because inorganic nitrogen compounds are necessary for the biosynthesis of the basic building blocks (e.g. nucleotides for DNA and RNA, enzymes co-factors, and amino acids for proteins) for all the living organisms. Some bacteria are able to remove atmospheric N_2 and convert it to ammonia (NH_3) through nitrogen fixation for use by plants and animals, and results in replenishment of soil nitrogen. Some bacteria fix nitrogen in both symbiotic associations with plants or in free-living form in soil and aquatic habitats. This process is crucial in some peculiar habitats, like the cactus community in the Sonoran Desert, where nitrogen-fixing bacteria are at the base of the food chain as the source of nitrogen for maintenance of cell material (Franklin et al., 2016).

Primary Production

In ecology, primary production is the synthesis of organic compounds from atmospheric or aqueous carbon dioxide (CO_2). Photosynthetic organisms are the primary producers that take up atmospheric CO_2 and fix it to organic (cellular) material. They form the basis of the food chain, as almost all life on Earth relies directly or indirectly on primary production. In terrestrial ecoregions, plants are mainly responsible for primary production, while in aquatic systems algae and microbes predominate in this role. Planktonic algae and cyanobacteria contribute to nearly half of the planet's primary production, and are "grass of the sea", from which the marine life is derived. For at least a billion years before plants evolved, oxygen on the earth was produced only by the microbes. The primary component of marine and freshwater plankton and microbial mats is represented by a cyanobacterium from the genus *Synechococcus*, which is involved in primary production, as well as nitrogen fixation and oxygenic photosynthesis, and thus has crucial roles in the cycles of carbon, nitrogen and oxygen (McDonald et al., 2011). It is among the most important photosynthetic bacteria in marine environments, estimated to account for nearly one-fourth of the primary production that occurs in typical marine habitats.

Microbes in Associations with Animals and Plants

Microbes are found in different kinds of association with all higher forms of organisms, including animals, plants, fish, insects, and invertebrates.

For example, bacteria and other microbes in the intestines of animals and insects digest nutrients and produce essential vitamins and growth factors. Leguminous plants (peas, beans, clover, alfalfa, etc.) have close associations with nitrogen fixing bacteria that extract atmospheric nitrogen and supply it to the plant for growth. Microbes in the rumen of the ruminant animals are responsible for the initial digestion of cellulose. Moreover, bacteria such as *Lactobacillus* and *Bifidobacterium*, normally live on human body surfaces as the normal flora, and protect their hosts from infections, as well as promote nutrition and health (Baron, 1996).

Production of Foods, Fuels, Drugs and Vaccines

In the home and in industry, microbes are used in the production of fermented foods like beer, wine, leavening of breads, yogurt, cheese, sour cream, buttermilk and other fermented milk products. While yeast is mainly involved in the production of ethanol, wine or gasohol, bacteria are the agents of many fermented foods, such as kombucha (fermented tea), kimchi (fermented vegetables), miso (fermented soy), yogurt and kefir (fermented milk), and sauerkraut (fermented cabbage).

Microorganisms and Oil-Spill Clean-Up

Each year, nearly more than 1 million liters crude oil (petroleum), a highly complex mixture of organic compounds, enters the environment, especially through oil-shipping disasters, in addition to deliberate discharge from processing sites. Around three million tons of oil enters the sea each year, through oil-pumping operations, transport, refining activities and non-tanker shipping, as well as natural seepages and illegal activities that include the dumping of ballast water and oil residues, and natural accidents. Illegal activities contribute to up the 55% of the oil present in the environment (Golyshin et al., 2003). In addition, hydrocarbons are also produced continuously by living cells as natural oils and fats.

Micoorganisms largely contribute to the breakdown of these organic compounds in the environment. An obvious indication of the activity of the hydrocarbon-degrading microorganisms is the observation that the oceans are not covered with an oily layer (Head et al., 2006). For some bacteria, hydrocarbons form an important source of food. These bacteria play an important role after an oil spill, in cleaning-up and forming the basis for the natural oil-degrading capacity of the ecosystem.

In addition to human activity-derived hydrocarbons, a significant part of the seawater hydrocarbon content originates from biological systems. Lipids and fatty acids from plants, animals and microbes, as well as their conversion products in anoxic zones are found everywhere. These systems 'attract' micro-organisms that are able to break down many of these aromatic compounds. Some of the obligate hydrocarbon degraders include the members of the genera *Oleispira*, *Oleiphilus*, *Thalassolituus*, *Alcanivorax* and *Cycloclasticus*.

Current interest in these hydrocarbon-degrading bacteria has spurred on various genome sequencing projects. The genome of the obligate hydrocarbon-degrading bacterium *Alcanivorax borkumensis* has been studied in detail (Schneiker et al., 2006). *Alcanivorax* is of particular interest because it is a cosmopolitan marine bacterium and thrives in oil-contaminated areas, making up to 80% of the bacterial population in these scenarios. This bacterium produces exopolysaccharides extensively, and possesses pili through which it can efficiently attach to the oil-water interface. Biosurfactants are massively produced, leading to an increase in the oil-water surface area by emulsifying the alkanes. Its genome contains several atypical GC stretches, one of which carries a complete set of genes for alkane degradation (*alkSB*$_1$*GJH*), and more genes that encode proteins that may be involved in alkane degradation, such as *alkK*, *alkL* and *alkN*. Some hydrocarbon decomposers have also been genetically modified to improve their degradation abilities. The first ever patent for a super-bug that can degrade oil was done for a genetically engineered *Pseudomonas* sp., that can break-down camphor, octane, salicylate and naphthalene (US Patent 4259444). However, as most of these bacteria fall prey to protozoa, their introduction into already colonized niches is very difficult (Cases and De Lorenzo, 2005).

The genomes of several sulfate-reducing bacteria have been completely sequenced and published in the last few years because of their ability to grow anaerobically and their significant ecological (and economic) impact. These include the *Desulfovibrio vulgaris Hildenborough*, the arctic *Desulfotalea psychrophila* and the archeal *Archaeoglobus fulgidus* (Brooijmans et al., 2009). The ability of these bacteria to anaerobically degrade oil may be useful in certain types of ecosystems for the natural oil-degrading capacity.

Microbes in Space

Microbiologists now have new opportunities with international plans already in place for exploration of solar system, either using robotic probes or with human crews. The search for signatures of life forms on another planet or moon in our solar system is one of the most prominent targets of these missions.

The key targets for the search for life are Mars and Jupiter's moon Europa. To understand the biology of microorganisms potentially able to colonize such extreme environments, potential extraterrestrial habitats simulating exo-planetary environments have been extensively studied. Extreme terrestrial regions, such as Antarctica, represent a model study because of their particular harsh conditions (cold and high UV irradiation) that allow the survival of microbial communities that well adapted to those environmental constraints (extremophiles). Therefore, microbiological studies under simulated planetary environments, and field studies with microbial communities in the extreme will provide valuable information to explore the life outside our Planet.

In the space exploration missions, other major role of microbiologists is the planetary protection. Unintentional introduction of terrestrial microorganisms to the planet or moon of concern through spacecraft, whether robotic orbiters, entry probes, or landers, may destroy the opportunity to examine these bodies in their niches. Thus, the concept of planetary protection has been introduced (COSPAR Planetary Protection Guidelines, COSPAR, 2005) has been initiated to prevent the introduction and probable proliferation of terrestrial microorganisms on the target body. Specific guidelines for biological load measurements, sterilization procedures, and planetary protection control represent additional upcoming paramount tasks for microbiologists. Before beginning any human exploratory mission in space, the issues concerning effects of microbes on human health and well-being need to be examined, in addition to establishing effective planetary protection protocols.

MICROBIAL ECOLOGY

Microbial Ecology is the study of interactions between microorganisms and their environment and also within microorganisms as inter- and intraspecific interactions and how they influence each other in terms of structural and functional diversity. This branch of ecology entails studies on microbial diversity as well as on environmental factors interacting and influencing this diversity. Structural diversity of a microbial community comes under the umbrella of Microbial Diversity (MD) studies. It refers to the isolation, identification/characterization, typing and quantification of microorganisms in different environments, and the microbial activity, or the functional aspects of what role they are playing in these environments_ in that is, what microorganisms are doing in their habitats and how their activities contribute to the observed microbial diversity and biogeochemical cycling. Microbial diversity in the environment can be measured by various indices such as phylogenetic diversity, species diversity, genotype diversity, and gene diversity.

Microbial Diversity indicates the number of species (Species Richness) and abundance (number of individuals for every species) comprising a microbial community in a given habitat at a given point of time. On a molecular level, microbial diversity is reflected by the sequence diversity for a given gene family in a community DNA sample (metagenome) isolated from an environment (Liesack et al., 1997; Garbeva et al., 2004). *Microbial community structure* is a qualitative term which indicates the number of individuals of diverse taxa which share a common habitat and perform key functions at a trophic level. The functions performed by different trophic levels of microbial communities (mainly bacterial and fungal) include decomposition of organic matter, mineralization and nutrients cycling (van Elsas and Trevors, 1997; Garbeva et al., 2004). Below ground soil communities also maintain essential ecosystem functions in both managed and unmanaged soils (Hairston et al., 1968; Doran et al., 1996; van Elsas and Trevors, 1997; Garbeva et al., 2004), by participating in soil structure formation and toxin removal (van Elsas and Trevors, 1997); by promoting plant growth, suppressing a number of soil borne pathogens and bringing about changes in above-ground vegetation (Nitta, 1991; Abawi and Thurston, 1992; Doran et al., 1996).

Interaction Among Microbial and Macrobial Communities-An Integral Concept in Microbial Ecology

Taking soil as a microbial habitat, it is seen that community structure, biomass and specific functions of soil microorganisms can indicate soil health and quality (Hooper et al., 2000; Wolters et al., 2000). Soil biota is assumed to be responsible for soil ecosystem processes, especially nutrient cycling and decomposition of soil organic matter (Wardle and Giller, 1996; Brussaard et al., 2004; Coleman et al., 2004). Soil organic matter, in turn, is an important attribute that shows positive correlation with above-ground plant diversity (Le Houerou, 1969; Lugo et al., 1986; Cesar, 1989; Aronson et al., 1993). This organic matter is mainly contributed by leaf litter, and animal biomass influencing microbial communities (Schaeffer and Whitford, 1981; Bardgett, 2005; Bardgett and Wardle, 2010). The composition of the soil microbial assemblage can (i) affect plant growth and competitive outcomes between plants, (ii) alter species composition, and (iii) affect the ability of a plant species to colonize an area (Turkington et al., 1988; Bever, 1994). In turn, microbial communities are influenced by site variables (seasonal variations, precipitation, temperature, relative humidity, soil pH, soil organic matter etc) and plant processes (e.g., root exudations, respiration and absorption of chemical ions) that affect bulk and rhizospheric soil chemistry at a particular site (Grayston et al., 1996; Paul and Clark, 1996a, b; Marschner et al., 2001). In fact, attempts for raising plant communities as restoration drives depends on the taxonomic and functional diversity of below ground communities (Wardle et al., 2002; 2004; Moore et al., 2004; Bardgett and Wardle, 2010; Miki et al., 2010). Microbial population genetics and Microbial biogeography entails spatial and temporal distributions. These distributions are seen in the context of several environmental factors viz. Temperature, pH, salinity, pressure, water and mineral nutrient availability, and carbon source. These factors play a crucial role in structure, composition as well as spatial and temporal changes occurring in microbiota in natural environments. As a result, microbial ecologist group microbiota on the basis of functional aspects i.e. on the basis of metabolic role played by different microbes e.g. Phototrophs, Chemotrophs, Heterotrophs, Mineralizers, Decomposers, Saprobes, Nitrogen fixers (free or associated forms). In fact, it has become obvious to us that the diversity of microbial metabolism is so large that they extend way beyond the metabolic capabilities of plant and animal systems taken together. The archaebacteria present a unique microbiota which shows metabolic capabilities to thrive in extreme environments e.g. high and low pressure, pH extremes, heavy metals, salinity, radiation, desiccation, oxidative stress, and temperature. For instance, *Pyrobolus fumarii,* a nitrate reducing chemolithotroph grows at temperature up to 113°C. Microorganisms in environment are like the higher organisms are organized in a hierarchical level of organization from individuals to population, population to metabolically related populations (guilds), set of interacting guilds or communities and then at ecosystem level. Microbes at ecosystem level have microbiota interacting with biotic (microbial and macrobial forms) and abiotic factors (mentioned above). The interactions range from being mutualistic to predatory to a number of other kinds. Thus, a microbial community is no different rather much more complex in terms of interactions and associations going on in between the biotic and abiotic forms.

Within a habitat, some microbes present themselves as indigenous ones, occupying functional niches in the ecosystem, carrying out ecosystem services, and these are called *Autochthonous*, while those which present themselves as random wanderer types surviving in the habitat for a period and then disappearing and thus neither filling any functional niche, nor sustaining any ecological services are called *Allochthonous*. The dynamic interaction between microbiota and their immediate environment is cause and consequence of productivity and environmental quality of a given ecosystem. As mentioned in the previous section, microbes are crucial for the waste management and bioremediation of contaminated polluted sites. At the same time, it is to be understood that though microbial ecology sounds pertaining to microbiota and environment only, but that is just not the case. As a matter of fact, there is a great deal of involvement of macrobial forms influencing the microbial community structure positively or negatively i.e. either enhancing the ability of populations to survive or limiting it or eliminating a population completely from a habitat as in the case of allelopathic weeds rendering the soil biologically sterile with microbial population due to allelochemicals.

REFERENCES

1. Joshi SR. *Microbes: redefined Personality*. 2007, APH publishing corporation, Ansari Road, Darya Ganj, New Delhi.
2. Lovley DR (2003). Cleaning up with genomics: applying molecular biology to bioremediation. *Nat Rev Microbiol.* 1 (1): 35–44.
3. Franklin KA, Sommers PN, Aslan CE, López BR, Bronstein JL, Bustamante E, et al. (March/April 2016). Plant Biotic Interactions

in the Sonoran Desert: Current Knowledge and Future Research Perspectives. *Int J Plant Sci.* 177(3): 217-234.

4. McDonald AE, Ivanov AG, Bode R, Maxwell DP, Steven R, Norman R, Hüner PA. (2011). Flexibility in photosynthetic electron transport: The physiological role of plastoquinol terminal oxidase (PTOX).
5. *Biochimica et Biophysica Acta-Bioenergetics.* In Medical Microbiology. 4th edition. Baron S, editor. University of Texas Medical Branch at Galveston, 1996.
6. Golyshin PN, Martins DSVA, Kaiser O, Ferrer M, Sabirova YS, Lunsdorf H. (2003). Genome sequence completed of *Alcanivorax borkumensis*, a hydrocarbon-degrading bacterium that plays a global role in oil removal from marine systems. *J Biotechnol.* 106:215–220.
7. Head IM, Jones DM, Roling WF. (2006). Marine microorganisms make a meal of oil. *Nat Rev Microbiol.* 4:173–182.
8. Schneiker S, Martins DSVA, Bartels D, Bekel T, Brecht M, Buhrmester J. (2006). Genome sequence of the ubiquitous hydrocarbon-degrading marine bacterium *Alcanivorax borkumensis*. *Nat Biotechnol.* 24:997–1004.
9. Cases I, De Lorenzo V. (2005). Genetically modified organisms for the environment: stories of success and failure and what we have learned from them. *Int Microbiol.* 8:213–222.
10. Brooijmans RJW, Pastink MI, Siezen RJ. (2009). Hydrocarbon-degrading bacteria: the oil-spill clean-up crew. *Microb Biotechnol.* 2(6): 587–594.
11. Horneck G, Klaus DM, Mancinelli RL. (2010) Space Microbiology. *Microbiol Mol Biol Rev.* 74(1):121-56. doi: 10.1128/MMBR.00016-09.
12. COSPAR Planetary Protection Policy (20 October 2002; Amended 24 March 2005)
13. US Patent 4259444. Microorganisms having multiple compatible degradative energy-generating plasmids and preparation thereof. 31 Mar 1981.
14. Liesack W, Janssen PH, Rainey FA, Ward-Rainey NL, Stackebrandt

E. (1997). *Microbial diversity in soil: the need for a combined approach using molecular and cultivation techniques.* In: Van Elsas, J.D., Trevors, J.T., Wellington, E.M.H., (Eds.), Modern Soil Microbiology. New York, N.Y: Marcel Dekker pp. 375-439.

15. Garbeva P, van Veen JA, van Elsas JD. (2004). Microbial diversity in soil: selection of microbial populations by plant and soil type and implications for disease suppressiveness. *Ann Rev Phytopathol.* 42, 243-270.
16. van Elsas JD, Trevors JT. (1997). Modern Soil Microbiology. New York, Marcel Dekker.
17. Hairston NG, Allan JD, Colwell RK, Futuyma DJ, Howell J, Lubin MD, Mathias J, Vandermeer JH. (1968). The relationship between species diversity and stability: an experimental approach with protozoa and bacteria. *Ecology.* 49, 1091-1101.
18. Doran JW, Sarrantonio M, Liebig MA. (1996). Soil health and sustainability. *Adv Agron.* 56, 2-54.
19. Nitta T. (1991). Diversity of root fungal floras: its implications for soil-borne diseases and crop growth. *Jpn Agric Res.* 25: 6-11.
20. Abawi GS, Thurston HD. (1992). Effect of organic mulches, soil amendments, and cover crops on soil borne plant pathogens and their root diseases. Proceeding of International Workshop on Slash/ Mulch Production systems, Turrialba, Costa Rica.
21. Hooper DU, Bignell DE, Brown VK, Brussaard L, Dangerfield JM, Wall DH, et al. (2000). Interactions between aboveground and belowground biodiversity in terrestrial ecosystems: patterns, mechanisms, and feedbacks. *Bioscience.* 50: 1049-1061.
22. Wolters V, Silver WL, Bignell DE. (2000). Effects of global changes on above- and belowground biodiversity in terrestrial ecosystems:implications for ecosystem functioning. *Bioscience.* 50: 1089-1098.
23. Wardle DA, Giller KE. (1996). The quest for a contemporary ecological dimension to soil biology. *Soil Biol Biochem.* 28, 1549-1554.
24. Brussaard L, Kuyper TW, Didden WAM, de Goede RGM, Bloem J. (2004). *Biological soil quality from biomass to biodiversity-importance and resilience to management stress and disturbance.*

In: Schjønning, P., Elmholt, S., Christensen, B.T., (Eds.), Managing soil quality: Challenges in modern agriculture. CAB International, Wallingford, UK pp. 139-161.

25. Coleman DC, Crossley DA, Jr. Hendrix PF. (2004). *Fundamentals of soil ecology*. Second edition. Elsevier Academic Press, Inc., Amsterdam, Netherlands. pp. 386.
26. Le Houerou HN. (1969). La vegetation de la Tunisie steppique. Annals de l'Institut National de Recherche Agronomique de la Tunisie 42, 1-622.
27. Lugo AE, Sanchez MJ, Brown S. (1986). Land use and organic carbon content of some subtropical soils. *Plant Soil*. 96, 185-196.
28. Cesar J. (1989). L'influence de l'exploitation sur la perennite des paturages de savane. II. Role du systeme racinaire dans la degradation du pattirage. *Fourrages*. 120, 382-392.
29. Aronson J, Floret C, LeFloch E, Ovalle C, Pontanier R. (1993). Restoration and rehabilitation of degraded ecosystems in Arid and semi-arid lands. I. A view from the South. *Restor Ecol*. 8-16.
30. Schaeffer DA, Whitford WA. (1981). Nutrient cycling by the subterranean termite *Gnathamitermes tubifovmis* in a Chihuahuan desert ecosystem. *Oecologia*. 48: 277-283.
31. Bardgett RD. (2005). The Biology of Soil: A Community and Ecosystem Approach, Oxford University Press, Oxford.
32. Bardgett RD, Wardle DA. (2010). Aboveground-belowground linkages: biotic interactions, ecosystem processes, and global change. Oxford Series in Ecology and Evolution. Oxford University Press.
33. Turkington R, Holl FB, Chanway CP, Thompson JD. (1988). *The influence of microorganisms, particularly Rhizobium, on plant competition in grass-legume communities*. In: Davy, A.J., Hutchings, M.J., Watkinson, A.R., (Eds.), Plant Population Ecology. Oxford, UK, Blackwell Scientific Publications, pp. 346-366.
34. Bever JD. (1994). Feedback between plants and their soil communities in an old field community. *Ecology*. 75: 1965-1977.
35. Grayston SJ, Vaughan D, Jones D. (1996). Rhizosphere carbon

flow in trees, in comparison with annual plants: the importance of root exudation and its impact on microbial activity and nutrient availability. *Appl Soil Ecol.* 5: 29-56.

36. Paul EA, Clark FE. (1996a). Soil Microbiology and Biochemistry. Academic Press, San Diego, CA.
37. Paul EA, Clark FE. (1996b). Sulphur transformations in soil. Soil Microbiology and Biochemistry. Academic Press, NY. pp. 299-313.
38. Marschner P, Yang C-H, Lieberei R, Crowley DE. (2001a). Soil and plant specic effects on bacterial community composition in the rhizosphere. *Soil Biol Biochem.* 33: 1437-1445.
39. Wardle DA. (2002). Communities and ecosystems: linking the aboveground and belowground components. Princeton University Press, Princeton, New Jersey.
40. Wardle DA, Bardgett RD, Klironomos JN, Setala H, van der Putten WH, Wall DH. (2004). Ecological linkages between aboveground and belowground biota. *Science.* 304, 1629-1633.
41. Wardle DA, Bonner KI, Barker GM. (2002). Linkages between plant litter decomposition, litter quality, and vegetation responses to herbivores. *Funct Ecol.* 16: 585-595.
42. Moore JC, Berlow EL, Coleman DC. (2004). Detritus, trophic dynamics and biodiversity. *Ecol Lett.* 7: 584-600.
43. Miki T, Ushio M, Fukui S, Kondoh M. (2010). Functional diversity of microbial decomposers facilitates plant coexistence in a plant-microbe-soil feedback model. *Proc Natl Acad Sci, USA.* 107: 14251-14256.

CHAPTER 3

RIBOSOMAL RNA GENE SEQUENCING AND ITS APPLICATION IN MICROBIAL ECOLOGY

Jennifer Wood[1], Prashant Pant[2]

[1]Jennifer Wood is post-doctoral fellow in Dr. Ashley Franks' Applied and Environmental Microbiology Laboratory at La Trobe University, Australia. She has recently completed her Doctoral research in Microbial Ecology at the same university. Her broad research interest is in understanding the ecology of soil-microbial communities and their interactions with macro-communities. She is interested in developing trait-based approaches to interpret and predict how environmental and anthropogenic changes impact microbial community ecology and function.

[2]Dr. Prashant Pant, is an Assistant Professor of Botany in Dyal Singh College, University of Delhi, New Delhi, India. He has been teaching for more than 5 years now and his main areas have been Bioinformatics, Molecular Biology and Genomics at the college. His PhD was on understanding the Microbial Succession using a Metagenomic approach in a Barren mined-out site undergoing ecological restoration. For this, he deployed many molecular and computational Biology tool and techniques. He has several publications (articles and book

chapters) in International Journals and International Publications. Currently, he is also Chief Editor of Bioinformatics Review and a co-founder/Executive Chief Editor of a Science Journal in Hindi Language "Swadeshi Vigyan" for the native language speaking communities. Both the Journals are a service to the scientific community as they are fully open access.

INTRODUCTION

Unravelling the complexities of microbial communities, or species assemblages, is an integral step in both microbial genetics and microbial ecology research. Traditional culture based approaches to understanding environmental microbial communities are not only difficult and time consuming but fraught with biases, as only a small fraction of environmental isolates are able to be cultured. Moreover, it is difficult to fully characterize the species within microbial communities mainly because of their inherent diversity and heterogeneity along different environmental gradients (Ovreas and Torsvik, 1998; Horner-Devine et al., 2003; Becker et al., 2006; Fierer and Jackson, 2006; Kang and Mills, 2006). Ribosomal RNA (rRNA) sequencing, also known as community profiling, is a culture independent, PCR-based approach to studying environmental microbial communities that is time efficient manner and bypasses many of the limitations of conventional methods (Garbeva et al., 2004). Community profiling and NGS technologies have now established that majority of soil bacteria, archaea and fungi are un-cultivable, hence culture-independent methods of community analysis are preferred over traditional culturing strategies (Torsvik et al., 1990a, b, 1996, 1998; Hugenholtz et al., 1998; Amann, 2000; Roselló-Mora and Amann, 2001; Prosser, 2002; Torsvik and Ovreas, 2002; Zwolinski, 2007; Pevsner, 2009).

Microbial community rRNA profiling uses molecular biology-based methods to amplify specific regions of microbial genomes which serve as a model candidate for identification, characterization of microorganisms. These community profiles can be used to investigate microbial community diversity, taxonomic and phylogenetic structure. This chapter attempts to bring together the methods and analytical tools used in rRNA profiling in a cohesive manner to develop an understanding of ribosomal RNA gene sequences and how they can be used in microbial ecology.

USE OF RIBOSOMAL RNA GENES FOR STUDYING MICROBIAL DIVERSITY ANALYSIS

Ribosomes are present in all prokaryotic and eukaryotic cells. They are the smallest organelles in a cell and are responsible for the crucial role of translating mRNA into functional proteins. Structurally, ribosomes are composed of a combination of ribosomal Ribo-Nucleic Acid (rRNA) and nucleoprotein. These rRNA molecules are never translated into protein, but rather fold together with nucleoproteins to form the ribosomal subunits, which are then assembled in the cytoplasm to form functional ribosomes (Nierhaus, 1991; Dlakic, 2005; Kaczanowska and Ryden-Aulin, 2007; Shazani et al., 2011). All ribosomes are made up of two subunits: a large subunit (LSU) and small subunit (SSU). These subunits, and their rRNA components are denoted in Svedberg units (S) after their sedimentation rates. The 16S rRNA gene, which encodes the prokaryotic SSU rRNA component, (Lane et al., 1985; Schloss and Handelsman, 2004; Kim et al., 2011) and the 18S and 28S rRNA genes which encode eukaryotic SSU and LSU rRNAs respectively (Smit et al., 1999) are commonly used as phylogenetic markers and can be used to assign and identify organisms at various taxonomic levels.

Due to the importance of ribosomes and rRNA to the normal function of the cell, these sequences have remained highly conserved over the course of evolution and are thus ideal targets for capturing information about entire microbial communities. However, as well as being highly conserved, rRNA genes contain within their structure hyper-variable regions, a property which can be exploited to differentiate microbial taxa and reconstruct phylogenies. The 16S rRNA genes have nine hyper-variable (V) regions namely V1-V9 (Figure 1) which are distributed over the ~1.5Kb long 16S rRNA gene (Stackebrandt and Goebel, 1994; Neefs et al., 1990, 1991, 1993). Some of these 'V' regions are reportedly more variable than others (Youssef et al., 2009; Yu and Morrison, 2004). Similar to 16S rRNA genes, 28S rRNA and 18S rRNA genes also display a number of hyper-variable regions (Khot et al., 2009; Okubo and Sugiyama, 2009).

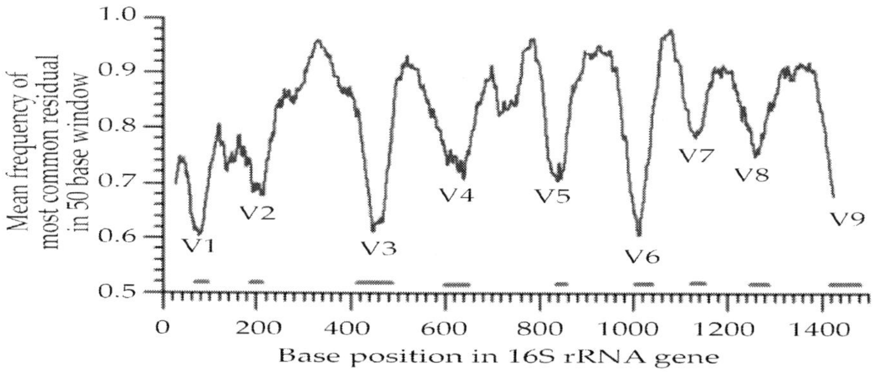

Figure 1: Nine hypervariable regions (V1-V9) distributed over 1.5 Kbp of the 16S rRNA gene, showing approximate positions of each variable region named V1-V9 (Neefs et al., 1990) *(source: http://www.bioinformatics-toolkit.org/ Help/Topics/ hypervariableRegions.html).*

RRNA COPY NUMBERS & ECOLOGICAL STRATEGIES OF MICROBES

Prokaryotic rRNA genes are typically found in operons, designated as 'rrn operons'. Each operon consists of a single copy of a 16S, 23S and 5S rRNA gene. The number of operons in a given microbial species is quite variable, ranging from 1-15 operons scattered through genome (Rainey et al., 1996). It is very interesting that genes encoding molecules that are absolutely essential to the function of the cell would exhibit such a tremendous range in copy number. This variable relationship between bacterial species and rRNA copy number confounds attempts to enumerate the total number of cells in a community (via methods such as qPCR) and is an important consideration when interpreting such data. Interestingly it has been established that the number of rRNA copy numbers in a cell appears to be related to the ecological strategy of the microbes. rRNA copy number generally fluctuates with nutrient availability and other environmental conditions (Stevenson and Schmidt, 1998; Klappenbach et al., 2000, 2001; Cavicchioli et al., 2003; Eichorst et al., 2007; Lee et al., 2009). High rRNA operon copy number is associated with better fitness in terms of ecological strategies including growth-related competition and the ability to utilize a complex or special substratum. On the other hand, low copy number is associated with commonly available source of nutrient where the genomes are not under selection pressure due to nutrient constraints and thus does not have to spend much energy to select for specific nutrient type (Klappenbach et al., 2000). In microorganism with multiple *rrn* operons, there exists a certain level (<3%) of sequence variation between the rRNA genes (Simon and Weiß, 2008). Thus, not only can a single cell be represented by multiple rRNA copies, but variation between rRNA sequences does not necessarily indicate that the genes originated from two different species. To deal with this heterogeneity as well as other useful aspects of rRNA gene operon number, Klappenbach et al., (2001) and Lee et al., (2009) describe "rrnDB" a curated online database that catalogues the maximum and minimum numbers of genes encoding for 16S, 23S and 5S ribosomal RNAs in Bacteria and Archaea (http://rrndb.mmg.msu.edu/index.php). This database can be used to determine the copy numbers associated with the diversity of bacteria observed in any sample.

PRIMER DESIGN FOR AMPLIFICATION OF RIBOSOMAL RNA GENES

When utilizing rRNA genes in microbial community analyses, almost all protocols incorporate Polymerase Chain Reaction (PCR) as a preliminary step. Following the extraction of community DNA from an environmental sample, PCR amplification of rRNA genes (or of the rRNA itself) is the next critical step in rRNA profiling. The choice of primers for this step is critical in ensuring that maximum information is captured from the community and bias is minimized. Weisburg et al., (1991) described several universal primer pairs which can be used to create robust rRNA profiles of bacterial communities associated with any environment. These "classical" primers are named FD1, FD2, FD3, FD4, RD1, RP2, RP3 and so forth, and are designed to amplify segments of 16S rRNA that collectively cover nearly the full length (~1.5 Kb) of the gene. Out of these, FD1 (forward primer) in combination with RD1 and RP2 primer (reverse primers) amplify almost all eubacteria as well as archaebacteria. The RP2 primer is nested in the full length FD1/RD1 amplicon and is located 17 bp upstream of RD1 primer region (Figure 2). While RD1 can amplify almost all eubacterial and cyanobacterial communities plus some selected archaebacteria, RP2 can amplify nearly all eubacteria, archaebacteria plus plant mitochondria and chloroplast.

Thus, it is regarded as the primer representing maximum diversity of bacterial communities (Weisburg et al., 1991).

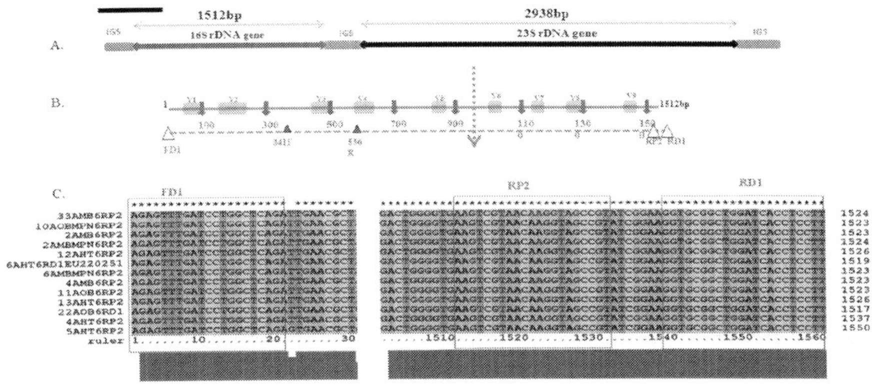

Figure 2: (A) shows prokaryotic rRNA operon depicting Intergenic spacer (IGS) regions (green), 16S (red) and 23S (black) rDNA regions. Respective positions of PCR primers for 16S rRNA amplification are shown by empty triangles belonging to FD1, RD1 and RP2 as described by Weisberg et al., 1991. These primer pairs amplify nearly a full length 16S rDNA amplicon and are capable of amplifying most eubacteria as well as archaebacterial groups (Weisberg et al., 1991). The position of primers used for qPCR analyses are shown by filled triangles. Bar in black indicates a span of 500bp approximately. (B) shows a full-length 1512 bp 16S rRNA gene with the distribution of the nine hypervariable regions highlighted. The *E. coli* 16S rRNA gene was used as a reference as described by Neefs et al., 1991, 1993. (C) A Multiple Sequence Alignment (MSA) showing conserved regions used as PCR primers to amplify nearly full length 16S rRNA gene sequences (Source: Pant, 2015)

Microbial communities also include eukaryotic members especially fungi, the largest phyla of which are the Ascomycota and Basidiomycota. In addition to fungi, Protozoans, Nematodes and microarthropods, which constitute the microfauna, are also important aspects of microbial communities. Ecology is the study of how organisms interact with each other and with their environmental, thus, in ecology a species cannot be studied in isolation. The same is true of microbial ecology. As such, most microbial analyses are designed for simultaneous assessment of both the bacterial and fungal members of a community. For detection and phylogenetic analysis of fungal communities, classical primers specific for 18S rRNA genes have been described by Smit et al., (1999).

Different primer combinations have different taxonomic affinities, for example EF4 forward primer when used with EF3 reverse primer has a bias for amplifying Basidiomycetes and Zygomycetous fungi, while the use of the classical nested reverse primer Fung5 has a slight bias for amplifying Ascomycetous fungi (Smit et al., 1999; Borneman and Hartin, 2000; Anderson and Cairney, 2004). Classical primer pairs targeting 3'conserved stretches of 28S rRNA genes from insect and terrestrial arthropods have been also used for studies on insect phylogeny (Burke et al., 1993; Werren et al., 1995a,b; de Leon et al., 2004). These primers detect insertions of ~150bp at retrotransposon insertion sites of the R1-R2 mobile elements that are conserved in various insect lineages (Burke et al., 1993). Until recently, the applicability of these primers in detection of rhizospheric soil fauna was largely unknown but recently these primers have been used extensively to type and characterize microfaunal communities from rhizospheric soil samples (Pant, 2015).

Figure 3 shows the conserved and hyper-variable regions of a typical eukaryotic rRNA operon consisting of the 18S rRNA gene, 5.8S rRNA gene, and 28S rRNA gene separated by inter-transcribed spacer (ITS) regions (Machida and Knowlton, 2012; Pant, 2015). The positions targeted by the "classical" EF4/EF3 primer pair and the nested primer pair EF4/fung5 after Smit et al., 1999 are shown. The 28SF/28SR primer pair targeting the D8-D10 variable region in the 28S rRNA genes of eukaryotes were originally used for typing terrestrial insects and invertebrates (Werren et al., 1995a, b; De Leon et al., 2004, 2006). These primer pairs have been incorporated in studying microbial communities associated with rhizospheric soil samples from a degraded mined out area undergoing ecological restoration (Pant, 2015). Primer pair 1TSF/5.8S target the variable regions of the ITS1 and 5.8S rRNA genes in fungi after Fierer et al., 2005; while the primer pair 28S1/28S5 D1-D2 target the variable regions of fungal 28S rRNA genes after Khot et al., 2009. These primer pairs have been used extensively for qPCR applications also.

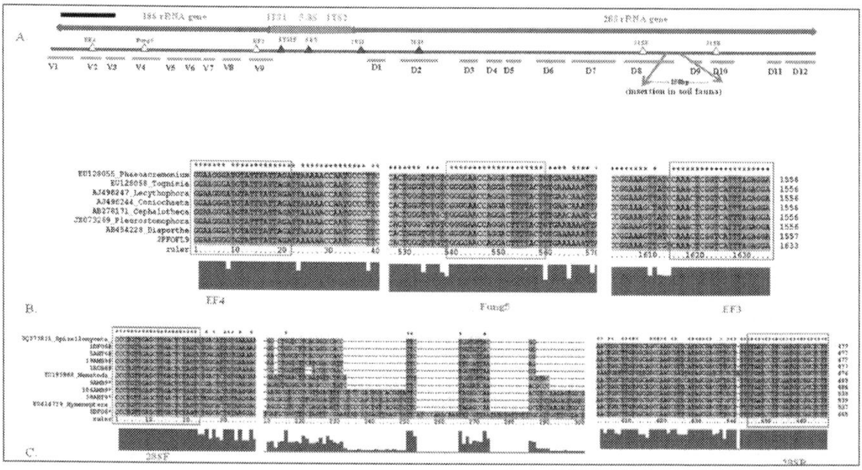

Figure 3: A diagrammatic representation of a eukaryotic rRNA operon showing (A) internal transcribed spacer regions (ITS1, ITS2) in green, the 18S rRNA gene in brown, the 5.8S rRNA gene in light blue and 28S rRNA gene in blue. The position of the classical PCR primers used for amplification of fungal 18S rRNA gene fragments (Smit et al., 1999) and insect 28S rRNA gene fragments (Werren et al., 1995) are shown by open triangles, while qPOPs are shown as closed triangles. Also, shown in orange are nine hyper-variable regions (V1-V9) in 18S rRNA gene (Okubo and Sugiyama, 2009) and expansion fragments D1-D12 within the 28S rRNA gene in blue (Zardoya and Meyer, 1996; Machida and Knowlton, 2012). An insertion of 150bp observed in this study between D8 and D9 regions is shown. The bar in black indicates 500bp. A MSA is shown for conserved regions of (B) fungal 18S rRNA genes built into the classical fungal primers (boxed in brown) and (C) fungal (#) and faunal (*) 28S rRNA sequences built into the classical insect-specific 28SF/28SR boxed in blue (Source: Pant, 2015).

PCR AND *rRNA* GENE TYPING OF MICROBIAL COMMUNITIES, QUANTITATIVE PCR ANALYSES

Universal and/or group-specific primers based on rRNA genes have facilitated the structural and taxonomic characterization of microbial communities from environmental samples via PCR-based techniques (Gardes and Bruns, 1993; Head et al., 1998; Borneman and Hartin, 2000; Schmalenberger et al., 2001; Theron and Cloete, 2000; Anderson et al., 2003). A range of community fingerprinting methods based on PCR generated rRNA amplicons are available to study soil microbial communities at different levels of resolution. These methods include: Denaturing Gradient Gel Electrophoresis (DGGE; Muyzer and Smalla, 1998), Temperature Gradient Gel Electrophoresis (TGGE; Heuer and Smalla,, 1997), Terminal Restriction Fragment Length Polymorphism (T-RFLP; Kuske et al., 2002), Single-Strand Conformational Polymorphism (SSCP; Schwieger and Tebbe, 1998), Amplified Ribosomal DNA Restriction Analysis (ARDRA; Massol-Deya et al., 1995), Automated Ribosomal Intergenic Spacer Analysis (ARISA; Borneman and Triplett, 1997; Fisher and Triplett, 1999; Ranjard et al., 2000; Yu and Mohn, 2001), DNA-DNA (Southern Blot or Chip based assays such as microarray or phylochip), mRNA-DNA hybridization, and PCR based cloning and sequencing (Kirk et al., 2004). The advantage of these methods is that they allow a direct comparative overview of the composition and diversity of environmental microbiota from a given sample (van Elsas and Boersma, 2011). However, many of these technologies developed prior to NGS have poor detection limits and are restricted to visualizing only the most abundant (~ top 10%) community members (Gobet et al., 2014).

Regular PCR based approaches help in selective amplification of target rRNA molecules from a sample. Further, to compare two or more different samples in terms of microbial load and its proportion of the kind viz. bacteria versus fungi or eubacteria versus archaebacteria and so on, a unique PCR based analytical tool called Quantitative PCRs (Q-PCR) is applied. Q-PCR detect the "quantum of amplicons" generated in real time during the exponential phase of a PCR (Higuchi et al., 1993; Heid et al., 1996; Lee et al., 1996b; Rincon-Florez et al., 2013). For this, a number of fluorescent dyes with varying chemistries (SYBR Green, TaqMan or Molecular Beacons) which bind to DNA are used and fluorometric detection of amplification is monitored in every PCR cycle. The cycle at which the detected fluorescence intensity reaches a threshold that is higher than the background fluorescence, is called cycle threshold (C_T). Similar to NGS, applications of Q-PCRs are immense and only limited by the questions asked (Metzker, 2010). In the rhizospheric soil microbial communities, Q-PCRs have been done using universal as well as group-specific primers (Bacchetti De Gregoris et al., 2011) to estimate relative abundances of taxonomic groups of microbial and fungal communities in soil (Fierer et al., 2005; Navarrete et al., 2013).

The use of PCR, cloning and DNA sequencing methods allows complex microbial communities from an environmental sample with many species and constituent genomes to be studied simultaneously. The collective community DNA is referred to as the metagenome or ecogenome or community genome or microbiome (Rondon et al., 2000; Galperin 2004; Handelsman,, 2004; Riesenfeld, 2004, Schloss and Handelsman, 2005a, b) and its study as metagenomics or community genomics. Conventionally, the PCR amplified rRNA gene amplicons are cloned and sequenced using Sanger sequencing approach. However, Sanger sequencing cannot efficiently sequence the massive genomic diversity contained within metagenomic samples. This is where high throughput Next Generation Sequencing (NGS) strategies comes into being. NGS methods have revolutionized the way sequence data is obtained from metagenomic samples (Metzker, 2010; Kim et al., 2013). NGS methods provide enormous quantities of sequence information from metagenomic samples. NGS has made contribution to the study of microbial diversity the forces that shape microbial community (see chapter 4).

LIMITATIONS OF CULTURE INDEPENDENT APPROACHES

Although they are considered preferable to culture-dependent methods, culture-independent approaches for studying microbial communities have their own limitations. Kirk et al., 2004 and Rincon-Florez et al., 2013 have reviewed these limitations in great detail, some of which are mentioned here.

Prior to PCR amplification, all community profiling methods (pre- and post NGS) are subject to potential bias due to soil DNA isolation protocols that may be more suited to a particular soil type (type and level of humic acid present, etc.). Some of these concerns can be addressed by using the same methods consistently among the samples being examined. Since, methods to study soil microbial diversity are continually evolving, no one method circumvents all bias or provides all answers adequately.

PRE-NGS rRNA PROFILING TECHNOLOGIES

The main criticism of community profiling techniques such as ARDRA and RFLP is that they are more suited to less complex environments. They require the digestion of amplicons with several restriction endonucleases, and share the problem of single species having different banding patterns (due to strain variations or PCR induced biases) or multiple species having common banding pattern (due to similar restriction sites) (Nocker et al., 2007; Liu et al., 1997; Tiedje et al., 1999). In T-RFLP and ARISA, PCR based artefacts can incorrectly inflate diversity estimates and common restriction sites shared by unrelated rRNA sequences will result is a loss of information as two microbial species will be 'binned' together (Nocker et al., 2007; Okubo and Sugiyama, 2009). DNA microarray 16S rRNA Phylochips (De Santis et al., 2007; Brodie et al., 2007) or functional gene Geochips (He et al., 2007) based approaches are costly; require expensive scanning facility and are limited by availability of probes as well as problems with specificity and cross hybridization (Hubert et al., 1999; Cho and Tiedje, 2001; Greene and Voordouw, 2003).

NGS rRNA Profiling Technologies

NGS approaches are much more expensive than PCR-based community fingerprinting, and require substantial computational and statistical treatment of large datasets. As mentioned previously, NGS methods are not fool-proof and have their own demerits and shortcomings viz. PCR biases, smaller reads length, over-estimations, Chimera formation (von Wintzingerode et al., 1997; Podar et al., 2007; Manichanh et al., 2008;).

On Chimera Detection

Detecting chimeric (or recombinant) sequences from a sequence dataset is a very important part of sequence analysis. In sequences from community DNA analysis, this becomes even more relevant. Recently, a number of chimera detecting softwares for 16S rRNA gene sequences have been launched namely Pintail and Mallard (Ashelford et al., 2005, 2006; http://www.bioinformatics-toolkit.org). These programs detect chimeras and anomalies in the 16S rRNA genes based on extent of pair-wise percentage similarity between the query and related sequences. In chimera analysis by Pintail 1.0, the query sequences which could be putative recombinants are compared on a one (query)-on-one (subject) basis with a list of closely related sequences identified by BLAST searches. As Pintail is a one-on-one query-subject comparison, it is highly stringent. In Mallard, one of the sequences from within a dataset of query sequences is randomly chosen as subject, while rest remain as query. A many (query)-on-one (subject) comparison follows, which is easy and completes in less time as compared to Pintail. However, Mallard is of limited use if the query sequences are too diverse in the first place.

Another software for detecting chimeras in 16S rRNA genes is Bellerophon ver 3.0 from Greengenes (http://greengenes.lbl.gov/cgi-bin/nph-index.cgi). Here, the sequences are submitted as NAST (Nearest Alignment Space Termination Tool) formatted file (De Santis et al., 2006a, b). The NAST alignment server at Greengenes has nearly 1049116 16S rRNA sequence records. Upon submission of the formatted file, the server launches a localized BLAST search for each query sequence with the 16S rRNA gene sequence library on its server. The program then checks for potential chimeras in the respective query-subject alignment, one-on-one. It is user-friendly and is particularly good for large datasets with high amount of sequence diversity. The only demerit of the software is that if there is a relatively novel sequence in the query batch, it receives a low score being highly unrelated with the existing records and thus stands at a risk of getting omitted. Hence there has to be careful scrutiny of the output results.

STRUCTURAL ANALYSIS OF *RRNA* COMMUNITY PROFILES

NGS rRNA profiling can be used to track the response of the microbial community to biotic and abiotic stimuli, characterize core microbiome and understand biogeographic patterns of microbial diversity. Although, pre-NGS community profiling strategies are expensive but the cost of materials needed for sequencing to create rRNA community profiles is steadily decreasing and is now within the realms of affordability for many researchers. Additionally, software associated with data quality control, taxonomic assignment and diversity analysis have been streamlined by platforms such as QIIME (quantitative insights into microbial ecology), USEARCH and the RDP (ribosomal database project) to enable rapid visualization of the taxonomic structure of microbial communities (Edgar, 2010; Caporaso et al., 2010; Cole et al., 2014.

PHYLOGENETIC ANALYSIS OF *RRNA* SEQUENCE BASED COMMUNITY PROFILES

An important aspect of studying microbial communities is to visualize the composition and structure of the community from sequence relatedness or phylogenetic point of interest. This helps the understanding of the communities in many ways, assessment of the predominant OTUs (at the rank of Genera or family or above), as well as understanding the spatial and temporal dynamics being the major ones. Especially in long term microbial succession based studies such as the one undertaken in ecological restoration programs, assessing the changes on community structure over time and the possible reasons (functional diversity) for the changes in incidence (presence/absence), abundance and predominance of members of the community. This is one way to establish initiation of ecosystem services, augmentation of mineral cycles, fixation of Nitrogen over time and space. From a taxonomic point of view, it is of great interest to work with environmental microbiomes and look at sequence relatedness, conserved residues, variation as the basis of differences between OTUs, as well as having a firsthand look at the general diversity which may have some novel sequences which stand out as unrelated or distantly related increasing the possibility of discovery of new strains or species or groups. It is because of these reasons that one need to analyze the sequence diversity that is obtained using NGS or Sanger Sequencing approach so as to extract maximum interpretation of the data.

Multiple Sequence Alignments and BLAST Results

In a Multiple Sequence Alignment (MSA), sequences are aligned by bringing the greatest number of characters to register in the same column of an alignment (Higgins and Sharp, 1988). MSA provide information as to the most similar regions in a set of sequences. In proteins, such regions may represent conserved functional or structural domains. Software generally use a progressive global alignment of the sequences involved (using threading algorithms), starting with a local alignment. An example of this approach is CLUSTAL (Higgins and Sharp, 1988). Sequences to be included in an alignment are usually obtained by BLAST based searches.

BLAST or Basic Local Alignment Search Tool (Altschul et al., 1990, 1997; Zhang et al., 2000) is the most popular tool for determining homologous sequences from a database. BLAST results provide bit scores and expect values (E-value) to compare a query with subject sequence alignment. These scores indicate quality of each alignment between the query and all the subjects (hits) in a database found significantly similar. The better the alignment, the higher is the score. This score takes into account, the penalties incurred due to gaps and mismatches. Selected sequences can be aligned using ClustalW (Higgins and Sharp, 1988; Thompson et al., 1994, 1997) and ClustalX (Jeanmougin et al., 1998). These software have been used to align DNA as well as amino acid sequences. They are versatile and can provide an alignment of large number of sequences and the structure of domains in the sequences (Mount, 2002). The steps of ClustalW and ClustalX include: (i) pair-wise alignment of the sequences, (ii) use the alignment score to produce a phylogenetic tree, and (iii) align the sequences sequentially, guided by the phylogenetic relationships indicated by a guide tree. A guide tree is a rooted dendrogram obtained with neighbor joining method (Young et al., 1999).

A tree showing relatedness of sequences reveals similarity and extent of diversity within a dataset. There exist several methods for constructing trees. Some were developed for interpreting morphological data, others for interpreting molecular data such as binary or multi-state character data and sequence data (an alignment of homologous DNA/ Protein sequence dataset). Some can be used with either kind of data. Some methods currently in use are either parametric or non-parametric. Distance based Neighbor Joining method (NJ), Maximum Parsimony (MP), and Maximum Likelihood (ML) are non-parametric methods while Bayesian methods are parametric ones.

Maximum likelihood Trees

An established approach to tree reconstruction is the maximum likelihood (ML) method (Felsenstein, 1981). This method seeks to identify the most likely tree, given a suitable evolutionary model, which would make it possible to estimate the probability of each possible individual change occurring in the selected dataset at the DNA or amino acid level (Yang, 1997; Posada and Crandall, 1998). The best tree is the one with the highest probability (or maximum likelihood) among all possible trees. Maximum likelihood method is computationally expensive when the number of sequences is large, because the number of possible trees grows exponentially with the number of sequences. In such situations, the NJ method is frequently used because of its computational speed. This method was initially proposed by Saitou and Nei (1987), and later modified by Studier and Kepler, (1988). The principle of NJ method is to find pairs of operational taxonomic units (OTUs=neighbors) that minimize the total branch length at each stage of clustering of OTUs starting with a star-like tree (Mount, 2008). A number of studies have supported NJ's performance in reconstructing correct evolutionary trees (Saitou and Imanishi, 1989; Kuhner and Felsenstein, 1994; Huelsenbeck, 1995). Maximum Parsimony (MP) infers a phylogenetic tree by minimizing the total number of evolutionary steps required to generate the observed variation in the sequences from common ancestral sequences. Parsimony refers to preference of the shortest path over all longer ones. The steps are base or amino-acid substitutions for sequence data. This method is best suited for sequences that are quite similar or conserved and limited to small datasets (Mount, 2008).

There are many software available free of cost for phylogenetic analysis and tree construction. One such very popular phylogenetic analysis software is known as MEGA (Molecular Evolutionary Genetics Analysis Software). It can be most favored for generating NJ, MP, ML based trees with options to test substitution models, and bootstrapping the trees for statistical robustness (Kumar et al., 1993, 1994, 2001; Tamura et al., 2011, 2013). The software also provides tree editing and annotation options. Another popular software has recently surfaced called RaxML or Randomized A(x)ccelerated Maximum Likelihood (Stamatakis et al., 2005, Stamatakis, 2006a,b). It is extremely fast in handling large datasets. The ML approach is computationally exhaustive and therefore there is a limit to how rapidly the results come when dealing with large datasets generated by metagenomic datasets or NGS platforms (Chor and Tuller, 2005). The algorithm of RAxML is so designed that it optimizes a likelihood tree which is initially generated by the parsimony method (which is comparatively faster than ML). The subtrees are subsequently rearranged into neighboring branches to obtain the best possible tree (Stamatakis et al., 2005).

Bayesian Approach for Drawing Trees Depicting Sequence Relatedness

BEAST (Bayesian Evolutionary Analysis by Sampling Trees) is a software package that attempts to provide a general framework for hypothesis testing of evolutionary models from sequence data (Drummond and Rambaut, 2007). It is intended both as a method of reconstructing phylogenies and as a framework for testing evolutionary hypotheses without being restricted to a single tree topology. Bayesian estimation is not necessarily faster than ML methods (Drummond and Rambaut, 2007). The core algorithm implemented in BEAST is Metropolis-Hastings Markov-chain Monte Carlo (MH-MCMC) (Metropolis et al., 1953; Hastings, 1970). MCMC is a non-deterministic or random algorithm that produces sample-based estimates of a desired distribution. BEAST uses MCMC methods to average over tree space, so that each tree is weighted proportional to its posterior probability. In Bayesian statistics, the posterior probability is the conditional probability of an event assumed to be randomly occurring. Posterior probability is a distribution of any random event treated as a variable, subject to outcomes from an experiment or trial run (Lee, 2004). The MCMC approach is especially promising, as it allows the analysis of large data sets (Pybus et al., 2003; Shapiro et al., 2004). In microbial community analyses, Bayesian approaches have made the sequence relatedness as well as phylogenetic analyses more robust with greater degrees of confidence with which the interpretation could be made more reliably.

REFERENCES

1. Altschul, S.F., Gish, W., Miller, W., Myers, E.W., Lipman, D.J., 1990. Basic local alignment search tool. Journal of Molecular Biology 215, 403-410.
2. Altschul, S.F., Madden T.L., Schaffer, A.A., Zhang, J., Zhang, Z., Miller, W., Lipman, D.J., 1997. Gapped BLAST and PSI-BLAST: a new generation of protein database search programs. Nucleic Acids

Research 25, 3389-3402.

3. Amann, R., 2000. Who is out there? Microbial aspects of biodiversity. Systematic and applied microbiology 23, 1-8.
4. Anderson, I.C., Cairney, J.W.G., 2004. Diversity and ecology of soil fungal communities: increased understanding through the application of molecular techniques. Environmental Microbiology 6, 769-779.
5. Anderson, I.C., Campbell, C.D., Prosser, J.I., 2003. Potential bias of fungal 18S rDNA and internal transcribed spacer polymerase chain reaction primers for estimating fungal biodiversity in soil. Environmental Microbiology 5, 36-47.
6. Ashelford, K.E., Chuzhanova, N.A., Fry, J.C., Jones, A.J., Weightman, A.J., 2005. At least one in twenty 16S rRNA sequence records currently held in public repositories estimated to contain substantial anomalies. Applied and Environmental Microbiology 71, 7724-7736.
7. Ashelford, K.E., Chuzhanova, N.A., Fry, J.C., Jones, A.J., Weightman, A.J., 2006. New Screening software shows that most recent large 16S rRNA gene clone libraries contain chimeras. Applied and Environmental Microbiology 72, 5734-5741.
8. Bacchetti De Gregoris, T., Aldred, N., Clare, A.S., Burgess, J.G., 2011. Improvement of phylum- and class-specific primers for real-time PCR quantification of bacterial taxa. Journal of Microbiological Methods 86, 351-356.
9. Becker, J.M., Parkin, T., Nakatsu, C.H., Wilbur, J.D., Konopka, A., 2006. Bacterial activity, community structure and centimeter-scale spatial heterogeneity in contaminated soil. Microbial Ecology 51, 220-231.
10. Borneman, J., Hartin, R.J., 2000. PCR primers that amplify fungal rRNA genes from environmental samples. Applied and Enviromental Microbiology 66, 4356-4360.
11. Borneman, J., Hartin, R.J., 2000. PCR primers that amplify fungal rRNA genes from environmental samples. Applied and Enviromental Microbiology 66, 4356-4360.
12. Borneman, J., Triplett, E., 1997. Molecular microbial diversity in soils from eastern Amazonia: evidence for unusual microorganisms

and microbial shifts associated with deforestation. Applied and Environmental Microbiology 63, 2647-2653.

13. Brodie, E.L., De Santis, T.Z., Moberg, J.P., Zubietta, I.X., Piceno, Y.M., Andersen, G.L., 2007. Urban aerosols harbor diverse and dynamic bacterial populations. Proceedings of the National Academy of Sciences, USA 104, 299-304.
14. Burke, W.D., Eickbush, D.G., Xiong, Y., Jakubczak, J., Eickbush, T.H., 1993. Sequence relationship of retrotransposable elements R1 and R2 within and between divergent insect species. Molecular Biology and Evolution 10, 163-185.
15. Caporaso, J.G., Kuczynski, J., Stombaugh, J., Bittinger, K., Bushman, F.D., Costello, E.K., Fierer, N., Pẽa, A.G., Goodrich, J.K., Gordon, J.I., Huttley, G.A., Kelley, S.T., Knights, D., Koenig, J.E., Ley, R.E., Lozupone, C.A., McDonald, D., Muegge, B.D., Pirrung, M., Reeder, J., Sevinsky, J.R., Turnbaugh, P.J., Walters, W.A., Widmann, J., Yatsunenko, T., Zaneveld, J., Knight, R., 2010. QIIME allows analysis of high-throughput community sequencing data. Nature Methods 7, 335-336.
16. Cavicchioli, R., Ostrowski, M., Fegatella, F., Goodchild, A., Guixa-Boixereu, N., 2003. Life under nutrient limitation in oligotrophic marine environments: an eco/physiological perspective of Sphingopyxis alaskensis (formerly Sphingomonas alaskensis). Microbial Ecology 45, 203-217.
17. Cho, J.-C., Tiedje, J.M., 2001. Bacterial species determination from DNA-DNA hybridization by using genome fragments and DNA microarrays. Applied and Environmental Microbiology 67, 3677-3682.
18. Chor, B., Tuller, T., 2005. Maximum Likelihood of evolutionary trees: hardness and approximation. Bioinformatics 21, 97-106.
19. Cole, J.R., Wang, Q., Fish, J.A., Chai, B., McGarrell, D.M., Sun, Y., Brown, C.T., Porras-Alfaro, A., Kuske, C.R., Tiedje, J.M., 2014. Ribosomal Database Project: Data and tools for high throughput rRNA analysis. Nucleic Acids Research 42, D633-D642.
20. De León J.H., Jones, W.A., Morgan, D.J.W., 2004. Molecular distinction between populations of Gonatocerus morrilli, egg parasitoids of the glassy-winged sharpshooter from Texas and

California: Do cryptic species exist? Journal of Insect Science 4, 1-7.

21. De Santis, T.Z., Brodie, E.L., Moberg, J.P., Zubieta, I.X., Piceno, Y.M., Andersen, G.L., 2007. High-density universal 16S rRNA microarray analysis reveals broader diversity than typical clone library when sampling the environment. Microbial Ecology 53, 371-383.
22. De Santis, T.Z., Hugenholtz, P., Keller, K., Brodie, E.L., Larsen, N., Piceno, Y.M., Phan, R., Andersen, G.L., 2006a. NAST: a multiple sequence alignment server for comparative analysis of 16S rRNA genes. Nucleic Acids Research 34, W394-W399.
23. De Santis, T.Z., Hugenholtz, P., Larsen, N., Rojas, M., Brodie, E.L., Keller, K., Huber, T., Dalevi, D., Hu, P., Andersen, G.L., 2006b. Greengenes, a Chimera-Checked 16S rRNA Gene Database and Workbench Compatible with ARB. Applied and Environmental Microbiology 72, 5069-5072.
24. Dlakic, M., 2005. The ribosomal subunit assembly line. Genome Biology 6, 234.
25. Drummond, A.J., Rambaut, A., 2007. BEAST: Bayesian evolutionary analysis by sampling trees. BMC Evolutionary Biology 7, 214-221.
26. Edgar, R.C., 2010. Search and clustering orders of magnitude faster than BLAST. Bioinformatics 26, 2460-2461.
27. Eichorst, S.A., Breznak, J.A., Schmidt, T.M., 2007. Isolation and characterization of soil bacteria that define Terriglobus gen. nov., in the phylum Acidobacteria. Applied and Environmental Microbiology 73, 2708-2717.
28. Felsenstein, J., 1981. Evolutionary trees from DNA sequences: a maximum likelihood approach. Journal of Molecular Evolution 17, 368-376.
29. Felsenstein, J., 1993. PHYLIP (Phylogeny Inference Package) version 3.5c. Seattle: Department of Genetics, University of Washington.
30. Fierer , N., Jackson, R., 2006 . The diversity and biogeography of soil bacterial communities. Proceedings of the National Academy of Sciences, USA 103, 626-631.

31. Fierer, N., Jackson, J.A., Vilgalys, R., Jackson, R.B., 2005. Assessment of soil microbial community structure by use of taxon-specific quantitative PCR assays. Applied and Environmental Microbiology 71, 4117-4120.

32. Fisher, M.M., Triplett, E.W., 1999. Automated approach for ribosomal intergenic spacer analysis of microbial diversity and its application to freshwater bacterial communities. Applied and Environmental Microbiology 65, 4630-4636.

33. Galperin, M.Y., 2004. Metagenomics: From acid mine to shining sea. Environmental Microbiology 6, 543-545.

34. Garbeva, P., van Veen, J.A., van Elsas, J.D., 2004. Microbial diversity in soil: selection of microbial populations by plant and soil type and implications for disease suppressiveness. Annual Reviews in Phytopathology 42, 243-270.

35. Gardes, M., Bruns, T.D., 1993. ITS primers with enhanced specificity for basidiomycetes-application to the identification of mycorrhizae and rusts. Molecular Ecology 2, 113-118.

36. Gobet, A., Boetius, A., Ramette, A., 2014. "Ecological coherence of diversity patterns derived from classical fingerprinting and Next Generation Sequencing techniques." Environmental Microbiology 16, 2672-2681.

37. Greene, E.A., Voordouw, G., 2003. Analysis of environmental microbial communities by reverse sample genome probing. Journal of Microbiological Methods 53, 211-219.

38. Handelsman, J., 2004. Metagenomics: application of genomics to uncultured microorganisms. Microbiology and Molecular Biology Reviews 68, 669–685.

39. Hastings, W.K., 1970. Monte Carlo sampling methods using markov chains and their applications. Biometrika 57, 97-109.

40. He, Z., Gentry, T.J., Schadt, C.W., Wu, L., Liebich, J., Chong, S.C., Huang, Z., Wu, W., Gu, B., Jardine, P., Criddle, C., Zhou, J., 2007. GeoChip: a comprehensive microarray for investigating biogeochemical, ecological and environmental processes, ISME Journal 1, 67-77.

41. Head, I.M., Saunders, J.R., Pickup, R.W., 1998. Microbial evolution, diversity, and ecology: a decade of ribosomal RNA analysis of

uncultivated microorganisms. Microbial Ecology 35, 1-21.

42. Heid, C.A., Stevens, J., Livak, K.J., Williams, P.M., 1996. Real-time quantitative PCR. Genome Research 6, 986-994.
43. Heuer, H., Smalla, K., 1997. Application of denaturing gradient gel electrophoreses (DGGE) and temperature gradient gel electrophoreses (TGGE) for studying soil microbial communities. In: van Elsas, J.D., Wellington, E.M.H., Trevors, J.T., (Eds.), Modern Soil Microbiology. Marcel Dekker, New York, pp. 353-373.
44. Higgins, D.G., Sharp, P.M., 1988. CLUSTAL: A package for performing multiple sequence alignment on a microcomputer. Gene 73, 237-244.
45. Higuchi, R., Fockler, C., Dollinger, G., Watson, R., 1993. Kinetic PCR analysis: Real-time monitoring of DNA amplification reactions. Biotechnology 11, 1026-1030.
46. Horner-Devine, M.C., Carney, K.M., Bohannan, B.J.M., 2003. An ecological perspective on bacterial biodiversity. Proceedings of Royal Society of London 271, 113-122.
47. Hubert, C., Shen, Y., Voordouw, G., 1999. Composition of toluene-degrading microbial communities from soil at different concentrations of toluene. Applied and Environmental Microbiology 65, 3064-3070.
48. Huelsenbeck, J.P., 1995. The robustness of two phylogenetic methods: four-taxon simulations reveal a slight superiority of maximum likelihood over neighbor joining. Molecular Biology and Evolution 12, 843-849.
49. Huelsenbeck, J.P., Ronquist, F., 2001. MRBAYES: Bayesian inference of phylogenetic trees. Bioinformatics 17, 754-755.
50. Hugenholtz, P., Goebel, B.M., Pace, N.R., 1998. Impact of culture-independent studies on the emerging phylogenetic view of bacterial diversity. Journal of Bacteriology 180, 4765-4774.
51. Jeanmougin, F., Thompson, J.D., Gouy, M., Higgins, D.G., Gibson, T.J., 1998. Multiple sequence alignment with Clustal X. Trends in Biochemical Sciences 23, 403-405.
52. Kaczanowska, M., Ryden-Aulin, M., 2007. Ribosome biogenesis and the translation process in Escherichia coli. Microbiology and

Molecular Biology Reviews 71, 477-494.

53. Kang, S., Mills, A.L., 2006. The effect of sample size in studies of soil microbial community structure. Journal of Microbiological Methods 66, 242-250.
54. Khot, P.D., Ko, D.L., Fredricks, D.N., 2009. Sequencing and analysis of fungal rRNA operons for development of broad-range fungal PCR assays. Applied and Environmental Microbiology 75, 1559-1565.
55. Kim, M., Lee, K.-H., Yoon, S.-W., Kim, B, -S., Chun, J., Yi, H., 2013. Analytical tools and databases for metagenomics in the next-generation sequencing era. Genomics and Informatics 11, 102-113.
56. Kim, M., Morrison, M., Yu, Z., 2011. Evaluation of different partial 16S rRNA gene sequence regions for phylogenetic analysis of microbiomes. Journal of Microbiological Methods 84, 81-87.
57. Kirk, J.L., Beaudette, L.A., Hart, M., Moutoglis, P., Klironomos, J.N., Lee, H., Trevors, J.T., 2004. Methods of studying soil microbial diversity. Journal of Microbiological Methods 58, 169-188.
58. Klappenbach, J.A., Dunbar, J.M., Schmidt, T.M., 2000. rRNA operon copy number reflects ecological strategies of bacteria. Applied and Environmental Microbiology 66, 1328-1333.
59. Klappenbach, J.A., Saxman, P.R., Cole, J.R., Schmidt, T.M., 2001. rrnDB: the Ribosomal RNA Operon Copy Number Database. Nucleic Acids Research 29, 181-184.
60. Kuhner, M.K., Felsenstein, J., 1994. Simulation comparison of phylogeny algorithms under equal and unequal evolutionary rates. Molecular Biology and Evolution, 11, 459-468.
61. Kumar, S., Tamura, K., Jakobsen, I.B., Nei, M., 2001. MEGA2: Molecular Evolutionary Genetics Analysis Software. Bioinformatics 17, 1244-1245.
62. Kumar, S., Tamura, K., Nei, M., 1993. Manual for MEGA: Molecular Evolutionary Genetics Analysis Software, Pennsylvania State University, USA.
63. Kumar, S., Tamura, K., Nei, M., 1994. 'MEGA: Molecular Evolutionary Genetics Analysis software for microcomputers'. Computational Applied Biosciences 10, 189-191.

64. Kuske, C.R., Ticknor, L.O., Miller, M.E., Dunbar, J.M., Davis, J.A., Barns, S.M., Belnap, J., 2002. Comparison of soil bacterial communities in rhizospheres of three plant species and the interspaces in an arid grassland. Applied and Environmental Microbiology 68, 1854-1863.
65. Lane, D.J., Pace, B., Olsen, G.J., Stahl, D.A., Sogin, M.L., Pace, N.R., 1985. Rapid determination of 16S ribosomal RNA sequences for phylogenetic analyses. Proceedings of Natural Academy of Sciences, USA 82, 6955-6959.
66. Lee, D.-H., Kahng, H.-Y., Lee, Y.S., Jung, J.S., Kim, J.M., Chung, B.S., Park, S.K., Jeon, C.O., 2009. Jejuia pallidilutea gen. nov., sp. nov., a new member of the family Flavobacteriaceae isolated from seawater. International Journal of Systematic and Evolutionary Microbiology 59, 2148-2152.
67. Lee, D-H., Zo, Y.-G, Kim, S.-J., 1996a. Nonradioactive method to study genetic profiles of natural bacterial communities by PCR-single-strand conformation polymorphism. Applied and Environmental Microbiology 62, 3112-3120.
68. Lee, P.M., 2004. Bayesian Statistics, an introduction (3rd edition). Wiley. ISBN 978-0-340-81405-5.
69. Liu, W.T., Marsh, T.L., Cheng, H., Forney, L.J., 1997. Characterization of microbial diversity by determining terminal restriction fragments length polymorphisms of genes encoding 16S rRNA. Applied and Environmental Microbiology 63, 4516-4522.
70. Machida, R.J., Knowlton, N., 2012. PCR primers for metazoan nuclear 18S and 28S ribosomal DNA sequences. PloS One 7, e46180.
71. Manichanh, C., Chapple, C.E., Frangeul, L., Gloux, K., Guigo, R., Dore, J., 2008. A comparison of random sequence reads versus 16S rDNA sequences for estimating the biodiversity of a metagenomic library. Nucleic Acids Research 36, 5180-5188.
72. Massol-Deya, A.A., Odelson, D.A., Hickey, R.F., Tiedje, J.M., 1995. Bacterial community fingerprinting of amplified 16S and 16S-23S ribosomal DNA gene sequences and restriction endonuclease analysis (ARDRA). In: Akkermans, A.D.L., van-Elsas, J.D., de-Bruijn, F.J. (Eds.), Molecular Microbial Ecology Manual. Kluwer

Academic Publishing, Boston, pp. 3.3.2 1-3.3.2 8.

73. Metropolis, N., Rosenbluth, A.W., Rosenbluth, M.N., Teller, A.H., 1953. Equation of state calculations by fast computing machines. The Journal of Chemcal Physics 21, 1087-1093.

74. Metzker, M.L., 2010. Sequencing technologies-the next generation. Nature Reviews Genetics 10, 31-46.

75. Mount, D.W., 2002. Bioinformatics-sequence and genome analysis. Second Edition. Coldspring Harbour Laboratory Press, New York.

76. Mount, D.W., 2008. Maximum parsimony method for phylogenetic prediction. Cold Spring Harbor Protocols.

77. Muyzer, G., Smalla, K., 1998. Application of denaturing gradient gel electrophoresis (DGGE) and temperature gradient gel electrophoresis (TGGE) in microbial ecology. Antonie van Leeuwenhoek 73, 127-141.

78. Navarrete, A.A., Kuramae, E.E., Hollander, M., Pijl, A.S., van Veen, J.A., Tsai, S.M., 2013. Acidobacterial community responses to agricultural management of soybean in Amazon forest Soils. FEMS Microbiology Ecology 83, 607-621.

79. Neefs, J.-M., De Peer, Y.V., Rijk, P.D., Goris, A., Wachter, R.D., 1991. Compilation of small ribosomal subunit RNA sequences. Nucleic Acids Research 19, 1987-2015.

80. Neefs, J.-M., Van de Peer, Y., De Rijk, P., Chapelle, S., De Wachter, R., 1993. Compilation of small ribosomal subunit RNA structures. Nucleic Acids Research 21, 3025-3049.

81. Neefs, J.-M., Van de Peer, Y., Hendriks, L., De Wachter, R., 1990. Compilation of small ribosomal subunit RNA sequences. Nucleic Acids Research 18, 2237-2317.

82. Nierhaus, K.H., 1991. The assembly of prokaryotic ribosomes. Biochimie 73, 739-755.

83. Nocker, A., Burr, M., Camper, A.K., 2007. Genotypic microbial community profiling: a critical technical review. Microbial Ecology 54, 276-289.

84. Okubo, A., Sugiyama, S., 2009. Comparison of molecular fingerprinting methods for analysis of soil microbial community structure. Ecological Research 24, 1399-1405.

85. Okubo, A., Sugiyama, S., 2009. Comparison of molecular fingerprinting methods for analysis of soil microbial community structure. Ecological Research 24, 1399-1405.

86. Ovreas, L., Torsvik, V., 1998. Microbial diversity and community structure in two different agricultural soil communities. Microbial Ecology 36, 303-315.

87. Pant, P., 2015. Rhizospheric Microbial Communities of a limestone-dolomite mined-out areas undergoing eco-restoration and their ecological significance. PhD Thesis. University of Delhi, Department of Botany, 2015.

88. Pevsner, J., 2009. Chapter 15: Completed Genomes: Bacteria and Archaea. In: Bioinformatics and Functional genomics. 2nd Edition. Willey-Blackwell publication, pp. 611-612.

89. Podar, M., Abulencia, C.B., Walcher, M., Hutchison, D., Zengler, K., Garcia, J.A., Holland, T., Cotton, D., Hauser, L., Keller, M., 2007. Targeted access to the genomes of low-abundance organisms in complex microbial communities. Applied and Environmental Microbiology 73, 3205-3214.

90. Posada, D., Crandall, K., 1998. MODELTEST: testing the model of DNA substitution. Bioinformatics 14, 817-818.

91. Prosser, J.I., 2002. Molecular and functional diversity in soil microorganisms. Plant and Soil 244, 9-17.

92. Pybus, O.G., Drummond, A.J., Nakano, T., Robertson, B.H., Rambaut, A., 2003. The epidemiology and iatrogenic transmission of Hepatitis C Virus in Egypt: a Bayesian coalescent approach. Molecular Biology and Evolution 20, 381-387.

93. Rainey, F.A., Ward-Rainey, N.L., Janssen, P.H., Hippe, H., 1996. Clostridium paradoxum DSM 7308(T) contains multiple 16S rRNA genes with heterogeneous intervening sequences. Microbiology 142, 2087-2095.

94. Ranjard, L., Brothier, E., Nazaret, S., 2000. Sequencing bands of ribosomal intergenic spacer analysis fingerprints for characterization and microscale distribution of soil bacterium populations responding to mercury spiking. Applied and Environmental Microbiology 66, 5334-5339.

95. Riesenfeld, C.S., Schloss, P.D., Handelsman, J., 2004. Metagenomics:

Genomic analysis of microbial communities. Annual Reviews in Genetics 38, 525-552.

96. Rincon-Florez, V.A, Carvalhais, L.C., Schenk, P.M., 2013. Culture-independent molecular tools for soil and rhizosphere microbiology. Diversity 5, 581-612.

97. Rondon, M.R., August, P.R., Bettermann, A.D., Brady, S.F., Grossman, T.H., Liles, M.R., Loiacono, K.A., Lynch, B.A., MacNeil, I.A., Minor, C., Tiong, C.L., Gilman, M., Osburne, M.S., Clardy, J., Handelsman, J., Goodman, R.M., 2000. Cloning the soil metagenome: A strategy for accessing the genetic and functional diversity of uncultured microorganisms. Applied and Environmental Microbiology 66, 2541-2547.

98. Roselló-Mora, R., Amann, R., 2001. The species concept for prokaryotes. FEMS Microbiology Review 25, 39-67.

99. Saitou, N., Imanishi, T., 1989. Relative efficiencies of the Fitch-Margoliash, maximum-parsimony, maximum-likelihood, minimum-evolution, and neighbor-joining methods of phylogenetic tree construction in obtaining the corrcct tree. Molecular Biology and Evolution 6, 514-525.

100. Saitou, N., Nei, M., 1987. The Neighbor-joining Method: A New Method for Reconstructing Phylogenetic Trees. Molecular Biology and Evolution 4, 406-425.

101. Schloss, P.D., Handelsman, J., 2004. Status of the microbial census. Microbiological and Molecular Biological Reviews 68, 686-691.

102. Schloss, P.D., Handelsman, J., 2005a. Introducing DOTUR, a computer programme for defining operational taxonomic units and estimating species richness. Applied and Environmental Microbiology 71, 1501-1506.

103. Schloss, P.D., Handelsman, J., 2005b. Metagenomics for studying unculturable microorganisms: Cutting the Gordian knot. Genome Biology 6, 229.

104. Schmalenberger, A., Schwieger, F., Tebbe, C.C., 2001. Effects of primers hybridizing to different evolutionarily conserved regions of the small subunit rRNA gene in PCR-based microbial community analyses and genetic profiling. Applied and Environmental Microbiology 67, 3557-3563.

105. Schwieger, F., Tebbe, C.C., 1998. A new approach to utilize PCR-single-strand conformation polymorphism for 16S rRNA gene-based microbial community analysis. Applied and Environmental Microbiology 64, 4870-4876.

106. Shapiro, B., Drummond, A.J., Rambaut, A., Wilson, M.C., Matheus, P.E., Sher, A.V., Pybus, O.G., Gilbert, M.T.P., Barnes, I., Binladen, J., Willerslev, E., Hansen, A.J., Baryshnikov, G.F., Burns, J.A., Davydov, S., Driver, J.C., Froese, D.G., Harington, C.R., Keddie, G., Kosintsev, P., Kunz, M.L., Martin, L.D., Stephenson, R.O., Storer, J., Tedford, R., Zimov, S., Cooper, A., 2004. Rise and fall of the Beringian steppe bison. Science 306, 1561-1565.

107. Shazani, Z., Sykes, M.T., Willaimson, J.R., 2011. Assembly of bacterial ribosomes. Annual Review in Biochemistry 80, 501-526.

108. Simon, U.K., Weisß, M., 2008. Intragenomic variation of fungal ribosomal genes is higher than previously thought. Molecular Biology and Evolution 25, 2251-2254.

109. Smit, E., Leeflang, P., Glandorf, B., van Elsas, J.D., Wernars, K., 1999. Analysis of fungal diversity in the wheat rhizosphere by sequencing of cloned PCR-amplified genes encoding 18S rRNA and temperature gradient gel electrophoresis. Applied and Environmental Microbiology 65, 2614-2621.

110. Stackebrandt, E., Goebel, B.M., 1994. Taxonomic note: a place for DNA-DNA reassociation and 16S rRNA sequence analysis in the present species definition in bacteriology. International Journal of Systematic Bacteriology 44, 846-849.

111. Stamatakis, A., 2006a. Phylogenetic models of rate heterogeneity: a high performance computing perspective. In: Proceedings of the IPDPS2006, Rhodos, Greece.

112. Stamatakis, A., 2006b. RAxML-VI-HPC: maximum likelihood-based phylogenetic analyses with thousands of taxa and mixed models. Bioinformatics 22, 2688-2690.

113. Stamatakis, A., Ludwig, T., Meier, H., 2005. RAxml-III: a fast program for maximum likelihood-based inference of large phylogenetic trees. Bioinformatics 21, 456-463.

114. Stevenson, B.S., Schmidt, T.M., 1998. Growth rate-dependent accumulation of RNA from plasmid-borne rRNA operons in

Escherichia coli. Journal of Bacteriology 180, 1970-1972.

115. Studier, J.A., Kepler, K.J., 1988. A note on the neighbour-joining method of Saitou and Nei. Molecular Biology and Evolution 5, 729-731.

116. Tamura, K., Dudley, J., Nei, M., Kumar, S., 2007. MEGA4: Molecular Evolutionary Genetics Analysis (MEGA) software version 4.0. Molecular Biology and Evolution 24, 1596-1599.

117. Tamura, K., Nei, M., 1993. Estimation of the number of nucleotide substitutions in the control region of mitochondrial DNA in humans and chimpanzees. Molecular Biology and Evolution 10, 512-526.

118. Theron, J., Cloete, T.E., 2000. Molecular techniques for determining microbial diversity and community structure in natural environments. Critical reviews on Microbiology 26, 37-57.

119. Thompson, J.D., Gibson, T.J., Plewniak, F., Jeanmougin, F., Higgins, D.G., 1997. The CLUSTAL X window interface: flexible strategies for multiple sequence alignment aided by quality analysis tools. Nucleic Acids Research 25, 4876-4882.

120. Thompson, J.D., Higgins, D.G., Gibson, T.J., 1994. CLUSTALW: improving the sensitivity of progressive multiple sequence alignment through sequence weighting, positions-specific gap penalties and weight matrix choice. Nucleic Acid Research. 22, 4673-4680.

121. Tiedje, J.M., Asuming-Brempong, S., Nusslein, K., Marsh, T.L., Flynn, S.J., 1999. Opening the black box of soil microbial diversity. Applied Soil Ecology 13, 109-122.

122. Torsvik, V., Daae, F.L., Sandaa, R.-A., Ovreas, L., 1998. Review article: novel techniques for analysing microbial diversity in natural and perturbed environments. Journal of Biotechnology 64, 53-62.

123. Torsvik, V., Goksoyr, J., Daae, F.D., 1990a. High diversity in DNA of soil bacteria. Applied and Environmental Microbiology 56, 782-787.

124. Torsvik, V., Ovreas, L., 2002. Microbial diversity and function in soil: from genes to ecosystems. Current Opinion in Microbiology 5, 240-245.

125. Torsvik, V., Ovreas, L., 2002. Microbial diversity and function in soil: from genes to ecosystems. Current Opinion in Microbiology

5, 240-245.

126. Torsvik, V., Salte, K., Soerheim, R., Goksoeyr, J., 1990b. Comparison of phenotypic diversity and DNA heterogeneity in a population of soil bacteria. Applied and Environmental Microbiology 56, 776-781.

127. Torsvik, V., Sorheim, R., Goksoyr, J., 1996. Total bacterial diversity in soil and sediment communities-a review. Journal of Indian Microbiology 17, 170-178.

128. van Elsas, J.D., Boersma, F.G.H., 2011. A review of molecular methods to study the microbiota of soil and the mycosphere. European Journal of Soil Biology 47, 77-87.

129. von Wintzingerode, F., Gobel, U.B., Stackebrandt, E., 1997. Determination of microbial diversity in environmental samples: pitfalls of PCR-based rRNA analysis. FEMS Microbiology Reviews 21, 213-229.

130. Weisburg, W.G., Barns, S. M., Pelletier D.A., Lane, D.J., 1991. 16S ribosomal DNA amplification for phylogenetic study. Journal of Bacteriology 173, 697-703.

131. Werren, J. H., Guo, L. R., Zhang, W., 1995a. Evolution and phylogeny of Wolbachia: reproductive parasites of Arthropods. Procceding of Royal Society of London, Biological Sciences 251, 55-63.

132. Werren, J.H., Windsor, D., Guo, L., 1995b. Distribution of Wolbachia among neotropical arthropods. Procceding of Royal Society of London, Biological Sciences 262, 197-204.

133. Yang, Z., 1997. PAML: a program package for phylogenetic analysis by maximum likelihood. Computational and Applied Biosciences 13, 555-556.

134. Young, G.B., Jack, D.L., Smith, D.W., Saier, M.H. Jr., 1999. The amino acid/auxin:proton symport permease family. Biochimica et Biophysica Acta- Membranes 1415, 306-322

135. Youssef, N., Sheik, C.S., Krumholz, L.R., Najar, F.Z., Roe, B.A., Elshahed, M.S., 2009. Comparison of species richness estimates obtained using nearly complete fragments and simulated pyrosequencing-generated fragments in 16S rRNA gene-based environmental surveys. Applied and Environmental Microbiology 75, 5227-5236.

136. Yu, Z., Mohn, W.W., 2001. Bioaugmentation with resin-acid-degrading bacteria enhances resin acid removal in sequencing batch reactors treating pulp mill effluents. Water Research 35, 883-890.
137. Yu, Z., Morrison, M., 2004. Comparisons of different hypervariable regions of rrs genes for use in fingerprinting of microbial communities by PCR-denaturing gradient gel electrophoresis. Applied and Environmental Microbiology 70, 4800-4806.
138. Zhang, Z., Schwartz, S., Wagner, L., Miller, W., 2000. A greedy algorithm for aligning DNA sequences. Journal of Computational Biology 7, 203-214.
139. Zwolinski, M.D., 2007. DNA Sequencing: Strategies for Soil Microbiology. Soil Science Society of America 71, 592-600.

CHAPTER 4

THE WORLD OF NEXT GENERATION SEQUENCING AND ITS ROLE IN MICROBIAL ECOLOGY

Jennifer Wood[1], Prashant Pant[2]

[4]Jennifer Wood is post-doctoral fellow in Dr. Ashley Franks' Applied and Environmental Microbiology Laboratory at La Trobe University, Australia. She has recently completed her Doctoral research in Microbial Ecology at the same university. Her broad research interest is in understanding the ecology of soil-microbial communities and their interactions with macro-communities. She is interested in developing trait-based approaches to interpret and predict how environmental and anthropogenic changes impact microbial community ecology and function.

[2]Dr. Prashant Pant, is an Assistant Professor of Botany in Dyal Singh College, University of Delhi, New Delhi, India. He has been teaching for more than 5 years now and his main areas have been Bioinformatics, Molecular Biology and Genomics at the college. His PhD was on understanding the Microbial Succession using a Metagenomic approach in a Barren mined-out site undergoing ecological restoration. For this, he deployed many molecular and computational Biology tool and techniques. He has several publications (articles and book

chapters) in International Journals and International Publications. Currently, he is also Chief Editor of Bioinformatics Review and a co-founder/Executive Chief Editor of a Science Journal in Hindi Language "Swadeshi Vigyan" for the native language speaking communities. Both the Journals are a service to the scientific community as they are fully open access.

With the dawn of the era of Next Generation Sequencing (NGS) or High Throughput Sequencing upon us, there is a whole new way at which we look at the DNA and RNA sequence data that has quintessentially the "Catch-all" approach for these molecules of life. NGs has in a very short period convinced biologist and chemist to be crucial for future research by showing potential for a wide variety of applications in Biology and related disciplines viz. computational biology and bioinformatics based analyses and interpretation. Microbial ecology is one of the most influenced subjects by NGS. Targeted sequencing of rRNA genes (or RNA itself) can be used to profile entire communities (discussed in Chapter 3); shot-gun sequencing, which involves the fragmentation and then sequencing of genetic material, can be used to reconstruct bacterial and lower-eukaryote genomes, or even community metagenomes. By sequencing the mRNA transcriptome, as oppose to genomic DNA, NGS can yield insights into which metabolic processes are active under different environmental conditions. This chapter will highlight diverse ways in which NGS can be used to advance microbial ecology through the discussion of notable discoveries and major collaborative projects that were facilitated by NGS. We will then discuss the major goal of understanding the breadth of microbial diversity and the importance and challenges associated with the discovery of novel genomes.

HOW HAS NGS CHANGED MICROBIAL ECOLOGY?

The advent of next-generation sequencing (NGS) has yielded insights into microbial communities that other PCR-based technologies, such as DGGE, ARISA and TRFLP, have not. Where previously community ecologists largely dealt with taxonomically ambiguous operational taxonomic units (OTUs), NGS can be used to ascribe names to community members and possible functions they may have. Where fragment based PCR-technologies were limited to visualizing the most abundant community members, NGS enables inspection of the rarest community members (Gobet et al., 2014, Van Dorst et al. 2014).

A NEW UNDERSTANDING OF DIVERSITY

NGS has dispelled the myth that microorganisms isolated in pure culture from an environment represent the numerically dominant or functionally important species within a community. NGS has revealed that these 'culturable fractions' are, in fact, numerically rare in the communities from which they were obtained (Hugenholtz, 2002). Culture independent studies using rRNA profiling have revealed that the marine bacteria clade SAR11 represent as much as 50% of the microbial community in photic layers of the ocean waters and 25% of the sub-euphotic microbial community (Morris et al., 2002). However, despite its ubiquity, the SAR11 clade remain difficult to cultivate (Rappé et al., 2002).

NGS has had a major impact upon our understanding of microbial diversity across a range of biomes including terrestrial, marine and the human gut. 16S rRNA profiling has revealed the presence of novel bacterial and archaeal communities growing on hydrothermal vents at depths of 1520 m in the northeast Pacific Ocean (Huber et al., 2007). These deep-ocean vent communities were dominated ε-proteobacteria and are now thought to be one of the largest reservoirs for ε-proteobacteria diversity on Earth (Campbell et al., 2006). Although ε-proteobacteria are thought to have important roles in the cycling of carbon, nitrogen and sulphur in these systems, until the advent of culture independent techniques and NGS, our understanding of the ε-proteobacteria has been limited by the lack of cultured isolates (Campbell et al., 2006).

16S rRNA profiling has also been used to reveal that a diverse community of microorganisms reside in the seemingly uninhabitable upper atmosphere. The vast majority of the Operational Taxonomic Units (OTUs) were classified as *Alpha-* and *Betaproteobacteria* and the average concentration bacterial cells in the upper atmosphere estimated to be 5.1×10^3 cells m^{-3} with the fraction of viable cells per sample ranging from 60 to almost 100% (DeLeon-Rodriguez et al., 2013).

rRNA profiling of unique and rare communities, such as hydrothermal vents has highlighted just how limited our knowledge of microbial diversity is. Direct interrogation of microbial genomes based on comparisons of orthologous gene sequences have shown that the number of different kinds or "species" of microbes is at least 100 times greater than estimates based on cultivation-dependent surveys (Pace 1997, Sogin et al., 2006).

UNDERSTANDING THE ACTIVITY OF THE RARE BIOSPHERE

As our understanding of the breadth of microbial diversity increases, the importance of what has been dubbed the 'rare biosphere' is emerging. Community profiling has revealed that most microbial communities that exist in nature, contain a small number of abundant taxa and a seemingly inexhaustible supply of rare taxa (Campbell et al., 2011). This begs the question: what is the role of these rare taxa? Are they metabolically active or dormant within the community? The activity of a microorganism can be estimated by using 16S rRNA profiling to compare the amount of 16S copies derived from community genomic DNA, compared to the number of copies derived from RNA extract. The ratio of 16S rRNA:rDNA for each microorganism is an index of the growth rate for specific taxa in natural communities under examination (Kemp et al., 1993). Studies examining the activity of abundant and rare marine taxa over 3 years suggests that at least half of the rare biosphere is metabolically active (Campbell et al., 2011).

LINKING MICRO- TO MACRO-COMMUNITY ECOLOGY

The availability of affordable sequencing to a wide range of users who have traditionally lacked access, spurred the initiative to integrate new sequences into a consistent universal taxonomic framework. By standardizing NGS pipelines, the outputs from different studies can be concatenated to from large collaborative NGS projects with the scope to answer global scale questions.

The Human Microbiome Project (HMP) was launched in 2008 as a logical concept of the Human Genome Project (Turnbaugh et al., 2007).

The project aimed to understand the diversity of our microbiome across the body and between individuals (Costello et al., 2009) as well as elucidate the factors that influence the distribution and evolution of our associated microorganisms. In addressing these questions, the HMP has eroded the barrier between microbial ecology and medical microbiology. To many, the human body is now viewed as another ecosystem in which one can study the microbiome and its interactions with the human body as the immediate environment and related questions on ecological aspects of their possible function and influences (Costello et al., 2012).

The HMP has revealed that our microbiome is more than a source of potential pathogens, but is integral to our health and development (Cho and Blaser, 2012). Some have suggested that humans and their microbiome should not be considered separate entities but rather a single meta-organism. The HMP and other NS efforts have uncovered links between our microflora and a diverse range of pathologies including: diabetes, inflammatory bowel disease, obesity, asthma, rheumatoid arthritis, and susceptibility to infections and neurodegenerative disorders (Wen et al., 2008, Turnbaugh et al., 2009, Chow et al., 2011, Scher and Abramson, 2011, Honda and Littman 2012, Huang and Boushey, 2015, Hill-Burns et al., 2017)

Similar to the human microbiome project, the TerraGenome project (Vogel et al., 2009) and the Earth Microbiome project (EMP) (Gilbert et al., 2014) were launched in 2009 and 2010, respectively, with the aim of constructing a global catalogue of the uncultured microbial diversity of this planet. The affordability of generating large-scale datasets has enabled researchers to ask unique questions regarding the biogeography, dynamic dispersal, and ecology of the microbial planet (Fierer and Jackson, 2006, Lauber et al., 2008, Tedersoo et al., 2010, Kivlin et al., 2011, Tedersoo et al., 2014).

To date these NGS surveys have contributed much to our understanding of the importance of microbes within ecosystems and of the microbial communities themselves (Gilbert et al., 2014). Metagenomic approaches have contributed to the discovery of autotrophic ammonia oxidising archaea. Ammonia oxidation is the first step in the nitrification pathway and is an important part of plant nutrition and ecosystem function, which was previously thought to be carried out by bacteria. Shot-gun sequencing of marine water samples identified an ammonium monooxygenase gene on an archaeal-associated scaffold and was one of the first line of evidence to indicate archaeal involvement in nitrogen transformations of global ecological significance (Venter et al., 2004, Prosser and Nicol, 2008).

EMP data from deep soil samples from the Russian permafrost is helping elucidate the bacteria that are responsible for degrading the soil organic matter in these systems (Gittel et al., 2014). Similarly, EMP data has also been used to examine structure of microbial communities from prairie soils across the Midwest of the United States of America. Tallgrass prairie ecosystems which once covered nearly 10% of the United States (>65 million ha) have mostly replaced through agricultural land-use (Fierer et al., 2013). Shifts in the taxonomic composition of remnant prairie soils were found to be largely driven by the changing relative abundance of the phylum Verrucomicrobia and its influence on carbon dynamics (Fierer et al., 2013). In addition to assisting with prairie restoration efforts, the information generated via the shot-gun metagenomics approach has also helped unravel the ecology of the phylum Verrucomicrobia. The ecology of the Verrucomicrobia remains poorly understood because they are difficult to culture and study in the laboratory (Joseph et al., 2003). However, this study supports the hypothesis that they are relatively slow-growing taxa that thrive under conditions of limited nutrient availability. Out of both animal- and plant-centric microbiome surveys clear links between the microbiome and health (be it of plants, animals or entire ecosystems) has emerged. These findings have realized a role for microbial ecology in conservation efforts and research is not emerging the focuses on understanding the ecology of microbial community for the purpose of conserving macro-organisms and communities (Redford et al., 2012, Cheng et al., 2015, Jiménez and Sommer, 2017).

ON THE IMPORTANCE OF REFERENCE GENOMES

A reference genome is a well annotated candidate genome sequence and assembled with greater degree of confidence from the consensus sequence data. NGS analysis of new genomes or related genomes require reference sequences for assembly, annotation and comparison. NGS surveys have revealed that only a small fraction of the phylogenetic diversity of bacteria, archaea and fungi are represented by cultivated organisms (Albertsen et al., 2013). Phyla with few cultured representatives, such as the Verrucomicrobia, and candidate phyla, play important roles in the environment (Knittel and Boetius, 2009, Albertsen et al., 2013). Thus, in addition to examining community level patterns, the use on NGS to assemble new genomes is of critical importance in microbial ecology.

SEQUENCING NEW GENOMES

The sequencing and initial assembly of the genome for a organism is called *de novo* sequencing (Sangal *et al.* 2014). The reconstruction of these contigs into a single genome without making use of a reference genome is known as *de novo* assembly (Ekblom and Wolf, 2014). The assembly of novel genomes involves a multistep workflow that begins with raw sequence data, typically consisting of single or paired reads. Raw reads typically undergo a number of pre-processing steps including adapter trimming, quality filtering and error correction before overlapping reads are assembled into a continuous sequence known as a 'contig' (Baker, 2012). The final genome is created by assembling the contigs into larger scaffolds both of which undergo verification steps.

Genome sequencing projects were long confined to the study of model organisms and required the concerted collaborative effort, time and money. However, NGS is now fast and cost effective enough to be considered part of the toolbox for investigating bacterial and viral genomes, and there are thousands of bacterial genome sequences available for comparison in the public domain (Edwards and Holt, 2013). A critical stage in genome sequencing is the assembly of the short reads into a set of contiguous sequences representing the DNA in the sample. Despite the accessibility and associability of NGS technologies, current high-throughput sequencing technologies generate large numbers of relatively short and error-prone reads, making the *de novo* assembly problem challenging (Al-Okaily, 2016). As with all technology relating to NGS, the advancement of software for the assembly of genomes is rapidly evolving. There are three main approaches for *de novo* genome assembly, the greedy strategy, the string overlap graph, and the de Bruijn graph (Al-Okaily, 2016).

The third approach, based on the de Bruijn graph model is currently the most commonly used by assemblers for the reconstruction of NGS data (Pevzner et al., 2001). Such assemblers include ABySS (Simpson et al., 2009), ALLPATHS-LG (Gnerre et al., 2011), SPAdes (Bankevich et al., 2012), Velvet (Zerbino and Birney, 2008) Ray meta (Boisvert et al., 2012) and A5-miseq (Coil et al., 2015).

FINDING NEW SPECIES FROM THE METAGENOME

The assembly of novel genomes from complex communities presents an additional challenge and it remains difficult to assemble low abundance genomes from populations using current metagenome sequencing and assembly approaches (Albertsen et al., 2013). More typically, only the genomes of the most abundant community members can be reconstructed from a metagenome.

The reconstruction of microbial genomes requires a sufficient number of overlapping reads to cover the genome multiple times (known as the depth of coverage). Where a good reference genome is availablc, dcpth of coverage as low as 20 may be sufficient. Where de novo assembly is required as much as 50 times coverage is necessary. The depth of coverage required will also depend upon the sequencing technology used. In complex communities, such as soils, the amount of sequencing needed to sufficiently cover the genomes of rare community members can be prohibitive (Howe et al., 2014). Recently an approach using differential coverage binning to recover rare community genomes from metagenomes has been developed (Albertsen et al., 2013). The method begins by performing two different extraction methods on the same sample to generate two metagenomes in which, due to extraction biases, each bacterial system will have different abundance in each metagenome. The differential abundance information for each is then used to extract genomes.

PHYLOGENETIC RELATIONSHIPS

Assembled genome sequences have been integral in unravelling questions regarding the evolution and phylogenetic interrelationships among microorganisms. Complete genome sequences have helped unravel microbial evolution considering horizontal gene transfer, such as the transfer of antibiotic resistance genes, between microbial lineages. Although the occurrence of horizontal gene transfer has been known for many years the process impedes the study of the evolution of microbial species as different genes in the same organism can have different evolutionary histories (Fraser et al., 2000).

Sequencing of multiple complete genomes of bacteria and archaea has made it possible to perform systematic, genome-scale comparisons. Such comparisons have revealed evidence that suggests there has been massive gene exchange between thermophilic archaea and bacteria (Aravind et al., 1998). Phylogenetic comparisons revealed that as much as 10% of the genes present in the bacterial lineage *Aquifex* appear to have been horizontally transferred from the archaea.

Phylogenetic analyses of bacterial genomes have also shed light on eukaryotic evolution. Mitochondria are vital to the evolution of all known eukaryotic cells and are thought to be the results of an endosymbiotic event that resulted in the permanent acquisition of bacteria that eventually became modern day mitochondria. The addition of new environmental genome sequences, particularly the difficult to culture SAR11 clade, suggest that mitochondria evolved from a planktonic marine alphaproteobacterial lineage, that participated in multiple inter-specific cell colonization events (Thrash et al., 2011).

The advent of NGS has in many ways, helped overcome the issue of observing microorganisms and their diversity *in situ* and stimulated a new wave of rhizospheric microbial diversity research. However, just as a list of plants and animals in the forest is only a first step in ecology, so too are the NGS-generated lists of microbial species present in rhizospheric communities. Many challenges are faced by those who attempt to interpret microbial communities in ecologically meaningful ways that go beyond describing patterns of changing diversity. As such, the development of broad ecological frameworks for the classification and interpretation of microbial communities has become a priority for the advancement of microbial ecology (Prosser et al., 2007, Widder et al., 2016). Readers are recommended to go through some of the selected articles in this book under open access licenses for a detailed and case specific studies where NGS has played an important role in elucidating microbial forms which remained unnoticed through conventional approaches.

REFERENCES

1. Al-Okaily, A. A. (2016). "HGA: De novo genome assembly method for bacterial genomes using high coverage short sequencing reads." BMC Genomics 17(1).
2. Albertsen, M., Hugenholtz, P., Skarshewski, A., Nielsen, K. L., Tyson, G. W. and Nielsen, P. H. (2013). "Genome sequences of rare, uncultured bacteria obtained by differential coverage binning of multiple metagenomes." Nature Biotechnology 31(6): 533-538.
3. Aravind, L., Tatusov, R. L., Wolf, Y. I., Walker, D. R. and Koonin, E. V. (1998). "Evidence for massive gene exchange between archaeal and bacterial hyperthermophiles." Trends in Genetics 14(11): 442-444.
4. Baker, M. (2012). "De novo genome assembly: What every biologist should know." Nature Methods 9(4): 333-337.
5. Bankevich, A., Nurk, S., Antipov, D., Gurevich, A. A., Dvorkin, M., Kulikov, A. S., . . . Pevzner, P. A. (2012). "SPAdes: A new genome assembly algorithm and its applications to single-cell sequencing." Journal of Computational Biology 19(5): 455-477.
6. Boisvert, S., Raymond, F., Godzaridis, E., Laviolette, F. and Corbeil, J. (2012). "Ray Meta: Scalable de novo metagenome assembly and profiling." Genome Biology.
7. Campbell, B. J., Engel, A. S., Porter, M. L. and Takai, K. (2006). "The versatile ε-proteobacteria: Key players in sulphidic habitats." Nature Reviews Microbiology 4(6): 458-468.
8. Campbell, B. J., Yu, L., Heidelberg, J. F. and Kirchman, D. L. (2011). "Activity of abundant and rare bacteria in a coastal ocean." Proceedings of the National Academy of Sciences of the United States of America 108(31): 12776-12781.
9. Cheng, Y., Fox, S., Pemberton, D., Hogg, C., Papenfuss, A. T. and Belov, K. (2015). "The Tasmanian devil microbiome-implications for conservation and management." Microbiome 3: 76.
10. Cho, I. and Blaser, M. J. (2012). "The human microbiome: At the interface of health and disease." Nature Reviews Genetics 13(4): 260-270.
11. Chow, J., Tang, H. and Mazmanian, S. K. (2011). "Pathobionts of

the gastrointestinal microbiota and inflammatory disease." Current Opinion in Immunology 23(4): 473-480.

12. Coil, D., Jospin, G. and Darling, A. E. (2015). "A5-miseq: An updated pipeline to assemble microbial genomes from Illumina MiSeq data." Bioinformatics 31(4): 587-589.
13. Costello, E. K., Lauber, C. L., Hamady, M., Fierer, N., Gordon, J. I. and Knight, R. (2009). "Bacterial community variation in human body habitats across space and time." Science 326(5960): 1694-1697.
14. Costello, E. K., Stagaman, K., Dethlefsen, L., Bohannan, B. J. M. and Relman, D. A. (2012). "The application of ecological theory toward an understanding of the human microbiome." Science 336(6086): 1255-1262.
15. DeLeon-Rodriguez, N., Lathem, T. L., Rodriguez-R, L. M., Barazesh, J. M., Anderson, B. E., Beyersdorf, A. J., . . . Konstantinidis, K. T. (2013). "Microbiome of the upper troposphere: Species composition and prevalence, effects of tropical storms, and atmospheric implications." Proceedings of the National Academy of Sciences of the United States of America 110(7): 2575-2580.
16. Edwards, D. J. and Holt, K. E. (2013). "Beginner's guide to comparative bacterial genome analysis using next-generation sequence data." Microbial Informatics and Experimentation 3(1): 2.
17. Ekblom, R. and Wolf, J. B. W. (2014). "A field guide to whole-genome sequencing, assembly and annotation." Evolutionary Applications 7(9): 1026-1042.
18. Fierer, N. and Jackson, R. B. (2006). "The diversity and biogeography of soil bacterial communities." Proceedings of the National Academy of Sciences of the United States of America 103(3): 626-631.
19. Fierer, N., Ladau, J., Clemente, J. C., Leff, J. W., Owens, S. M., Pollard, K. S., . . . McCulley, R. L. (2013). "Reconstructing the microbial diversity and function of pre-agricultural tallgrass prairie soils in the United States." Science 342(6158): 621-624.
20. Fraser, C. M., Eisen, J. A. and Salzberg, S. L. (2000). "Microbial genome sequencing." Nature 406(6797): 799-803.
21. Gilbert, J. A., Jansson, J. K. and Knight, R. (2014). "The Earth Microbiome project: Successes and aspirations." BMC Biology

12(1).

22. Gittel, A., Bárta, J., Kohoutová, I., Mikutta, R., Owens, S., Gilbert, J., Urich, T. (2014). "Distinct microbial communities associated with buried soils in the siberian tundra." ISME Journal 8(4): 841-853.
23. Gnerre, S., MacCallum, I., Przybylski, D., Ribeiro, F. J., Burton, J. N., Walker, B. J., Jaffe, D. B. (2011). "High-quality draft assemblies of mammalian genomes from massively parallel sequence data." Proceedings of the National Academy of Sciences of the United States of America 108(4): 1513-1518.
24. Gobet, A., Boetius, A. and Ramette, A. (2014). "Ecological coherence of diversity patterns derived from classical fingerprinting and Next Generation Sequencing techniques." Environmental Microbiology 16(9): 2672-2681.
25. Hill-Burns, E. M., Debelius, J. W., Morton, J. T., Wissemann, W. T., Lewis, M. R., Wallen, Z. D., Payami, H. (2017). "Parkinson's disease and Parkinson's disease medications have distinct signatures of the gut microbiome." Movement Disorders 32(5): 739-749.
26. Honda, K. and Littman, D. R. (2012). The microbiome in infectious disease and inflammation. Annual Review of Immunology. 30: 759-795.
27. Howe, A. C., Jansson, J. K., Malfatti, S. A., Tringe, S. G., Tiedje, J. M. and Brown, C. T. (2014). "Tackling soil diversity with the assembly of large, complex metagenomes." Proceedings of the National Academy of Sciences, USA 111(13): 4904-4909.
28. Huang, Y. J. and Boushey, H. A. (2015). "The microbiome in asthma." Journal of Allergy and Clinical Immunology 135(1): 25-30.
29. Huber, J. A., Mark Welch, D. B., Morrison, H. G., Huse, S. M., Neal, P. R., Butterfield, D. A. and Sogin, M. L. (2007). "Microbial population structures in the deep marine biosphere." Science 318(5847): 97-100.
30. Hugenholtz, P. (2002). "Exploring prokaryotic diversity in the genomic era." Genome Biology 3(2): 0003.0001-0003.0008.
31. Jiménez, R. R. and Sommer, S. (2017). "The amphibian microbiome: natural range of variation, pathogenic dysbiosis, and role in

conservation." Biodiversity and Conservation 26(4): 763-786.

32. Joseph, S. J., Hugenholtz, P., Sangwan, P., Osborne, C. A. and Janssen, P. H. (2003). "Laboratory Cultivation of Widespread and Previously Uncultured Soil Bacteria." Applied and Environmental Microbiology 69(12): 7210-7215.
33. Kemp, P. F., Lee, S. and LaRoche, J. (1993). "Estimating the growth rate of slowly growing marine bacteria from RNA content." Applied and Environmental Microbiology 59(8): 2594-2601.
34. Kivlin, S. N., Hawkes, C. V. and Treseder, K. K. (2011). "Global diversity and distribution of arbuscular mycorrhizal fungi." Soil Biology and Biochemistry 43(11): 2294-2303.
35. Knittel, K. and Boetius, A. (2009). Anaerobic oxidation of methane: Progress with an unknown process. Annual Review of Microbiology. 63: 311-334.
36. Lauber, C. L., Strickland, M. S., Bradford, M. A. and Fierer, N. (2008). "The influence of soil properties on the structure of bacterial and fungal communities across land-use types." Soil Biology and Biochemistry 40(9): 2407-2415.
37. Morris, R. M., Rappé, M. S., Connon, S. A., Vergin, K. L., Siebold, W. A., Carlson, C. A. and Giovannoni, S. J. (2002). "SAR11 clade dominates ocean surface bacterioplankton communities." Nature 420(6917): 806-810.
38. Pace, N. R. (1997). "A molecular view of microbial diversity and the biosphere." Science 276(5313): 734-740.
39. Pevzner, P. A., Tang, H. and Waterman, M. S. (2001). "An Eulerian path approach to DNA fragment assembly." Proceedings of the National Academy of Sciences of the United States of America 98(17): 9748-9753.
40. Prosser, J. I., Bohannan, B. J. M., Curtis, T. P., Ellis, R. J., Firestone, M. K., Freckleton, R. P., Young, J. P. W. (2007). "The role of ecological theory in microbial ecology." Nature Reviews Microbiology 5(5): 384-392.
41. Prosser, J. I. and Nicol, G. W. (2008). "Relative contributions of archaea and bacteria to aerobic ammonia oxidation in the environment." Environmental Microbiology 10(11): 2931-2941.

42. Rappé, M. S., Connon, S. A., Vergin, K. L. and Giovannoni, S. J. (2002). "Cultivation of the ubiquitous SAR11 marine bacterioplankton clade." Nature 418(6898): 630-633.

43. Redford, K. H., Segre, J. A., Salafsky, N., Del Rio, C. M. and McAloose, D. (2012). "Conservation and the Microbiome." Conservation Biology 26(2): 195-197.

44. Sangal, V., Nieminen, L., Tucker, N. P. and Hoskisson, P. A. (2014). Revolutionizing prokaryotic systematics through next-generation sequencing. Methods in Microbiology. 41: 75-101.

45. Scher, J. U. and Abramson, S. B. (2011). "The microbiome and rheumatoid arthritis." Nature Reviews Rheumatology 7(10): 569-578.

46. Simpson, J. T., Wong, K., Jackman, S. D., Schein, J. E., Jones, S. J. M. and Birol, I. (2009). "ABySS: A parallel assembler for short read sequence data." Genome Research 19(6): 1117-1123.

47. Sogin, M. L., Morrison, H. G., Huber, J. A., Welch, D. M., Huse, S. M., Neal, P. R., Herndl, G. J. (2006). "Microbial diversity in the deep sea and the underexplored "rare biosphere"." Proceedings of the National Academy of Sciences of the United States of America 103(32): 12115-12120.

48. Tedersoo, L., Bahram, M., Põlme, S., Kõljalg, U., Yorou, N. S., Wijesundera, R., Abarenkov, K. (2014). "Fungal biogeography. Global diversity and geography of soil fungi." Science (New York, N.Y.) 346(6213): 1256688.

49. Tedersoo, L., May, T. W. and Smith, M. E. (2010). "Ectomycorrhizal lifestyle in fungi: Global diversity, distribution, and evolution of phylogenetic lineages." Mycorrhiza 20(4): 217-263.

50. Thrash, J. C., Boyd, A., Huggett, M. J., Grote, J., Carini, P., Yoder, R. J., Giovannoni, S. J. (2011). "Phylogenomic evidence for a common ancestor of mitochondria and the SAR11 clade." Scientific Reports 1.

51. Turnbaugh, P. J., Hamady, M., Yatsunenko, T., Cantarel, B. L., Duncan, A., Ley, R. E., . Gordon, J. I. (2009). "A core gut microbiome in obese and lean twins." Nature 457(7228): 480-484.

52. Turnbaugh, P. J., Ley, R. E., Hamady, M., Fraser-Liggett, C. M., Knight, R. and Gordon, J. I. (2007). "The Human Microbiome

Project." Nature 449(7164): 804-810.

53. Van Dorst, J., Bissett, A., Palmer, A. S., Brown, M., Snape, I., Stark, J. S., . Ferrari, B. C. (2014). "Community fingerprinting in a sequencing world." FEMS Microbiology Ecology 89(2): 316-330.
54. Venter, J. C., Remington, K., Heidelberg, J. F., Halpern, A. L., Rusch, D., Eisen, J. A., Smith, H. O. (2004). "Environmental Genome Shotgun Sequencing of the Sargasso Sea." Science 304(5667): 66-74.
55. Vogel, T. M., Simonet, P., Jansson, J. K., Hirsch, P. R., Tiedje, J. M., van Elsas, J. D., Philippot, L. (2009). "TerraGenome: A consortium for the sequencing of a soil metagenome." Nature Reviews Microbiology 7(4): 252.
56. Wen, L., Ley, R. E., Volchkov, P. Y., Stranges, P. B., Avanesyan, L., Stonebraker, A. C., Chervonsky, A. V. (2008). "Innate immunity and intestinal microbiota in the development of Type 1 diabetes." Nature 455(7216): 1109-1113.
57. Widder, S., Allen, R. J., Pfeiffer, T., Curtis, T. P., Wiuf, C., Sloan, W. T., Soyer, O. S. (2016). "Challenges in microbial ecology: Building predictive understanding of community function and dynamics." ISME Journal 10(11): 2557-2568.
58. Zerbino, D. R. and Birney, E. (2008). "Velvet: Algorithms for de novo short read assembly using de Bruijn graphs." Genome Research 18(5): 821-829.

CHAPTER 5

METAGENOMIC ANALYSES FOR UNDERSTANDING STRUCTURAL AND FUNCTIONAL ASPECTS OF MICROBIAL COMMUNITIES AND THEIR INTERACTIONS

Jennifer Wood[1], Prashant Pant[2]

[1]Jennifer Wood is post-doctoral fellow in Dr. Ashley Franks' Applied and Environmental Microbiology Laboratory at La Trobe University, Australia. She has recently completed her Doctoral research in Microbial Ecology at the same university. Her broad research interest is in understanding the ecology of soil-microbial communities and their interactions with macro-communities. She is interested in developing trait-based approaches to interpret and predict how environmental and anthropogenic changes impact microbial community ecology and function.

[2]Dr. Prashant Pant, is an Assistant Professor of Botany in Dyal Singh College, University of Delhi, New Delhi, India. He has been teaching for more than 5 years now and his main areas have been Bioinformatics, Molecular Biology and Genomics at the college. His PhD was on understanding the Microbial Succession using a Metagenomic approach in a Barren mined-out site undergoing ecological restoration. For this, he deployed many molecular and computational Biology tool and techniques. He has several publications (articles and book

chapters) in International Journals and International Publications. Currently, he is also Chief Editor of Bioinformatics Review and a co-founder/Executive Chief Editor of a Science Journal in Hindi Language "Swadeshi Vigyan" for the native language speaking communities. Both the Journals are a service to the scientific community as they are fully open access.

Ecology being the study of how living organisms interact with one another and with their environment, there is a critical need for strategies that link community-level taxonomic changes to ecological and functional changes (Ganong, 1902; Prosser et al., 2007). While NGS has increased the ability to successfully capture the richness and composition of microbial communities, researchers are often limited in their ability to ecologically interpret taxonomic changes, or assign specific community members into ecologically meaningful categories. The use of microbial functions, as opposed to taxonomic structure, presents an opportunity for describing microbial communities in terms of ecologically relevant processes. This chapter describes current approaches to infer and interpret microbial community functions from both rRNA community profiling and from shot-gun metagenomics data.

FUNCTIONAL TRAITS

The development of broad ecological frameworks for the classification and interpretation of microbial communities is considered a priority for the advancement of microbial ecology and may provide a strategy for understanding community assembly (Prosser, 2015; Widder et al., 2016). Emerging trends such as the use of community aggregated traits, which utilize functional information derived from a community metagenome, provide opportunities to describe microbial communities in terms of ecologically relevant functional traits (Fierer et al., 2014). In ocean communities, metagenomic approaches have been used to show a separation in microbial-community functional traits between subsurface-phototropic zones in the water column and near-ocean-floor depths (DeLong et al., 2006). Traits enriched close to the surface were related to high rates of carbon capture (high rates of metabolic activity and foraging abilities) whilst traits related to persistence in unproductive environments (such as polysaccharide production) were enriched in deep waters (DeLong et al., 2006). Similarly, there exist in all micro and macro ecosystems a diverse range of functionalities which are coherent with their microclimatic conditions and available nutrients/resources.

THE KEGG DATABASE FOR ACCESSING FUNCTIONAL TRAITS FROM METAGENOMIC DATA

The advent of NGS technologies has increased the ease with which novel genome sequences from organisms or environmental samples, that contains multiple organisms, can be reconstructed (see Chapter 4). However, unravelling the biological meaning encoded in the genome (i.e. predicted gene functions and their ecological relevance) remains a challenging task. As part of the Japanese Human Genome Program, the KEGG (Kyoto Encyclopedia of Genes and Genomes) database project was launched in 1995 in anticipation of the need for a reference knowledge base for biological interpretation of genome sequence data (Kanehisa and Goto, 2000). The main objective of KEGG has been to establish links among collective sets of genes and high-level functions of the cell or organism (metabolic pathways and processses). Although, the KEGG database was originally restricted to complete genome sequences, as of 2015, genetic information from viruses and plasmids were added as was protein sequence data (Kanehisa et al., 2016).

With the arrival of high-throughput sequence data, KEGG has become one of the most widely used biological databases in the world. Unlike many genome annotation procedures, KEGG annotation is done using ortholog annotation to enable extension of experimental evidence from a specific organism to other organisms. Annotation of individual genes within the database is performed by creating links to the KEGG orthology database by assigning genes KO entry identifiers called K numbers. Ortholog groups are then defined in the context of KEGG pathway maps and other molecular networks, which are networks of K number nodes. Thus, the genome annotation procedure automatically leads to the reconstruction of KEGG pathways and other networks, enabling interpretation of high-level functions (Kanehisa et al., 2016). The KEGG function BlastKOALA is suitable for annotating fully sequenced genomes, while GhostKOALA, runs 100 times faster, is suitable for annotating large data sets such as metagenomes.

LINKING RRNA PROFILES TO FUNCTIONAL GENOMIC INFORMATION

For highly diverse microbial communities, such as soil-microbial communities (bulk or rhizospheric), construction of a metagenome that adequately samples community diversity requires large volumes of sequencing data which can be prohibitive for many researchers (Howe et al., 2014). For such communities, rRNA profiling has provided a useful strategy for capturing the taxonomic diversity of the community, but at the expense of functional information. Recently, predictive metagenomics profiling (PMP) has been added to the microbial ecologist's arsenal of strategies for probing microbial communities and can be used to associate 16S rRNA profile with KEGG functional annotations. Currently there are two platforms available for PMP; PICRUSt and tax4fun (Langille et al., 2013; Aßhauer et al., 2015).

PICRUSt and tax4fun are described in detail in their respective publications. Briefly, both programs generate a database of organisms with known gene content. For tax4fun, this is achieved using KEGG organisms with sequenced genomes. For PICRUSt, a reference phylogenetic tree with GreenGenes identifiers (see Chapter 3) is created. The PICRUSt reference phylogenetic tree is first populated by gene content complied directly from databases containing the annotated sequence genomes. Then, where reference 16S sequences have not associated genomic information, gene content is inferred using an ancestral state reconstruction algorithm. The underlying assumption is that taxonomically similar organisms will have functionally similar capabilities. Indeed, the authors of PICRUSt highlight that phylogeny and biomolecular function are highly correlated (Langille et al., 2013).

Gene family predictions for experimental data are then made by associating the taxonomic identification of OTUs in experimental data to organisms in the reference database with precomputed gene content. For PICRUSt, GreenGenes taxonomic identifiers are used to create the phylogenetic reference tree, thus to map the experimental 16S rRNA data to this tree, taxonomic assignments of OTUs must also be performed using

GreenGenes. Conversely, tax4fun requires SILVA assigned taxonomies which are then mapped to the reference database of KEGG organisms. Both programs use 16S rRNA copy number information to normalize the abundance of identified OTUs.

The predictive nature of these programs is a potential limitation as the software relies on taxonomic similarities between environmental 16S rRNA sequences and published metagenomes to make robust predictions (Langille et al., 2013). However, for environments with moderate amounts of reference genomes, including soils and the human gut, Langille et al. (2013) have demonstrated that PICRUSt can make accurate metagenome predictions.

Whilst PMP is well suited to understanding prokaryotic communities, PICUSt in particular, cannot be used to examine eukaryotic communities. These limitations stem from the fact that, relative to prokaryotes, fungi have large genomes and considerably fewer fully-sequenced species (Hibbett et al., 2013).

FUNGuild, is an open access annotation tool that attempts to relate fungal rRNA profiles to functional information by parsing fungal OTUs into functional guilds based on their taxonomic assignments (Nguyen et al., 2016). FUNGuild is a two-component tool consisting of a community-annotated database and a bioinformatics script that parses. Importantly, FUNGuild output should be manually inspected as unassigned sequences, which often constitute the bulk of the output, are unlikely to be assigned to a guild. Through manual inspection of unassigned sequences against the UNITE database, Fernandez et al. (2017) were able to almost double the amount of sequences assigned to functional guilds (with 1.1 million annotated sequences increasing to nearly 2 million sequences in the final dataset).

STATISTICAL TOOLS FOR ANALYZING FUNCTIONAL TRAITS

Once functional traits have been assigned to a metagenome, it remains to be asked, "which traits are (statistically) significantly different between communities?". Statistically robust methods for the comparison of microbial community functions have been developed that overcome this issue. The software 'statistical analysis of taxonomic and functional profiles (STAMP)' and 'BugBase' allow the user to compare communities based on a predetermined suite of functional traits and determine whether there are significant differences in trait abundances (Parks et al., 2014; Ward et al., 2017). STAMP v2 and BugBase are not limited to pair-wise comparisons but facilitate the analysis of trait abundances across multiple samples. Both metagenomics data, and rRNA generated predictive metagenomics data can be used as inputs. Whilst STAMP specialized in testing for differences between functional pathways defined by KEGG or COG databases, BugBase, provides additional comparisons of biologically interpretable phenotypes such as oxygen tolerance, Gram staining and pathogenic potential. The emergence of tools for the functional analysis of microbial communities is an encouraging sign that the importance of functional differences, as opposed to taxonomic differences, is being recognized by microbial ecologists.

INTERPRETING FUNCTIONAL TRAITS: NETWORK ANALYSIS

Whilst rRNA profiling can answer 'who is present' in a microbial community, and sequenced or predicted metagenomic data can help elucidate important community functions, these are inherently static ways to view a microbial community. Understanding the ecology of microbial communities requires a strategy to interpret interactions occurring within the community, and interactions that occur between community members and the environment. An emerging approach to investigate community interactions is to create networks that visualize statistically significant interactions that occur within and amongst microbial communities. Broadly, in network analysis, microorganisms (or operational taxonomic units; OTUs) are called features and are represented as nodes and feature–feature pairs are represented as edges (lines connecting the nodes; Weiss et al., 2016). An edge implies an ecologically meaningful relationship between features and the strength of type of relationship (i.e. positive of negative) can be visualized using line width or color.

Studies using network analysis have been used to uncover interactions between species within microbial communities and to reveal potential keystone species (Ruan et al., 2006; Steele et al., 2011; Eiler et al., 2012). Network analysis can also be used to deduce cross-kingdom relationships such as plant-bacterial associations which were observed to vary across gradients of environmental harshness, with fewer interactions reported in more extreme environments (King et al., 2012).

Two major strategies have emerged to describe microbial networks: Co-occurrence networks and correlation networks. Co-occurrence networks are based on the pair-wise co-occurrence of OTUs across environments, whilst Correlation networks use a more complex regression-based approach to determine how the abundances of two OTUs correlate across space or time (Faust and Raes, 2012; Weiss et al., 2016). In both co-occurrence and correlation networks, interactions within microbial communities are classified in one of the three broad ways: interactions can have a positive effect (such as mutualisms), a negative impact (parasitism or predation) or no impact on the OTUs involved (neutralism). Using abundance data, collected using rRNA profiling or shout-gun metagenomics, it is possible to infer the interaction between any two species within a community and to classify that interaction as one of positive, negative and neutral. The underlying assumption being that strongly non-random distribution patterns are mostly due to ecological reasons, although this assumption has been debated as abundance distribution can also be driven by random processes (Faust and Raes, 2012).

In principal, predicting the relationships between two OTUs is straightforward: positive relationships are inferred when two OTUs co-occur or exhibit similar abundances over multiple samples; conversely when two OTUs fail to co-occur or exhibit negatively correlated abundances, it is assumed that they have a negative relationship. In practice, generating networks using microbial abundance data faces multiple complications arising from the way abundance data is normalized, biases of different similarity measures, the choice of null models and multiple testing issues all of which need to be considered by the user to ensure best practices used (Faust and Raes, 2012; Kurtz et al., 2015).

Interpreting the ecological relevance of relationships described by co-occurrence and correlation networks is also far from straightforward because both positive and negative relationship can arise from a diversity of interactions. For instance, a positive relationship can be due to cross-feeding (i.e. metabolites output by one bacterium being consumed by another; Falony et al., 2006), syntrophic interactions such as direct interspecies election transport which can promote co-aggregation (Summers et al., 2010), co-colonization and niche overlap. In contrast, negative relationship may result from a prey–predator relationship, competition for nutrients, or space and so on. Additionally, the outcome of these interactions can be time delayed, thus modelling abundance correlations over time-series data may be more informative than temporal comparisons that represent a single snapshot in time (Faust and Raes, 2012).

Several programs are available to generate microbial networks. The popular QIIME workflow incorporates a network analysis procedure using Cytoscape (Shannon et al., 2003; Caporaso et al., 2010). For correlation networks, software such as CoNet can be used to detect significant non-random patterns of co-occurrence in incidence and abundance data (Detti et al., 2011). As not all correlation techniques are equal, CoNet also incorporates a strategy to optimally detect different functional relationships (Weiss et al., 2016). SparCC is a python module for computing correlations in compositional data (i.e. 16S rRNA profiles) (Friedman and Alm, 2012). Molecular Ecological Network Analysis (MENA) is a pipeline that adopts random matrix theory to generate correlation networks and that offers a number of network analysis options once the network is constructed (Deng et al., 2012). Finally, Local Similarity Analysis (LSA) is designed to analyze correlations across time series data and address the issue of time delayed relationships (Ruan et al., 2006).

In addition to OTU abundance information, network inference can also incorporate measurements of environmental traits predict relationships between OTUs and environmental trait parameters. In co-occurrence networks, environmental traits are incorporated by treating them as additional 'species'. In correlation networks, they are incorporated as additional factors (Faust and Raes, 2012).

PERSISTENT BARRIERS TO INTERPRETING MICROBIAL COMMUNITY ECOLOGY

Microbial communities form the microscopic foundation of all ecosystems. The importance of understanding microbial communities is recognized across a range of applied and theoretical fields including: bioremediation, agriculture, plant-pathogen interactions, community ecology and invasion ecology (Lugtenberg and Kamilova, 2009; Bhattacharyya and Jha, 2012; Wood et al., 2016, 2017). Many of these fields share the common goal of elucidating strategies for manipulating microbial communities to optimize desirable processes yet despite the advances afforded by NGS there is still major practical and theoretical barriers to translating our understanding of microbial community function, into strategies to manipulate community function.

Practical barriers include our still limited knowledge of the relationship between genes and their ecological function. Roughly half of the genes identified in metagenomics studies encode products of unknown function, and existing functional annotations are often incomplete or inaccurate (Alivisatos et al., 2015). Additionally, the ecological roles of many well characterized microbial functions are based on assumptions from *in vitro* studies, which may not be environmentally relevant. For example, there is some debate as to whether antimicrobial secondary metabolites have a competitive function *in situ* – a role that has been assumed from decade of laboratory research (Hibbing et al., 2010). Recent studies have shown that secondary metabolites, previously recognized for their antibiotic activity, influence biofilm formation in *Bacillus subtilis* and may actually function in cell-cell communication (López et al., 2009). Integrating multi-omic datasets also presents ongoing challenge. Studies that integrate metagenomics, transcriptomics, proteomics, and metabolomics have been reported, but they are limited by coarse temporal and spatial scales (Alivisatos et al., 2015). Moreover, in a research landscape dominated by NGS and omics, the complimenting of *in situ* observations from NGS surveys with controlled experiments designed to test specific questions and hypotheses is often overlooked. Although, they cannot reproduce the full diversity and complexity of natural biotic interactions, controlled microcosms experiments allow for experimental replication as well as the control of environmental parameters such as substrate concentration, system size or temperature, which would be otherwise unmanageable in natural conditions (Wood et al., 2017). Such controlled experimental approaches are necessary for the systematic unravelling of many ecological processes (Widder et al., 2016).

The ability to assign trophic strategies to individual OTUs remains a broad challenge in the field of microbial ecology despite the emerging power of meta-transcriptomics (Filiatrault, 2011). The realization that microbial ecologists generally lack an ecological framework for describing community-level changes is steadily gaining momentum (Prosser et al., 2007; Powell et al., 2013; Krause et al., 2014; Litchman et al., 2015; Thijs et al., 2016; Wood et al., 2017). The potential of couching functional trait-based analyses of microbial communities within a broader ecological framework is a largely overlooked strategy that could help bridge gap between community-level observations and ecological interpretation.

Macro-ecological theories such as the copiotrophic or oligotrophic R/*k* strategist theory have been used to describe and predict the composition of microbial communities with relative success (Fierer et al., 2007). The use of functional traits (either at the species or the community level) to understand processes that govern the assembly of communities is a major goal of the Competitor/Stress-tolerator/Ruderal (CSR) theory (Grime and Pierce, 2012). The CSR theory predicts that organisms face a three-way resource trade-off between the investment in competitive, stress-tolerant or ruderal traits that is governed by the levels of stress due to resource availability and the levels of disturbance present in an environment (Grime and Pierce, 2012). Using functional traits, an organism can be ascribed a C, S or R classification, with intermediate (e.g. CS, CR, CSR etc.) classifications also being recognized (Hodgson et al., 1999). Levels of environmental stress, disturbance and competition collectively govern which class of CSR traits can persist in a community, thus providing a link between community function and environmental constraints.

The CSR theory was originally developed to explain plant-community assembly, however, as with any good theory, its principals are generalizable and have been recognized as being theoretically applicable to explaining microbial community assembly as well (Prosser et al., 2007; Krause et al., 2014; Thijs et al., 2016). A number of trait trade-offs akin to those proposed in the CSR theory have been described in microorganisms (Lipson et al., 2009; Wallenstein and Hall, 2012; Litchman et al., 2015), including trade-offs between stress tolerance and competitive ability which are likely to affect community assembly in contaminated environments (Ferenci and Spira, 2007).

Although they are broad in nature, ecological theories such as R/k and CSR theory are essential for classifying, interpreting and predicting interactions that occur in the natural world. In the absence of a coherent ecological hypothesis, microbial ecology is merely an accumulation of situation-bound statements gathered together from disparate microbial investigations.

REFERENCES

1. Alivisatos, A.P., Blaser, M.J., Brodie, E.L., Chun, M., Dangl, J.L., Donohue, T.J., et al. (2015). A unified initiative to harness Earth's microbiomes: Transition from description to causality and engineering. *Science* 350(6260), 507-508. doi: 10.1126/science.aac8480.
2. Aßhauer, K.P., Wemheuer, B., Daniel, R., and Meinicke, P. (2015). Tax4Fun: Predicting functional profiles from metagenomic 16S rRNA data. *Bioinformatics* 31(17), 2882-2884. doi: 10.1093/bioinformatics/btv287.
3. Bhattacharyya, P.N., and Jha, D.K. (2012). Plant growth-promoting rhizobacteria (PGPR): Emergence in agriculture. *World Journal of Microbiology and Biotechnology* 28(4), 1327-1350.
4. Caporaso, J.G., Kuczynski, J., Stombaugh, J., Bittinger, K., Bushman, F.D., Costello, E.K., et al. (2010). QIIME allows analysis of high-throughput community sequencing data. *Nat. Methods.* 7(5), 335-336. doi: 10.1038/nmeth.f.303.
5. DeLong, E.F., Preston, C.M., Mincer, T., Rich, V., Hallam, S.J., Frigaard, N.U., et al. (2006). Community genomics among stratified microbial assemblages in the ocean's interior. *Science* 311(5760), 496-503. doi: 10.1126/science.1120250.
6. Deng, Y., Jiang, Y.H., Yang, Y., He, Z., Luo, F., and Zhou, J. (2012). Molecular ecological network analyses. *BMC Bioinformatics* 13(1). doi: 10.1186/1471-2105-13-113.
7. Detti, A., Blefari Melazzi, N., Salsano, S., and Pomposini, M. (Year). "CONET: A content centric inter-networking architecture", in: *Proceedings of the ACM SIGCOMM Workshop on Information-Centric Networking, ICN'11*), 50-55.
8. Eiler, A., Heinrich, F., and Bertilsson, S. (2012). Coherent dynamics

and association networks among lake bacterioplankton taxa. *ISME J.* 6(2), 330-342. doi: 10.1038/ismej.2011.113.

9. Falony, G., Vlachou, A., Verbrugghe, K., and De Vuyst, L. (2006). Cross-feeding between Bifidobacterium longum BB536 and acetate-converting, butyrate-producing colon bacteria during growth on oligofructose. *Appl. Environ. Microbiol.* 72(12), 7835-7841. doi: 10.1128/AEM.01296-06.
10. Faust, K., and Raes, J. (2012). Microbial interactions: From networks to models. *Nature Rev. Microbiol.* 10(8), 538-550. doi: 10.1038/nrmicro2832.
11. Ferenci, T., and Spira, B. (2007). Variation in stress responses within a bacterial species and the indirect costs of stress resistance. *Annals of the New York Academy of Sciences* 1113, 105-113. doi: 10.1196/annals.1391.003.
12. Fernandez, C.W., Nguyen, N.H., Stefanski, A., Han, Y., Hobbie, S.E., Montgomery, R.A., et al. (2017). Ectomycorrhizal fungal response to warming is linked to poor host performance at the boreal-temperate ecotone. *Global Change Biology* 23(4), 1598-1609. doi: 10.1111/gcb.13510.
13. Fierer, N., Barberán, A., and Laughlin, D.C. (2014). Seeing the forest for the genes: Using metagenomics to infer the aggregated traits of microbial communities. *Front. Microbiol.* 5(NOV). doi: 10.3389/fmicb.2014.00614.
14. Fierer, N., Bradford, M.A., and Jackson, R.B. (2007). Toward an ecological classification of soil bacteria. *Ecology* 88(6), 1354-1364. doi: 10.1890/05-1839.
15. Filiatrault, M.J. (2011). Progress in prokaryotic transcriptomics. *Current Opinion in Microbiology* 14(5), 579-586. doi: http://dx.doi.org/10.1016/j.mib.2011.07.023.
16. Friedman, J., and Alm, E.J. (2012). Inferring Correlation Networks from Genomic Survey Data. *PLoS Comput. Biol.* 8(9). doi: 10.1371/journal.pcbi.1002687.
17. Ganong, W.F. (1902). The word 'ecology' [5]. *Science* 15(380), 593-594.
18. Grime, J.P., and Pierce, S. (2012). *The Evolutionary Strategies that Shape Ecosystems.* Oxford: Wiley-Blackwell.

19. Hibbett, D.S., Stajich, J.E., and Spatafora, J.W. (2013). Toward genome-enabled mycology. *Mycologia* 105(6), 1339-1349.
20. Hibbing, M.E., Fuqua, C., Parsek, M.R., and Peterson, S.B. (2010). Bacterial competition: Surviving and thriving in the microbial jungle. *Nature Rev. Microbiol.* 8(1), 15-25. doi: 10.1038/nrmicro2259.
21. Hodgson, J.G., Wilson, P.J., Hunt, R., Grime, J.P., and Thompson, K. (1999). Allocating C-S-R plant functional types: A soft approach to a hard problem. *Oikos* 85(2), 282-294.
22. Howe, A.C., Jansson, J.K., Malfatti, S.A., Tringe, S.G., Tiedje, J.M., and Brown, C.T. (2014). Tackling soil diversity with the assembly of large, complex metagenomes. *Proc. Natl. Acad. Sci. U.S.A.* 111(13), 4904-4909. doi: 10.1073/pnas.1402564111.
23. Kanehisa, M., and Goto, S. (2000). KEGG: Kyoto Encyclopedia of Genes and Genomes. *Nucleic Acids Res.* 28(1), 27-30.
24. Kanehisa, M., Sato, Y., Kawashima, M., Furumichi, M., and Tanabe, M. (2016). KEGG as a reference resource for gene and protein annotation. *Nucleic Acids Res.* 44(D1), D457-D462. doi: 10.1093/nar/gkv1070.
25. King, A.J., Farrer, E.C., Suding, K.N., and Schmidt, S.K. (2012). Co-occurrence patterns of plants and soil bacteria in the high-alpine subnival zone track environmental harshness. *Front. Microbiol.* 3(OCT). doi: 10.3389/fmicb.2012.00347.
26. Krause, S., Le Roux, X., Niklaus, P.A., Van Bodegom, P.M., Lennon, J.T., Bertilsson, S., et al. (2014). Trait-based approaches for understanding microbial biodiversity and ecosystem functioning. *Front. Microbiol.* 5, 251. doi: 10.3389/fmicb.2014.00251.
27. Kurtz, Z.D., Müller, C.L., Miraldi, E.R., Littman, D.R., Blaser, M.J., and Bonneau, R.A. (2015). Sparse and Compositionally Robust Inference of Microbial Ecological Networks. *PLoS Comput. Biol.* 11(5), e1004226. doi: 10.1371/journal.pcbi.1004226.
28. Langille, M.G.I., Zaneveld, J., Caporaso, J.G., McDonald, D., Knights, D., Reyes, J.A., et al. (2013). Predictive functional profiling of microbial communities using 16S rRNA marker gene sequences. *Nat Biotechnol* 31(9), 814-821. doi: 10.1038/nbt.2676.
29. Lipson, D.A., Monson, R.K., Schmidt, S.K., and Weintraub, M.N.

(2009). The trade-off between growth rate and yield in microbial communities and the consequences for under-snow soil respiration in a high elevation coniferous forest. *Biogeochemistry* 95(1), 23-35. doi: 10.1007/s10533-008-9252-1.

30. Litchman, E., Edwards, K.F., and Klausmeier, C.A. (2015). Microbial resource utilization traits and trade-offs: implications for community structure, functioning, and biogeochemical impacts at present and in the future. *Front. Microbiol.* 6(254). doi: 10.3389/fmicb.2015.00254.
31. López, D., Fischbach, M.A., Chu, F., Losick, R., and Kolter, R. (2009). Structurally diverse natural products that cause potassium leakage trigger multicellularity in Bacillus subtilis. *Proceedings of the National Academy of Sciences of the United States of America* 106(1), 280-285. doi: 10.1073/pnas.0810940106.
32. Lugtenberg, B., and Kamilova, F. (2009). Plant-growth-promoting rhizobacteria. *Annu. Rev. Microbiol.* 63, 541-556.
33. Nguyen, N.H., Song, Z., Bates, S.T., Branco, S., Tedersoo, L., Menke, J., et al. (2016). FUNGuild: An open annotation tool for parsing fungal community datasets by ecological guild. *Fungal Ecology* 20, 241-248. doi: 10.1016/j.funeco.2015.06.006.
34. Parks, D.H., Tyson, G.W., Hugenholtz, P., and Beiko, R.G. (2014). STAMP: Statistical analysis of taxonomic and functional profiles. *Bioinformatics* 30(21), 3123-3124. doi: 10.1093/bioinformatics/btu494.
35. Powell, J.R., Anderson, I.C., and Rillig, M.C. (2013). A new tool of the trade: Plant-trait based approaches in microbial ecology. *Plant. Soil.* 365(1-2), 35-40. doi: 10.1007/s11104-012-1581-0.
36. Prosser, J.I. (2015). Dispersing misconceptions and identifying opportunities for the use of 'omics' in soil microbial ecology. *Nature Rev. Microbiol.* 13(7), 439-446. doi: 10.1038/nrmicro3468.
37. Prosser, J.I., Bohannan, B.J.M., Curtis, T.P., Ellis, R.J., Firestone, M.K., Freckleton, R.P., et al. (2007). The role of ecological theory in microbial ecology. *Nature Rev. Microbiol.* 5(5), 384-392. doi: 10.1038/nrmicro1643.
38. Ruan, Q., Dutta, D., Schwalbach, M.S., Steele, J.A., Fuhrman, J.A., and Sun, F. (2006). Local similarity analysis reveals

unique associations among marine bacterioplankton species and environmental factors. *Bioinformatics* 22(20), 2532-2538. doi: 10.1093/bioinformatics/btl417.

39. Shannon, P., Markiel, A., Ozier, O., Baliga, N.S., Wang, J.T., Ramage, D., et al. (2003). Cytoscape: A software Environment for integrated models of biomolecular interaction networks. *Genome Research* 13(11), 2498-2504. doi: 10.1101/gr.1239303.
40. Steele, J.A., Countway, P.D., Xia, L., Vigil, P.D., Beman, J.M., Kim, D.Y., et al. (2011). Marine bacterial, archaeal and protistan association networks reveal ecological linkages. *ISME J.* 5(9), 1414-1425. doi: 10.1038/ismej.2011.24.
41. Summers, Z.M., Fogarty, H.E., Leang, C., Franks, A.E., Malvankar, N.S., and Lovley, D.R. (2010). Direct exchange of electrons within aggregates of an evolved syntrophic coculture of anaerobic bacteria. *Science* 330(6009), 1413-1415. doi: 10.1126/science.1196526.
42. Thijs, S., Sillen, W., Rineau, F., Weyens, N., and Vangronsveld, J. (2016). Towards an Enhanced Understanding of Plant–Microbiome Interactions to Improve Phytoremediation: Engineering the Metaorganism. *Front. Microbiol.* 7, 341. doi: 10.3389/fmicb.2016.00341.
43. Wallenstein, M.D., and Hall, E.K. (2012). A trait-based framework for predicting when and where microbial adaptation to climate change will affect ecosystem functioning. *Biogeochemistry* 109(1), 35-47. doi: 10.1007/s10533-011-9641-8.
44. Ward, T., Larson, J., Meulemans, J., Hillmann, B., Lynch, J., Sidiropoulos, D., et al. (2017). BugBase Predicts Organism Level Microbiome Phenotypes. *bioRxiv*.
45. Weiss, S., Van Treuren, W., Lozupone, C., Faust, K., Friedman, J., Deng, Y., et al. (2016). Correlation detection strategies in microbial data sets vary widely in sensitivity and precision. *ISME J.* 10(7), 1669-1681. doi: 10.1038/ismej.2015.235.
46. Widder, S., Allen, R.J., Pfeiffer, T., Curtis, T.P., Wiuf, C., Sloan, W.T., et al. (2016). Challenges in microbial ecology: Building predictive understanding of community function and dynamics. *ISME J.* 10(11), 2557-2568. doi: 10.1038/ismej.2016.45.
47. Wood, J.L., Franks, A.E., and Egidi, E. (2017). Linking microscopic

interactions with macroscopic effects. *J. Veg. Sci.* 28(3), 462-463. doi: 10.1111/jvs.12532.

48. Wood, J.L., Tang, C., and Franks, A.E. (2016). Microbial associated plant growth and heavy metal accumulation to improve phytoextraction of contaminated soils. *Soil. Biol. Biochem.* 103, 131-137. doi: http://dx.doi.org/10.1016/j.soilbio.2016.08.021.

CHAPTER 6

MICROBIAL ECOLOGY IN THE ERA OF NEXT GENERATION SEQUENCING

Surajit De Mandal[1], Amrita Kumari Panda[2], Satpal Singh Bisht[2], and Nachimuthu Senthil Kumar[1]

[1]Department of Biotechnology, Mizoram University, Aizawl-796004, Mizoram, India

[2]Department of Zoology, Kumaun University, Nainital-263002, Uttarakhand, India

ABSTRACT

Microorganisms are the major part of earth's biological diversity. Due to most of the microbes being nonculturable, it is necessary to use culture independent techniques to study the uncultured microbes. Metagenomics is the molecular tool which helps to understand the genetic makeup of the wide variety of uncultivable microorganisms. Currently, Next Generation Sequencing (NGS) is one of the most advanced technologies used in metagenomics studies and different computational tools have been developed for the analysis of large metagenomic dataset. This review demonstrates the different tools used in NGS analysis and its applications in microbial ecology.

Keywords: Biological diversity, Metagenomics, Next generation sequencing

INTRODUCTION

Natural environments harbor a large number of microorganisms representing the major reservoir of undescribed biodiversity [1] and occupy diverse habitats from deep sea sediments to high up in the atmosphere [2]. But, most of the microbes are unculturable by standard techniques [3,4]. Thus, their ecology and functional roles are unknown and our understanding of the composition of the natural microbial world is therefore rudimentary. The reason for uncultivable nature may be the fact that they depend on other organisms for critical processes, fail to grow in vitro or have even become extinct in fossil records [5]. Some are metabolically active, even though they were unculturable in laboratory [6]. Culturing these microbes in artificial environment needs in-depth knowledge of the nutrient and growth requirements. Although some work has attempted to culture these uncultivable bacteria [7,8], the technique is time consuming and laborious and therefore needs some alternative methods to study them.

DNA sequence-based methods can overcome these problems since we can isolate the genetic material directly from live or dead cells from various environmental samples, allowing a new emerging field called metagenomics. Metagenomics involves the culture independent approach to study the unculturable microorganisms directly from an environmental samples such as soil [9,10], seawater [11], ground water [12], antarctic desert soil [13] etc. In the mid-1980s, Norman Pace [14] suggested a culture-independent approach which was later named as "metagenomics" by Handelsman and first appeared in publication in 1998. This field is also called as environmental genomics, ecogenomics or community genomics. In this technique, the isolation of environmental DNA directly resulted in the isolation of a number of novel genes and was designed with several practical gains such as the discovery of new genes and gene products that would lead to agricultural innovations, medicinal chemistry and industrial processes. It can also be used to reveal diversity patterns of microorganisms, horizontal gene transfer analysis, identification of novel metabolic pathways and examining genes/operons for desirable enzyme candidates (e.g., cellulases, chitinases, lipases, antibiotics) and other natural products [3,15,16].

Ribosomal RNA for the Identification of Uncultured Bacteria and its Community

Ribosomal RNA is largely used for the study of uncultured bacteria as it has several advantages. The 16S gene occurs in all living organisms with the notable exception of viruses and represents more than 80% of total bacterial RNA found in every living cell and generally stable and long enough (approx. 1500 nt), contains independently evolving domains, i.e. variable regions. They are shorter than 23s (approx. 2300 nt) and easier to sequence. The 16S rRNA gene structure consists of interspersed conserved and variable regions which is suitable for PCR amplification and sequencing. As a molecular chronometer, these molecules have preserved their evolutionary history. Most of the studies have been carried out with 16S rRNA and has been used as a phylogenetic marker as it is present in all prokaryotes and contains different hypervariable regions separated by highly conserved gene segments. They have highly conserved portion which carries the information on early evolutionary events and more recent changes are documented within less conserved position or stretches. The degree of divergence of present day rRNA sequence gives an estimate of their phylogenetic distance. They donot have any ortholog - paralog issues or horizontal gene transfer. It is easy to PCR amplify and they have a very slow rate of mutation. The new era of metagenomics was ushered in by studies using 16S rRNA as a phylogenetic marker of microbial taxa. Focusing on a small part of the microbial genome brings down the sequencing costs dramatically [17].

It is important to choose the appropriate variable region of the 16S rRNA for microbial ecological studies mainly for biogeography, metacommunity theory or microbiome analysis. Previous studies selected the hypervariable region which can exhibit sufficient phylogenetic signals to classify at the genus level [18]. Liu et al. [19] suggested for the 250 bp region between the V2 and V3 hypervariable regions for microbial ecological studies. The use of tags of the V3 and V6 regions in 454 pyrosequencing also provides sufficient taxonomic assignments [20]. Analysis by Wang et al. [21] proved that V2 and V4 regions can results best taxonomic assignment at the genus level, where as for the study of microbial community complex ecosystem it is recommended to use V7-V8 region [22,23].

Analysis of Metagenome using DGGE and TGGE

Sanger sequencing (first generation sequencing) was widely used for molecular biology studies with no or minimum errors. Based on the Sanger sequencing approach, different PCR based methods have been developed to study the microbial communities. Denaturing Gradient Gel Electrophoresis (DGGE) or Terminal Restriction Fragment Length Polymorphism (T-RFLP) has been developed to analyze the uncultured microbial communities in diverse environments [24,25]. These methods allow differentiating gene molecules (in polyacrylamide gels) based on the decreased electrophoresis mobility of a partially melted double stranded DNA molecule at the level of a single base containing a linear gradient of DNA denaturants (a mixture of urea and formamide) or a linear temperature gradient. However, these techniques present several drawbacks as the dominant populations are better revealed and bands from more than one species may be hidden behind a single band resulting in underestimation of bacterial diversity. The same isolate can have different identifiable bands because multiple copies of 16S rRNA gene are present in a single genome of species. The Sanger sequencing based methods present several drawbacks since it is expensive and cannot detect rare microorganisms [26,27].

Next Generation Sequencing Technology (NGS)

NGS allows massively parallel sequencing with thousands to millions of sequences in one experiment at considerably low cost compared to Sanger method. There are no requirements of bacterial cloning of DNA fragments and electrophoresis, since NGS libraries preparation process are carried out in a cell-free system and the sequence output is directly detected. However, it produces shorter reads with a higher error rate than Sanger sequencing [28-30]. Moreover, secondary structure and thermal stability of the genome affects the efficiency of PCR amplification of the genomic fragments. There are different NGS platforms (e.g. Illumina, Pacific Biosciences, Ion Torrent, SOLiD) with different principle and divergent features such as run times, yields and read lengths [31] (Table 1).

Advances in NGS have revolutionized the field of microbial ecology. Any NGS study follow some common steps such as sample and metadata collection, DNA extraction, library construction, sequencing and read preprocessing followed by quantitative analysis and functional binning [32,33]. Collection of environment sample is the first step in any metagenomic study. But, the problem faced in microbial ecology studies are that no information is available about the amount of required sampling that will comprise entire population in an environment. Usually rarefaction curve, a plot of number of species versus number of individuals sampled, which typically defines the population richness and evenness is used to estimate the fraction of species sequenced.

Another important aspect for any successful metagenomic study is the extraction of high quality DNA from the environmental samples [34]. Environmental samples contain DNA in a variety of packages like free DNA, virus particle, and prokaryotic and eukaryotic cells. These can be suspended in water, bound to a solid matrix like soil, or encased in a biofilm or tissue. So extraction methods must be chosen carefully based on the medium and the DNA population of interest (Susannah Green Tringe) and DNA should come from all the microbial representative present in the sample without any shearing or contamination [35]. Although many commercial kits are available for metagenome isolation, it is required to develop our own method to optimize extraction and reducing bias caused by unequal lysis of different members of the soil microbial community [36,37]. Beat beating, sonication, detergents or enzymatic lysis have been used for the isolation purpose. Since some samples contain high humic acid, additional purification is required. Polyphenol compound often co-purified along with the DNA can enzymatically modify the isolated DNA [38]. One strategy may be the use of Skimmed milk which can aid in DNA extraction process from soil [39]. Multiple displacement amplification can be used to increase template DNA for samples with very less amount of DNA [40,41].

Two kind of NGS strategy is applied in microbial community analysis: 1) deep amplicon sequencing and 2) complete metagenome or metatranscriptome analysis. Shotgun sequencing is applied to know the function of the community. The analysis procedure involves assembly (merging overlapping short reads into contigs), binning (grouping reads or contigs into individual genomes and assigning the groups to specific species, subspecies, or genus), functional annotation (for the prediction of CDS, CRISPR, noncoding RNAs) and functional assignment to the predicted protein coding sequences. On the other hand, amplicon sequencing is used for community profiling using marker genes (e.g., 16S rRNA gene) in different ecosystems. Procedure involves denoising (filtered the 'noise' sequences from the raw data), Chimera detection (recombinants which are formed when prematurely terminated fragments reanneal to other template DNA during PCR), OTU clustering (alignment is done by either aligning query sequences against pre-aligned reference sequences (42) or using pairwise and multiple sequence alignments), Taxonomic classification (taxonomic assignment of OTUs) and Statistical analysis, (alpha and betadiversity analysis, multidimensional scaling and analysis of similarities (ANOSIM), hypothesis testing etc). A flow diagram for the analysis of metagenomic data is shown in Figure 1 and Table 2 represents metagenomics tools available for Microbial Ecology Studies.

Table 1: Summary of the major next-generation sequencing platform

Technology (company)	*Amplification*	Chemistry	Sequencing method	Yield (Gb/run)	*Highest Average Read Length*	Error rate	Output file	Disadvantage	Advantage	Website
Illumina	Bridge amplification	Reversible dye terminator (seq-by-synthesis)	incorporation of fluorescent nucleotides	1- 60	300 bp (overlapping paired-end sequencing available	≥0.1	Fastq (Phred +64 & 33, Illumina +1.8)	Short reads and long runtime	High throughput and low cost	www.illumina.com
SOLiD (Life Technologies)	Emulsion PCR	Sequencing by ligation (SBL)	fluorescent short linkers	3	75 bp (paired-end sequencing available)	>0.06	Fastq (Phred +33)	Long run-time	Low error rate	www.applied-biosystems.com
454 (Roche)	Emulsion PCR	Pyrosequencing (seq-by-synthesis)	incorporation of normal nucleotides	0.7	700 bp	1	SFF, fasta, fastq	High error rate in homopolymer	Long read	www.454.com
SMAT (Pacific Bio)	N/A (single molecule)	Single Molecule Real Time (SMRT™)	incorporation of fluorescent nucleotides	0.3-0.5	5,000 bp average; maximum read length ~22,000 bases	16	Fastq (Phred +33)	No PCR longest reads	High error	www.pacificbio.com
Ion Torrent (Life Technology)	N/A	Proton detection (seq-by-synthesis)	measuring pH change	1	400 bp (bidirectional sequencing available)	1	Fastq (Phred +33)	new	Short read	www.iontorrent. com

Third-Generation Sequencing

Third-generation sequencing (SMRT sequencing) is based on the sequencing by synthesis approach and allows the detection of single molecules, but also enables real-time sequencing. The biggest advantage is that it allows 20,000 reads or more, with average read lengths of 5 kilobases. Moreover, it can also directly detect epigenetic modifications such as 4-methylctosine (mC), 5-mC and 6-methyladenine [42].

Application of NGS in Microbial Ecology

Analysis of microbial community composition and its function is a

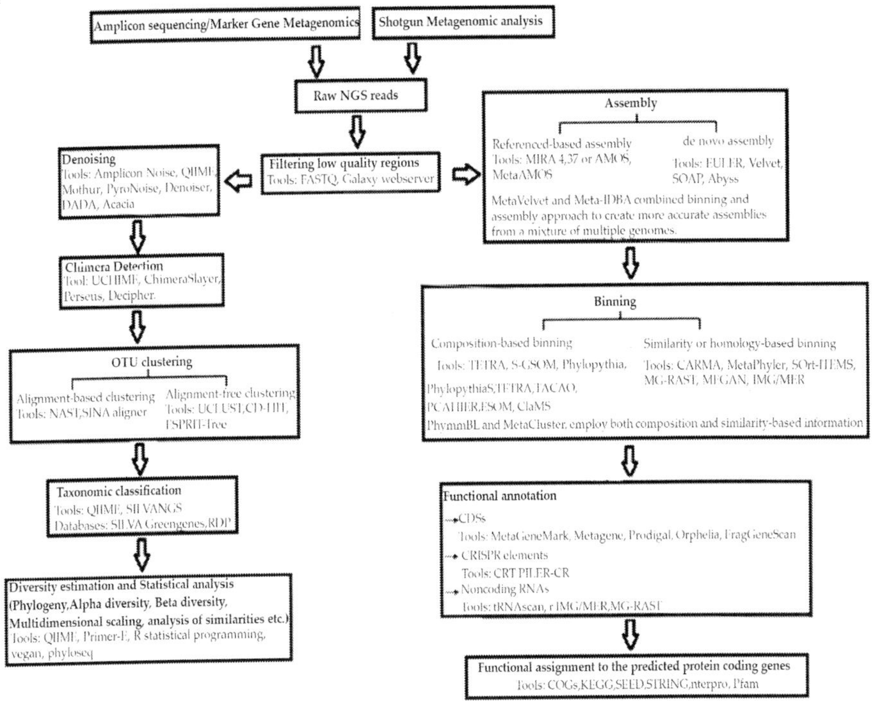

Figure 1: Flow diagram for the analysis of Metagenomics data.

Table 2: Metagenomics tools available for microbial ecology studies

Metagenomics tools	website
General Software package	
STAMP	http://kiwi.cs.dal.ca/Software/STAMP
CD-HIT-OUT	http://weizhong-lab.ucsd.edu/cd-hit-otu/
GAAS	http://sourceforge.net/projects/gaas/
Megan	http://www-ab.informatik.unituebingen.de/software/megan/ welcome.html
MetaPhlAn	http://huttenhower.sph.harvard.edu/metaphlan
MetaSim	http://www-ab.informatik.uni-tuebingen.de/software/metasim
Mocat	http://vm-lux.embl.de/~kultima/MOCAT/index.htm
Livermore Metagenomics Analysis Toolkit	https://computation-rnd.llnl.gov/lmat
Straine	http://www.hsls.pitt.edu/obrc/index.php?page=URL1221230034
Metagenome assembly	
Velvet	http://www.ebi.ac.uk/~zerbino/velvet/
Celera	http://www.cbcb.umd.edu/research/assembly.shtml#software
Metasim	http://ab.inf.uni-tuebingen.de/software/metasim/welcome.html #Download
Euler	http://nbcr.sdsc.edu/euler/ JAZZ
Mapping to reference genome	
Bowtie	http://bowtie-bio.sourceforge.net/index.shtml
BWA	http://bio-bwa.sourceforge.net/
MAPLE:	http://www.genome.jp/maple-bin/mapleSubmit.cgi?aType=sDirect
Quality analysis	
FastQC	http://www.bioinformatics.babraham.ac.uk/projects/download.html#fastqc
Prinseq	http://edwards.sdsu.edu/cgi-bin/prinseq/prinseq.cgi
Gene calling	
Genemark.	hmm http://exon.gatech.edu/GeneMark/metagenome/Prediction/
Microbial diversity Analysis	
MLST	http://www. mlst.net) http://www.mlst.net/
MOTHUR	http://www.mothur.org/
EstimateS	http://viceroy.eeb.uconn.edu/EstimateS/
QIIME	http://qiime.org/install/virtual_box.html
PHACCS	http://phaccs.sourceforge.net/
Binning	
TETRA	http://www.megx.net/tetra/index.html
Phylopathia	http://cbcsrv.watson.ibm.com/phylopythia.html
MEGAN	http://ab.inf.uni-tuebingen.de/software/megan/
CARMA	http://www.cebitec.uni-bielefeld.de/brf/carma/carma.html
Phymm	http://www.cbcb.umd.edu/software/phymm/
Functional Annotation	
MEX (Motif Extraction)	http://adios.tau.ac.il/SPMatch/
RAMMCAP	http://weizhong-lab.ucsd.edu/rammcap/cgi-bin/rammcap.cgi
Analysis of quantitative metagenomics data.	
FANTOM	http://www.sysbio.se/Fantom/
Comparitive Metagenomics	
MEGAN	http://metagenomics.anl.gov/
MG-RAST	http://metagenomics.anl.gov
Camera	http://camera.calit2.net/
Shotgun FunctionalizeR	http://shotgun.math.chalmers.se/
MetaStats	http://metastats.cbcb.umd.edu/detection.html
UniFrac	http://bmf.colorado.edu/unifrac/
Galaxy	https://main.g2.bx.psu.edu/u/aun1/w/metagenomic-analysis
MetaMine	http://www.megx.net/metamine/
MetaLook	http://www.megx.net/metalook/index.php
IMG/M	http://img.jgi.doe.gov/cgi-bin/m/main.cgi
Online tools for NGS analysis	
Parallel Meta	http://www.computationalbioenergy.org/parallel-meta.html
SOrt-ITEMS	http://metagenomics.atc.tcs.com/binning/SOrt-ITEMS/
PANGEA	http://www.ohloh.net/p/pangea-plus
Genohub:	https://genohub.com/bioinformatics/17/metagenomic-analysis
Additional Metagenome analysis links	
EBI Metagenomics	https://www.ebi.ac.uk/metagenomics
MetaPhlAn:	http://huttenhower.sph.harvard.edu/metaphlan
Amphora	Net http://pitgroup.org/amphoranet

major objective in modern microbial ecology studies. NGS has already been applied in such studies in diverse ecosystem such as caves, forests, hot springs, deserts etc [43-47]. A pyrosequencing based approach revealed that the soil stratification and resource availability can impact the taxonomic and functional diversity of bacterial communities [48]. NGS based studies have proved that the agriculture management system can significantly affect the microbial diversity [49,50]. Chhabra et al. [51] identified genes and operons responsible for mineral phosphate solubilization in the barley rhizosphere using pyrosequencing. A NGS based study of soil bacterial communities showed that the conversion of primary Amazonian forest to pasture increases bacterial alphadiversity, but decreases beta diversity resulting in the homogenization of communities across space. This homogenization is driven by the loss of forest soil bacteria with restricted ranges (endemics) and results in a net loss of diversity [52]. Analysis using 227 million Illumina shotgun sequences from forest soil and 246 million from deforested soils revealed that bacterial taxonomic and functional adaptations at the bacterial community level is due to an increase in nutrient availability from slash and burn clearing of Amazon forest [53]. Analysis of 62081 pyrosequencing reads from 10 soil samples revealed that bacterial diversity and soil physicochemical properties did not show consistent changes along the elevation gradients in southwestern highlands of Saudi Arabia [54]. An NGS study carried out in temperate deciduous forest of Northeast Ohio, USA found that the vernal pool microbial communities may rely on their metabolic plasticity for growth and survival during limited resources [55]. Analysis of NGS reads shown that microbial functional activity increased throughout decomposition in spring, summer and winter, while it decreased in autumn in a Midwest temperate forest in Morris Bean Reserve of Greene County, Ohio, USA [56]. Pyrosequencing of V3-V4 regions described the bacterioplankton communities in a coastal Antarctic lake which is under long-term environmental change [57]. Fierer et al. [47] used Illumina shotgun sequencing in different ecosystem and found that desert microbial communities had high relative abundance of genes associated with osmoregulation and dormancy, whereas genes linked with nutrient cycling and catabolism of plant-derived organic compounds was less. This study also revealed that abiotic factors are more important in harboring the desert microbial communities. Another shotgun study detected identification of genes related to phytic acid utilization [58]. Analysis of total of 1,294,216 raw 16S rRNAV6 sequences revealed that Actinobacteria, Acidobacteria, Nitrospirae, and Verrucomicrobia were abundant in nutrient-rich inner mangrove sediments, while Proteobacteria and Deferribacterias were high in outer mangrove sediments [59]. Based on pyrosequencing of the V2-V3 16S rRNA gene regions, Nacke et al. [60] found that soil bacterial community composition and diversity of the six analyzed management types showed significant differences between the land use patterns in grassland and forest ecosystems and more over bacterial community structure was largely driven by tree species and soil pH.

NGS technologies are improving day by day with more accuracy and longer sequence reads which are applied in many microbial ecology studies. Due to the continuous decrease in NGS cost, a huge number of environmental samples has been sequenced to solve research problems in microbial ecology. This paper will be helpful in understanding the basics of next generation sequencing as well as for designing new metagenomics projects. Future research with the help of NGS technologies will further unzip the unseen microbial diversity and its function useful in solving ecological problems.

ACKNOWLEDGEMENTS

The authors thank DBT, New Delhi for supporting Bioinformatics Infrastructure Facility, Mizoram University which has provided the essential facilities to carry out the work. The first author is thankful to DBT-State Biotech Hub, Mizoram University for the fellowship.

REFERENCES

1. Nichols DS, Sanderson K, Buia A, Kamp J, Halloway P et al. (2002) Bioprospecting and biotechnology in Antarctica. In: Jabour-Green J, Haward M (eds). The Antarctic: Past, Present and Future. Antarctic CRC. Res. Report, Hobart, 85-103.
2. Torsvik VL, Øvreås L (2011) DNA Reassociation Yields Broad-Scale Information on Metagenome Complexity and Microbial Diversity. In: Handbook of Molecular Microbial Ecology I. John Wiley & Sons, Inc., 3-16.
3. Handelsman J (2004) Metagenomics: application of genomics to uncultured microorganisms. Microbiol Mol Biol Rev 68: 669-685.
4. Kimura N (2006) Metagenomics: Access to unculturable microbes

in the environment. Microb Environ 21: 201-215.

5. Tringe SG and Rubin EM (2005) Metagenomics: DNA sequencing of environmental samples. Nat Rev Genet 6: 805-814.
6. Rappe MS, Connon SA, Vergin KL, Giovannoni SJ (2002) Cultivation of the ubiquitous SAR11 marine bacterioplankton clade. Nature 418: 630-633.
7. Sait M, Hugenholtz P, Janssen PH (2002) Cultivation of globally distributed soil bacteria from phylogenetic lineages previously only detected in cultivation independent surveys. Environ Microbiol 4: 654-666.
8. Zwirglmaier K, Ludwig W, Schleifer KH (2004) Recognition of individual genes in a single bacterial cell by fluorescence in situ hybridization-RING-FISH. Mol Microbiol 51: 89-96.
9. Voget S, Steele HL, Streit WR (2006) Characterization of a metagenomederived halotolerant cellulose. J Biotechnol 126: 26-36.
10. Waschkowitz T, Rockstroh S, Daniel R (2009) Isolation and characterization of metalloproteases with a novel domain structure by construction and screening of metagenomic libraries. Appl Environ Microbiol 75: 2506-2516.
11. Stein JL, Marsh TL, Wu KY, Shizuya H, DeLong EF (1996) Characterization of uncultivated prokaryotes: isolation and analysis of a 40-kilobase- pair genome fragment from a planktonic marine archeon J Bacteriol 178: 591-599.
12. Uchiyama T, Abe T, Ikemura T, Watanable K (2005) Substrate induced geneexpression screening of environmental metagenome libraries for isolation of catabolic genes. Nat Biotechnol 23: 88-93.
13. Heath C, Hu XP, Cary SC, Cowan D (2009) Identification of a novel alkaliphilic esterase active at low temperatures by screening a metagenomic library from Antarctic desert soil. Appl. Environ. Microbiol 75: 4657-4659.
14. Pace N, Stahl D, Lane D, Olsen G (1985) Analysing natural microbial populations by rRNA sequences. ASM News 51: 4-12.
15. Elend C, Schmeisser C, Leggewie C, Babiak P, Carballeira JD, et al. (2006) Isolation and biochemical characterization of two novel

metagenome derived esterases. Appl Environ Microbiol 72: 3637-3645.

16. Vakhlu J, Kour A, Johri B (2008) Metagenomics: future of microbial gene mining. Indian J Microbiol 48: 202-215.
17. Pace NR (1997) A molecular view of microbial diversity and the biosphere. Science 276: 734-740.
18. Birtel J, Walser JC, Pichon S, Bürgmann H, Matthews B (2015) Estimating Bacterial Diversity for Ecological Studies: Methods, Metrics, and Assumptions. PLoS ONE 10- e0125356.
19. Liu Z, DeSantis TZ, Andersen GL, Knight R (2008) Accurate taxonomy assignments from 16S rRNA sequences produced by highly parallel pyrosequencers. Nucleic Acids Res 36: e120.
20. Huse SM, Dethlefsen L, Huber JA, Mark Welch D, Welch DM, et al. (2008) Exploring microbial diversity and taxonomy using SSU rRNA hypervariable tag sequencing. PLoS Genet 4: e1000255.
21. Wang Q, Garrity GM, Tiedje JM, Cole JR (2007) Naive Bayesian classifier for rapid assignment of rRNA sequences into the new bacterial taxonomy. Applied and Environment Microbiology 73: 5261-5267.
22. Kumar PS, Brooker MR, Dowd SE, Camerlengo T (2011) Target region selection is a critical determinant of community fingerprints generated by 16S Pyrosequencing. PLoS One 6: e20956.
23. Nikolaki S, Tsiamis G (2013) Microbial Diversity in the Era of Omic Technologies. BioMed Research International 2013: 1-15.
24. Muyzer G, de Waal EC, Uitterlinden AG (1993) Profiling of complex microbial populations by denaturing gradient gel electrophoresis analysis of polymerase chain reaction amplified genes encoding for 16S rRNA. Appl Environ Microbiol 59: 695-700.
25. Cancilla MR, Powell IB, Hillier AJ, Davidson BE (1992) Rapid genomic fingerprinting of Lactococcus lactis strains by arbitrarily primed polymerase chain-reaction with P-32 and fluorescent labels. Appl Environ Microbiol 58: 1772-1775.
26. Moyer CL, Dobbs FC, Karl DM (1995) Phylogenetic diversity of the bacterial community from a microbial mat at an active, hydrothermal vent system, Loihi Seamount, Hawaii. Appl Environ

Microbiol 61: 1555-1562.

27. Sheffield VC, Beck JS, Stone EM, Myers RM (1992) A simple and efficient method for attachment of a 40-base pair, GC-rich sequence to PCR-amplified DNA. BioTechniques 12: 386-387.
28. Rastogi G, Sani R (2011) Molecular Techniques to Assess Microbial Community Structure, Function, and Dynamics in the Environment. In: Springer New York 29-57.
29. Pop M, Salzberg SL (2008) Bioinformatics challenges of new sequencing technology. Trends Genet 24: 142-149.
30. Kircher M, Kelso J (2010) High-throughput DNA sequencing: Concepts and limitations. Bioessays 32: 524-536.
31. Quail MA, Smith M, Coupland P et al. (2012) A tale of three next generation sequencing platforms: comparison of Ion Torrent, Pacific Biosciences and Illumina MiSeq sequencers. BMC Genomics 13: 341.
32. Metzker ML (2010) Sequencing technologies-the next generation. Nat Rev Genet 11: 31-46.
33. Richter DC, Ott F, Auch AF, Schmid R, Huson DH (2008) MetaSim-A Sequencing Simulator for Genomics and Metagenomics. PLoS ONE 3: e3373.
34. Kennedy J, Marchesi, JR (2007) Metagenomic approach to exploit the biotechnological potential of the microbial consortia of marine sponges. Appl Microbiol Biotechnol 75: 11-20.
35. Schmeisser C, Helen S, Wolfgang RS (2007) Metagenomics, biotechnology with non culturable microbes. Appl Microbiol Biotechnol 75: 955-962.
36. Krsek M, Wellington EM (1999) Comparison of different methods for the isolation and purification of total community DNA from soil. J Microbiol Methods 39: 287-294.
37. Miller DN, Bryant JE, Madsen EL, Ghiorse WC (1999) Evaluation and optimization of DNA extraction and purification procedure for soil and sediment samples. Appl Environ Microbiol 65: 4715-4724.
38. Tsai YL, Olson BH (1992) Rapid method for separation of bacterial DNA from humic substances in sediments for polymerase chain reaction. Appl Environ Microbiol 58: 2292-2295.

39. Hoshino YT, Matsumoto N (2004) An improved DNA extraction method using skim milk from soils that strongly adsorb DNA. Microbes Environ 19: 13-19.

40. Eisen JA (2007) Environmental shotgun sequencing: its potential and challenges for studying the hidden world of microbes. PLoS Biol 5: e82.

41. Lasken RS (2009) Genomic DNA amplification by the multiple displacement amplification (MDA) method. Biochem Soc Trans 37: 450-453.

42. Murray IA, Clark TA, Morgan RD, Boitano M, Anton BP, et al. (2012) The methylomes of six bacteria Nucleic Acids Res 40: 11450-11462.

43. De Mandal S, Sanga Z, Senthil Kumar N (2015) Metagenome sequencing reveals Rhodococcus dominance in Farpuk Cave, Mizoram, India, an Eastern Himalayan biodiversity hot spot region. Genome Announc 3: e00610-e00615.

44. De Mandal S, Panda AK, Lalnunmawii E, Bisht SS, Senthil Kumar N (2015) Illumina-based analysis of bacterial community in Khuangcherapuk cave of Mizoram, Northeast India. Genomics Data 5: 13-14.

45. De Mandal S, Sanga Z, Lalremsanga HT, Senthil Kumar N (2015) Bacterial diversity of Murlen National Park located in Indo-Burman Biodiversity hotspot region: A metagenomic approach. Genomics Data 5: 25-26.

46. Panda AK, Bisht SS, Senthil Kumar N, De Mandal S (2015) Investigations on microbial diversity of Jakrem hot spring, Meghalaya, India using cultivation independent approach. Genomics Data 4: 156-157.

47. Fierer N, Leff JW, Adams BJ, Nielsen UN, Scott TB, et al. (2012) Cross-biome metagenomic analyses of soil microbial communities and their functional attributes. PNAS 109: 21390-21395.

48. Uroz S, Ioannidis P, Lengelle J, Cébron A, Morin E, Buée M, et al. (2013) Functional Assays and Metagenomic Analyses Reveals Differences between the Microbial Communities Inhabiting the Soil Horizons of a Norway Spruce Plantation. PLoS ONE 8: e55929.

49. Roesch LF, Fulthorpe RR, Riva A, Casella G, Hadwin AK, et al.

(2007) Pyrosequencing enumerates and contrasts soil microbial diversity. ISME Journal 1: 283-290.

50. Acosta-Martínez V, Dowd S, Sun Y, Allen V (2008) Tag-encoded pyrosequencing analysis of bacterial diversity in a single soil type as affected by management and land use. Soil Biology and Biochemistry 40: 2762-2770.
51. Chhabra S, Brazil D, Morrissey J, Burke JI, O'Gara F, et al. (2013) Characterization of mineral phosphate solubilization traits from a barley rhizosphere soil functional metagenome. Microbiologyopen 2: 717-724.
52. Rodrigues JLM, Pellizari VH, Mueller R, Baek K, Jesus EC, et al. (2013) Conversion of the Amazon rainforest to agriculture results in biotic homogenization of soil bacterial communities. Proc Natl Acad Sci USA 110: 988-993.
53. Navarrete AA, Tsai SM, Mendes LW, Faust K, de Hollander M, et al. (2015) Soil microbiome responses to the short-term effects of Amazonian deforestation. Mol Ecol 24: 2433-2448.
54. Yasir M, Azhar E, Khan I, Bibi F, Baabdullah, et al. (2015) Composition of soil microbiome along elevation gradients in southwestern highlands of Saudi Arabia. BMC Microbiol 15: 65.
55. Pechal JL, Crippen TL, Tarone AM, Lewis AJ, Tomberlin JK, et al. (2013) Microbial Community Functional Change during Vertebrate Carrion Decomposition. PLoS ONE 8: e79035.
56. Carrino-Kyker S, Smemo K, Burke D (2013) Shotgun metagenomic analysis of metabolic diversity and microbial community structure in experimental vernal pools subjected to nitrate pulse. BMC Microbiol 13: 78.
57. Logares R, Lindström ES, Langenheder S, Logue JB, Paterson H et al., (2013) Biogeography of bacterial communities exposed to progressive long-term environmental change. The ISME Journal 7: 937-948.
58. Unno Y, Shinano T (2013) Metagenomic analysis of the rhizosphere soil microbiome with respect to phytic acid utilization. Microbes Environ 28: 120-127.
59. Jiang XT, Peng X, Deng GH, Sheng HF, Wang Y, et al. (2013) Illumina sequencing of 16S rRNA tag revealed spatial variations

of bacterial communities in a mangrove wetland. Microb Ecol 66: 96-104.

60. Nacke H, Thürmer A, Wollherr A, Will C, Hodac L, et al. (2011) Pyrosequencingbased assessment of bacterial community structure along different management types in german forest and grassland soils. PLoS ONE 6: e17.

CHAPTER 7

METASIM—A SEQUENCING SIMULATOR FOR GENOMICS AND METAGENOMICS

Daniel C. Richter[1], Felix Ott[1], Alexander F. Auch[1], Ramona Schmid[2], Daniel H. Huson[1]

[1] ZBIT- Center for Bioinformatics Tu¨bingen, University of Tu¨bingen, Tu¨bingen, Germany

[2] Boehringer Ingelheim Pharma GmbH & Co. KG, Biberach, Germany

ABSTRACT

Background

The new research field of metagenomics is providing exciting insights into various, previously unclassified ecological systems. Next-generation sequencing technologies are producing a rapid increase of environmental data in public databases. There is great need for specialized software solutions and statistical methods for dealing with complex metagenome data sets.

Methodology/Principal Findings

To facilitate the development and improvement of metagenomic tools and the planning of metagenomic projects, we introduce a sequencing simulator called MetaSim. Our software can be used to generate collections of synthetic reads that reflect the diverse taxonomical composition of typical metagenome data sets. Based on a database of given genomes, the program allows the user to design a metagenome by specifying the number of genomes present at different levels of the NCBI taxonomy, and then to collect reads from the metagenome using a simulation of a number of different sequencing technologies. A population sampler optionally produces evolved sequences based on source genomes and a given evolutionary tree.

Conclusions/Significance

MetaSim allows the user to simulate individual read datasets that can be used as standardized test scenarios for planning sequencing projects or for benchmarking metagenomic software.

INTRODUCTION

Metagenomics is based on the isolation and characterization of DNA from environmental samples without the need for prior cultivation of microorganisms. In contrast to single genome studies, analyses are applied to entire communities of microbes instead of only few isolated organisms. It has already led to exciting insights into the ecology of different habitats such as ocean [1], soil [2], acid mine [3], human and mouse gut [4], [5] and even into ancient DNA [6].

The research field of Metagenomics is spurred by the recent development and improvement of next-generation sequencing technologies like Roche›s 454 pyrosequencing [7]. Although these high through-put technologies promise faster and relatively inexpensive generation of reads, Sanger sequencing still has been used in environmental genome projects [5] to avoid the drawbacks of shorter read lengths.

In general, studies show that algorithms developed for single-genome assembly are only suitable for environmental sequences under special conditions, for example in low complexity populations [2], [8]. In particular, it is very difficult to assemble reads from highly diverse ecologic systems [9]. The problem is that the arrangement of reads into contigs fails or is misleading because contigs are put together from reads from many different genomes.

Currently, the primary goals of metagenomic studies are the investigation of the phylogenetic composition of the sample (taxonomical binning, "Who is out there"), the quantitive analysis ("How many are there?") and the prediction of genes and their functions (functional binning, "What are they doing"). Since the amount of comparable environmental data is rapidly growing, comparative studies of multiple metagenomic data sets are of great interest as wells. As of September 2008, 44 metagenome studies have already been conducted whereas 86 projects still are on-going [10].

Common strategies for taxonomical binning are for example: (1) detecting phylogenetic markers like rRNA, RecA, heat shock protein (HSP70) and elongation factors (EF-Tu, EF-G) [11], (2) comparing reads against a reference database such as NCBI-nr [12] and then analyzing the matches to place the reads in the NCBI taxonomy [13] and (3) measuring the oligonucleotide frequency caused by codon usage or restriction-site frequency [14]–[18].

When it comes to functional binning, sequences are compared to known protein functions, families and pathways provided by several databases, for example COG, KEGG, PFAM, SEED, STRING and TIGRFAM [19]–[24]. A de novo search for (unknown) functional units is only feasible if either long reads or contigs are available for the detection of open reading frames.

Another challenge in metagenomic studies is the development of robust statistical techniques [25]. Particularly with regard to comparative metagenomics dealing with highly variable data, these techniques are considered as indispensable for a well-founded analysis. Despite the enormous amount of sequence data that was generated and analyzed in the past few years, the number of publicly available software specialized in metagenomic data analysis is surprisingly low. Hence, many studies still make use of classic methods, software or web services that originally were not intended for metagenomic data analysis and have to be adapted or pipelined to produce the desired results [8].

Thus, there is a great demand for specialized metagenomic software supporting the analysis process. Because of the complexity of metagenomic data, it is crucial to benchmark new and existing software with standardized test cases using simulated and verifiable data. A first study [9] provides three data sets with varying complexity by selecting original sequence reads from 113 isolated genomes. In their paper, the authors anticipate that these data sets will be used as standard test cases for software testing.

Some other publications already applied the software ReadSim (pre-version of MetaSim, unpublished) to generate simulated read data sets for testing their software [18], [26].

Description of MetaSim

MetaSim takes as input a set of known genome sequences and an abundance profile. This profile determines which genome sequences are selected for the simulation and the relative abundance of each genome sequence in the dataset.

MetaSim integrates an "induced tree view" of the NCBI taxonomy [27] that can be used to interactively select taxa and inner nodes of the taxonomy to configure their relative abundances. Additionally, the user is able to simulate an "evolved" population of a single genome sequence, using a population simulator. This feature is aimed at simulating the common real world situation that many different, but closely related strains of a lineage coexist in the same habitat.

Finally, for the construction of a realistic read data set, MetaSim includes a versatile read sequencing simulator. The user is able to choose from different (adaptable) error models of current sequencing technologies (e.g. Sanger [28], [29], Roche›s 454 [7] and Illumina (former Solexa) [30]).

MetaSim allows one to construct verifiable read data sets, and additionally, metagenomes variable in size, taxonomical composition and abundance to reflect the diverse and complex output of real metagenomic studies. The resulting data sets can be used to plan and design metagenomic studies and for evaluation and improvement of metagenomics software tools, statistical methods or assembly algorithms.

Availability

MetaSim is written in Java and can be run with a graphical user interface or in command line mode. Installers for Linux/Unix, MacOS X and Windows are freely available from our website at: http://www-ab.informatik.uni-tuebingen.de/software/metasim.

METHODS

MetaSim's processing pipeline consists of several phases:

- Selection of source genome sequences from the internal database
- Configuration of the species abundance profile by setting the relative copy number of the genome sequences
- Sampling sequencing of fragments according to the species abundance profiles
- Application of technology-specific error models to the fragments to create sequencing reads

Configuration of Species Abundance Profiles

At the beginning, whole genome sequences available from public database can be stored locally as source sequences in an integrated database. The user specifies the relative abundance of each genome sequence in a text-based profile file. An interesting feature of MetaSim is the possibility of assigning frequency values not only at the species level but also at higher taxonomical levels. For example, if the genus Escherichia is assigned a certain amount of genome copies, this amount is split and applied uniformly to all descendant species whose sequences are available from the internal database.

To facilitate this data composition process in GUI mode, MetaSim provides an interactive taxonomy editor that visualizes the induced NCBI taxonomy, i.e. the genome sequences listed in the profile file are displayed as nodes in a rooted tree (Figure 1). Node sizes reflect the relative number of genome copies for each given taxon.

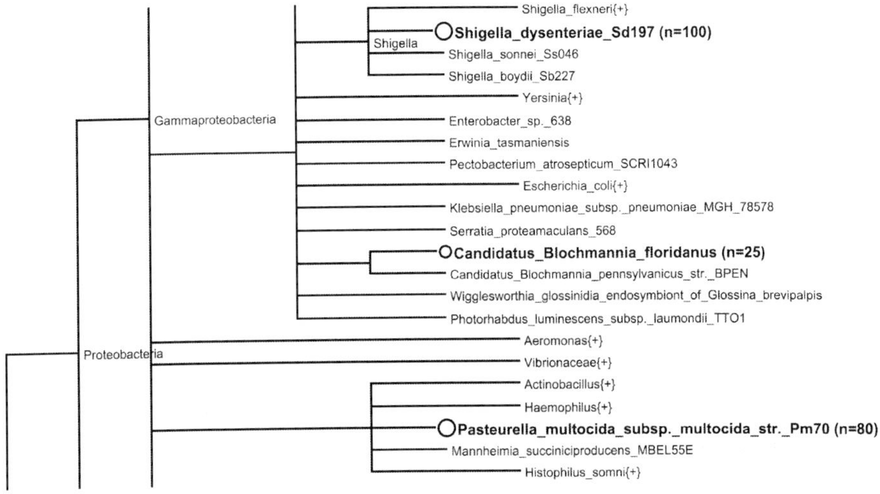

Figure 1: Taxonomy Editor.

A clipping of the taxonomy editor view is shown. Three taxa are assigned an abundance value (number in parenthesis). These settings can be either determined in a text-based abundance profile file or directly in the taxonomy editor by right-clicking on a node.

https://doi.org/10.1371/journal.pone.0003373.g001

Population Sampling

The current genome databases reflect only a small part of earth's still unexplored microbial diversity. Thus, a simulated metagenome only based on known genome sequences does not adequately reflect the complexity of realistic data sets.

MetaSim therefore includes a population sampler that optionally generates a set of evolved (mutated) offsprings derived from single source genomes, using a given evolutionary tree. This tree describes how the offspring sequences descend from the source sequence. By default, a random pyholgenetic tree is generated under the Yule-Harding model [31], [32], but alternatively, user-defined trees can also be loaded. As a simple model of DNA evolution, the Jukes-Cantor formula [33] is applied to estimate a probability of change for each base pair, with a customizable transition rate α (0.001 by default) and time t based on the edge weights. MetaSim then generates the designated number of evolved genomes and then adds them to the internal genome database. As an example, a fragment recruitment plot (according to [1]) shows 10000 sampled Sanger reads of 100 evolved offspring sequences (α=0.004) mapped to the source genome (Escherichia coli K-12 substr. MG1655) using blastn (Figure 2). Read sequences sampled directly from the source genome show a significantly higher identity compared to the mutated sequencing reads.

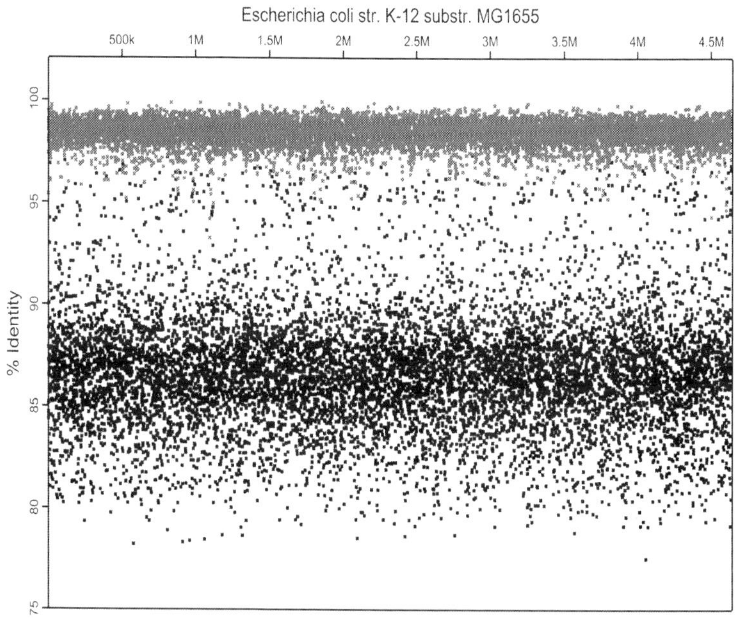

Figure 2: Fragment Recruitment Plot.

Black dots represent 10,000 sequencing reads (Sanger technology, ≈800 bp) drawn from 100 evolved offsprings (α=0.004) of the source genome Escherichia coli K-12 substr. MG1655. Their sequence identity is lower compared to the mapped reads sampled directly form the source genome (red dots).

https://doi.org/10.1371/journal.pone.0003373.g002

Read Sampling

MetaSim simulates both Sanger sequencing and Roche's 454 (sequencing-by-synthesis) approach. Additionally, it provides a flexible, empirical error model usable to simulate Illumina's ultra-short reads. For the simulation of read sequences, statistical approaches are adopted to simulate the distribution of read lengths, its frequency rate and the use of error models depending on the chosen sequencing technology.

To be able to model mate-pairs as well, MetaSim first extracts large fragments called clonesfrom the set of genomes with normally or uniformly distributed lengths. For example, clones with a length of 1000 bp and a standard deviation of 100 bp are modelled with a normal distribution N(1000,100) (Figure 3). The overall number of clones is determined by the number of reads or mate-pairs the user desires to generate.

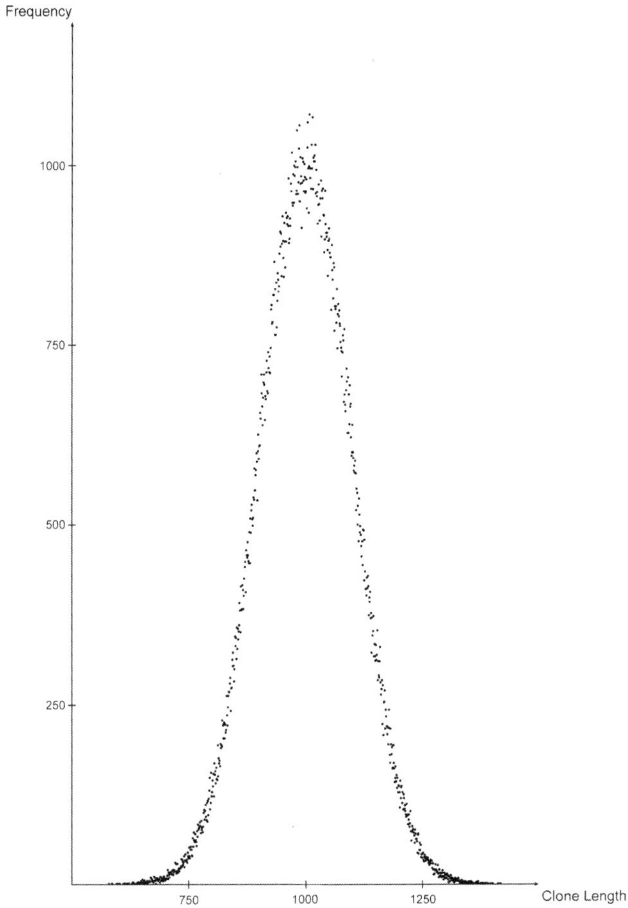

Figure 3: Frequency distribution of clone lengths.

As an example, 250,000 clones with mean length 1000 bp and standard deviation of 100 bp were modelled with a normal distribution.

https://doi.org/10.1371/journal.pone.0003373.g003

If only one source genome is present in the given profile, the clones are randomly extracted from this single sequence. In contrast, in a typical metagenome simulation, the clones have to be sampled from many genomes of varying length, copy number (e.g. to model the abundance of plasmids versus the organsim genomes) and abundancies.

So, each genome sequence s is assigned a weight

$$w_s = l_s \times c_s \times a_s \qquad (1)$$

where l_s is the length, c_s is the copy number and a_s is the specified relative abundance of the genome sequence s as determined in the profile.

For each length of the clone length distribution, the weights of all sequences are summed up to receive the summarized weight w_{sum} that is used to compute a sequence probability $p_s=w_s/w_{sum}$. Considering the overall lengths distribution, a frequency value for each source sequence is then obtained.

After the clone sampling, the ends of the clones are the basis for the subsequent sampling of the reads or mate-pairs, respectively. Again, read lengths can be either uniformly or normally distributed. Finally, read sequences are processed and modified by applying the selected error model.

Simulation of Sanger Sequencing

A widely-used approach to sequencing large DNA molecules is Sanger sequencing, using a shotgun approach that involves cloning small pieces (or inserts) of DNA and then determining their sequence using fluorescent dideoxynucleotides for termination and capillary electrophoresis.

To simulate Sanger sequencing, we closely followed the implementation of celsim reported in [34]. Each read is subjected to a linearly increasing error rate. We model fixed percentages of deletion errors, insertion errors and substitutions. Further, the simulator is capable of modeling mate-pairs and one can specify the length distribution of inserts.

Simulation of Sequencing-by-Synthesis

In pyrosequencing, the intensity of emitted light is used to estimate the length of homopolymers, i.e. runs of identical nucleotides in a sequence. During sequencing, the four DNA composing nucleotides are periodically flowed over the inserts to be sequenced. Within each flow, the intensity of the signal emitted (which is linear up to 8 bp) reflects the number of nucleotides incorporated. Thus, the addition of a single base or even homopolymer stretches of multiple bases in a single flow can be detected.. For chemical and technical reasons, this signal is subject to fluctuations that lead to sequencing errors. In [7], an error rate of about 3% is reported.

Let r denote the length of a given homopolymer. We model the emitted light intensity using a normal distribution N(μ, σ), with mean μ=r and standard deviation $\sigma = k \cdot \sqrt{r}$, where k is a fixed proportionality factor. Following [7], by default we use k=0.15. Although basic statistics implies that the standard deviation should grow with the square root of r, in [7] the standard deviation of the light intensity emitted during 454-sequencing is reported to be σ=k·r. Both variants of the calculation are implemented in our software.

A negative flow is a flow of nucleotides in which the sequence to synthesize is not elongated. Light intensities of negative flows follow a lognormal distribution, with mean μ=0.23, and standard deviation σ=0.15, see [7]. A random variable X is said to be lognormally distributed, if the random variable ln(X) is normally distributed.

We simulate base-calling intensities of negative flows and model the misinterpretation of null-mers as homopolymers of length=1 (insertion). Our algorithm takes the order of the sequencing flows into account. Since the nucleotides are cyclically flowed in the order T,A,C,G, after a given base only two specific negative flows in a specific order are allowed.

During base-calling the algorithm looks at each homopolymer, generates a N(μ, σ) distributed random variable and, according to this variable, decides which length to set for the observed homopolymer.

The signal intensity space is separated into disjoint intervals by probability density functions of all homopolymer signals. The intersections of the density functions $f_{N(r_1,\sigma)}(x)$, and $f_{N(r_2,\sigma)}(x)$ of the normal distributions for different homopolymer lengths r_1 and r_2 are calculated and stored. In each interval one probability density function is maximized. These values are used to decide which homopolymer length is called given a certain signal intensity.

Simulation of Reads with Empirical Models

As an additional feature, MetaSim includes an empirical error model that allows the incorporation of user-defined error statistics.

The probability of an occurrence of a sequencing error often depends on the position of the erroneous base and its surrounding bases. The program GenFrag [35] originally developed an error model which incorporates different error types (deletion, insertion, substitution) at certain positions with empirical error probabilities.

MetaSim adopts this general approach. The error model is based on mappings (error curves) that assign error rates to specific base positions. Each mapping has three parameters (the last two are optional):

- type of error,
- base at the position where the error occurs and
- base preceding the position where the error occurs.

In this way, 48 independent mappings can be individually determined. In addition, in case of a substitution error the user can specify the probability of integrating a particular base depending on the type of the base at the error position and the preceding base.

Using empirical data, MetaSim provides an error model for the short reads of the Illumina technology.

RESULTS

MetaSim can be controlled either by a graphical user interface (Figure 4) or by command line. The command line mode provides access to most of the functions needed for automatic simulation runs. A simulation run generating 400,000 454 reads of length ≈250 bp (a total of 100 Mbp) takes less than 80 seconds on a 2,13 GHz single processor computer.

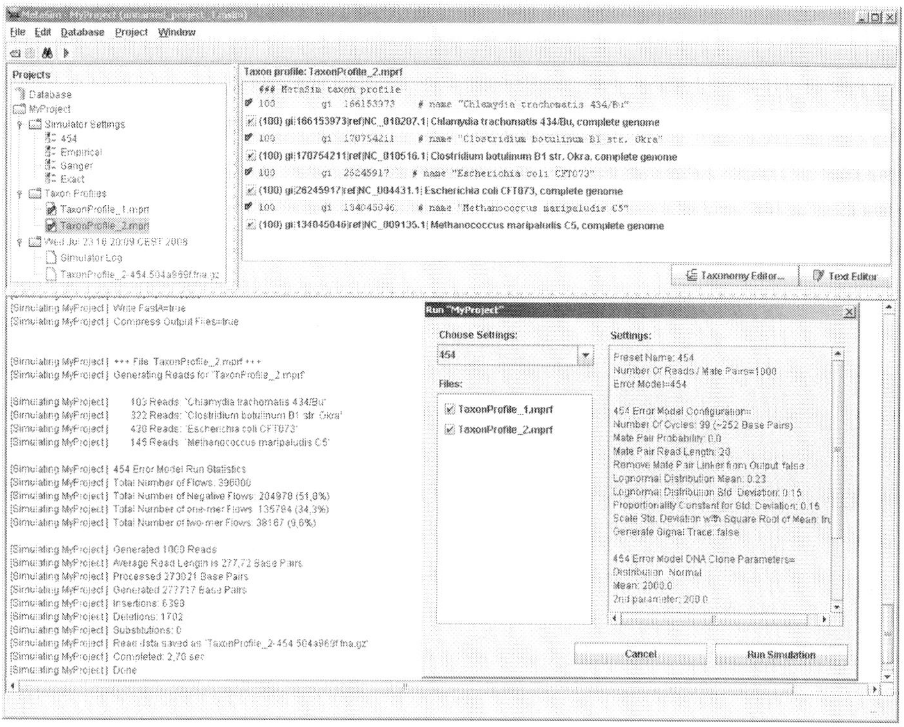

Figure 4: The graphical user interface of MetaSim is divided into three panels: a project tree on the left containing all simulation settings and taxon profiles, an overview and edit panel on the right and a message panel at the bottom.

Additionally, a configuration window is shown.

https://doi.org/10.1371/journal.pone.0003373.g004

To show the utility of this simulation software, for example, in the benchmarking of new software, we generated 9 data sets using a range of parameters and used them to test how well the MEGAN software succeeded in successfully binning sequences based on taxonomic classification by homology.

Simulation Study

MetaSim was used to generate three simulated data sets of different species compositions. These data sets were designed and named simLC, simMC and simHC representing low, medium and high complexity communities, respectively (in correspondence to Mavromatis et al.[9]). For each data set, three runs were conducted with different sequencing error models and read lengths. Also, each resulting read data set comprises of approximately 15 Mbp, so the number of reads differ accordingly: for the 454 sequencing technology with read length ≈100 bp (≈250 bp), 150,000 (60,000) read sequences were generated. The third data set consists of Sanger reads (≈800 bp) and consisted of 18,750 reads.

Table 1–3 show the relative abundancies and the resulting number of sampled reads for each selected taxon. The simLC data set (Table 1) consists of two taxonomically distant microbes whose relative abundance values differ significantly. The simMC data set (Table 2) is composed of nine microbial species (all from the phylum γ-Proteobacteria) with two dominant populations. The simHC data set (Table 3) consists of 11 diverse microbial species covering several phyla from the superkingdom bacteria. All species in the simHC data set are sampled with the same relative abundance.

Table 1: Species abundance and percentage of sampled reads of the simLC dataset

https://doi.org/10.1371/journal.pone.0003373.t001

Abdce	Species	Mbp	454-100[a]	454-250[b]	S-800[c]
90	Methanoculleus marisnigri JR1	2.5	82,70	82,61	82,71
10	Escherichia coli str. K-12 substr. MG1655	4.6	17,30	17,39	17,29

[a]454 technology, 150000 reads (length: 100 bp).
[b]454 technology, 60000 reads, (length: 250 bp).
[c]Sanger technology, 18750 reads, (length: 800 bp).
doi:10.1371/journal.pone.0003373.t001

Table 2: Species abundance and percentage of sampled reads of the simMC dataset

https://doi.org/10.1371/journal.pone.0003373.t002

Abdce	Species	Mbp[a]	454-100[b]	454-250[c]	S-800[d]
100	Pseudomonas fluorescens PfO-1	6.4	38,42	38,39	38,13
100	Shigella dysenteriae Sd197	4.6	27,07	27,47	27,25
80	Pasteurella multocida subsp. multocida str. Pm70	2.3	10,82	10,87	10,81
50	Buchnera aphidicola str. APS	6.6	1,97	1,94	1,78
50	Francisella tularensis subsp. tularensis Schu 4	1.9	5,69	5,60	5,56
25	Alcanivorax borkumensis SK2	3.1	4,68	4,57	4,60
25	Candidatus Blochmannia floridanus	7.1	1,03	1,08	1,21
25	Pseudomonas entomophila L48	5.9	8,89	8,65	9,26
5	Escherichia coli str. K-12 substr. MG1655	4.6	1,43	1,43	1,40

[a]The length of plasmids is considered as well.
[b]454 technology, 150,000 reads (length: 100 bp).
[c]454 technology, 60,000 reads, (length: 250 bp).
[d]Sanger technology, 18,750 reads, (length: 800 bp).
doi:10.1371/journal.pone.0003373.t002

Table 3: Species abundance and percentage of sampled reads of the sim-HC dataset.

https://doi.org/10.1371/journal.pone.0003373.t003

Abdce	Species	Mbp[a]	454-100[b]	454-250[c]	S-800[d]
100	Agrobacterium tumefaciens str. C58	5.7	11,7	11,7	11,3
100	Anabaena variabilis ATCC 29413	7.1	14,7	14,9	14,6
100	Archaeoglobus fulgidus DSM 4304	2.2	4,54	4,41	4,55
100	Bdellovibrio bacteriovorus HD100	3.8	7,84	7,79	7,83
100	Campylobacter jejuni subsp. jejuni 81-176	1.7	3,52	3,6	3,57
100	Clostridium acetobutylicum ATCC 824	4.1	8.6	8.56	8,49
100	Lactococcus lactis subsp. cremoris SK11	2.6	5,38	5,32	5,54
100	Nitrosomonas europaea ATCC 19718	2.8	5,81	5,66	5,59
100	Pseudomonas aeruginosa PA7	6.6	13,6	13,6	14
100	Streptomyces coelicolor A3(2)	9.1	18,7	18,9	18,9
100	Sulfolobus tokodaii str. 7	2.7	5,64	5,59	5,6

[a]The length of plasmids is considered as well.
[b]454 technology, 150,000 reads (length: 100 bp).
[c]454 technology, 60,000 reads, (length: 250 bp).
[d]Sanger technology, 18,750 reads, (length: 800 bp).
doi:10.1371/journal.pone.0003373.t003

Obviously, the amount and ratio of sampled reads in each simulation result reflects the configured abundances and genome sizes. For example, in the simHC data set both Campylobacter jejuni subsp. jejuni 81-176 and Streptomyces coelicolor A3(2) are assigned the same relative abundance of 100. The percentage of sampled reads differs about 15% which can be explained by thc difference in genome size (length of contained plasmids are considered as well). Considering all simulations runs, the ratio of sampled reads is almost equal for each species. For example in Table 1, 17,30%, 17,39% and 17,29% reads were sampled for Escherichia coli str. K12 substr. MG1655 when simulating 454 sequencing technology with read length 100 bp and 250 bp and Sanger sequencing with read length of 800 bp, respectively.

Taxonomical Classification using MEGAN

Following the generation of the nine data sets, a taxonomical assignment of the reads with the MEGAN software was conducted to test its binning functionality. Therefore, all generated read sets were blasted against the NCBI-nr database (downloaded March 2008). MEGAN then assigns these reads to taxa in such a way that the taxonomical level of the assigned taxon reflects the level of conservation of the sampled sequence (Parameter settings of MEGAN: minscore: 0.0, toppercent: 1.0, minsupport: 2, winscore: 0.0).

The result for each simulation run is shown in Figure 5A–C. Summarizing all simulations, one observation can be made: The amount of total assignments of reads to taxa correlates with the read length. The longer the read sequence, the more assignments for a taxon are found. This is also true for the correct assignments (true positives).

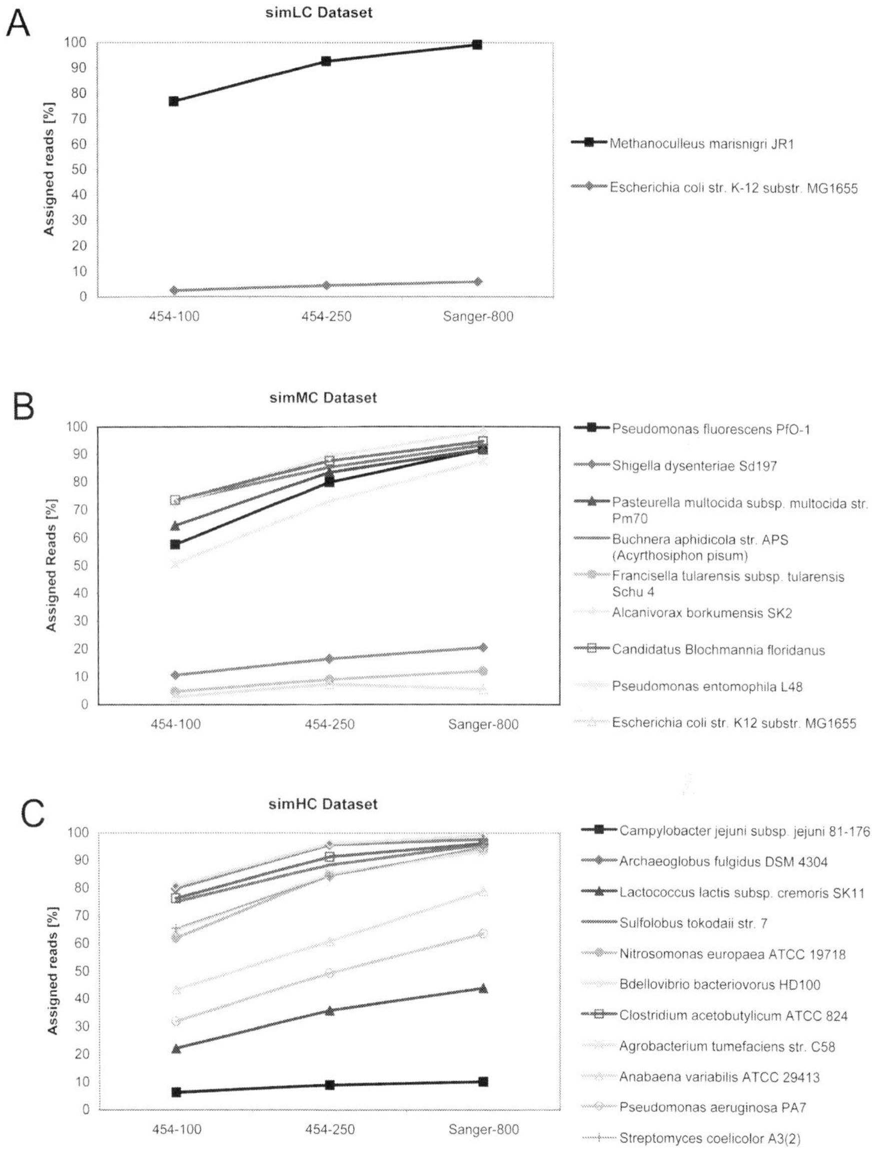

Figure 5: Assignment curves of reads taxonomically classified by MEGAN.

The precentage values refer to the number of sampled reads generated for each organism. (A) The simLC dataset consists of only two organisms. The number of assigned reads to M. marisnigri JR1 almost equals the number of its sampled reads whereas E. coli str. K-12 substr. MG1655 has only few assignmentss. (B) In the simMC dataset, the number of assigned reads increases significantly with longer read lengths (except for Shigella dysenteriae Sd197, Francisella tularensis subsp. tularensis Schu 4 and E. coli str. K12 substr. MG1655). (C) In the simHC dataset, the fraction of assigned reads to Campylobacter jejuni subsp. jejuni 81-176, Lactococcus lactis subsp. cremoris SK11 and Pseudomonas aeruginosa PA7 is rather low compared to the other organisms.

https://doi.org/10.1371/journal.pone.0003373.g005

Accordingly, the fraction of reads that did not match anything in the NCBI-nr database (% no hits) (Table 4) decreases significantly in case of longer read sequences. These findings were expected since longer read sequences generally give rise to more significant high scoring pairs using BLAST.

Table 4: Percentage of assigned, unassigned and "No Hits" reads for all simulation runs.

https://doi.org/10.1371/journal.pone.0003373.t004

	Total Reads	%Assigned Reads	%Unassigned Reads	%No Hits[a]
simLC-454-100	150000	83,14	0,46	16,40
simLC-454-250	60000	98,58	0,85	0,57
simLC-S-800	18750	99,45	0,55	0,00
simMC-454-100	150000	81,71	0,52	17,76
simMC-454-250	60000	98,08	1,02	0,91
simMC-S-800	18750	99,28	0,71	0,01
simHC-454-100	150000	81,68	0,51	17,81
simHC-454-250	60000	97,55	0,93	1,52
simHC-S-800	18750	99,08	0,87	0,05

[a]Reads that did not match anything in the NCBI-nr database.
doi:10.1371/journal.pone.0003373.t004

Independent of the read length, MEGAN is able to assign reads with a high true positive rate. Except for E. coli str. K12 substr. MG1655 and Shigella dysenteriae Sd197 in simMC, virtually all reads (98–100%) are correctly classified. It is important to mention that the composition of the simulated data sets was explicitly intended as an example for a benchmarking study, not for modelling real ecological environments. Thus, simLC consists of only two microbial species that derive from two distinct superkingdoms of the taxonomy (Bacteria and Archae) (Table 1 and Figure 5A). The majority of the sampled reads of E. coli str. K12 substr. MG1655 were assigned to other taxa and clades in the subtree of Bacteria (Figure 6) leading to false positive hits in MEGAN. However, many closely related strains in the subtree of genus Escherichia were hit supporting the evidence that a high number of genetic functions are shared among them.

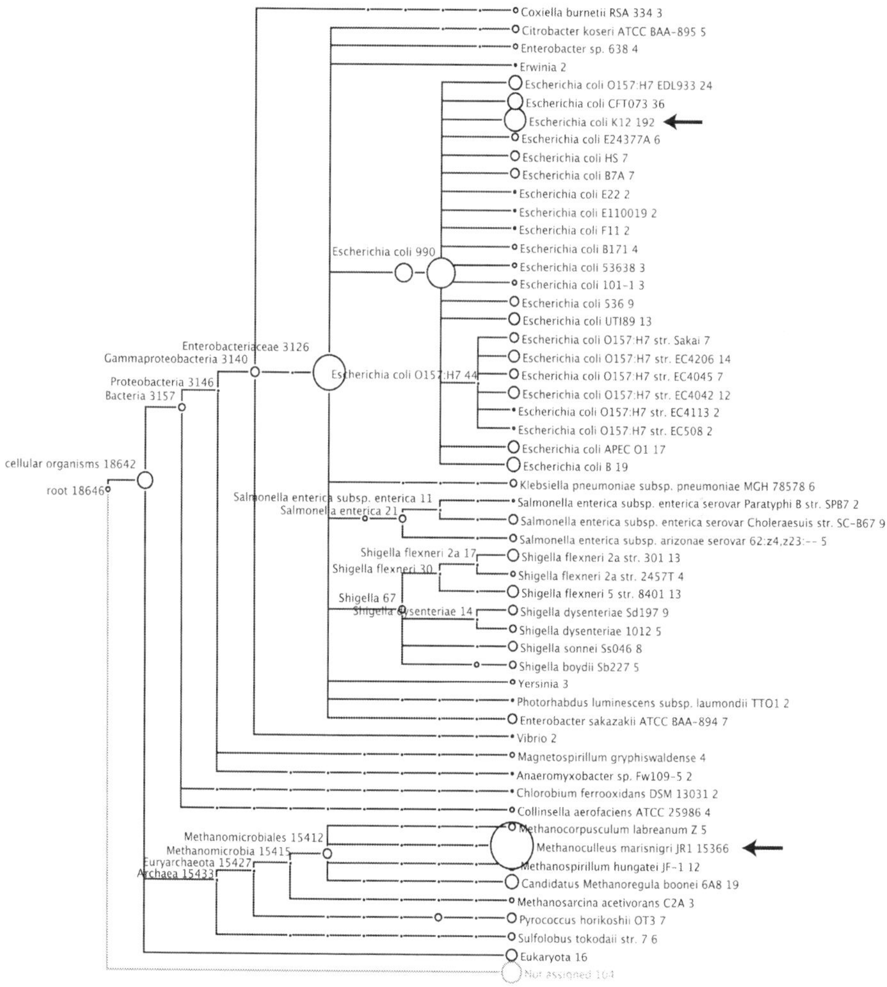

Figure 6: MEGAN visualization of the simLC data set (Sanger technology, read length ≈800 bp).

Two arrows point out the two source genomes of the simulation run. The number of assigned reads to E.coli K12 (192) is rather small compared to the number of sampled reads from the genome of E. coli str. K12 substr. MG1655 (192 assigned vs. 3214 sampled reads). Many reads have BLAST hits in multiple strains and clades, so that MEGAN assigns them to an high-order level in the tree e.g. node Bacteria (3157 reads). M. marisnigri JR1 has only few related strains. In this case, the assignment of reads is more specific (15,366 assigned vs. 15,509 sampled reads).

https://doi.org/10.1371/journal.pone.0003373.g006

A similar problem can be observed in the simMC data set (Figure 5B). To complicate the distinct classification of the reads, all species were taken from the class γ-Proteobacteria. In this example, reads of S. dysenteriae Sd197, Francisella tularensis subsp. tularensis Schu 4and E. coli str. K12 substr. MG1655 could not be completely assigned to the correct taxa. However, the high assignment accuracy (correct hits, %TP) means that practically all assigned reads were correctly classified by MEGAN.

Data set simHC contains sequences of many phyla assigned with the same relative abundance of 100 (Table 3). This implies that the amount of sampled reads depends mainly on the length of the source genome sequences. At first sight, the fraction of assigned reads for the three genomes Campylobacter jejuni subsp. jejuni 81-176, Lactococcus lactis subsp. cremoris SK11and Pseudomonas aeruginosa PA7 is quite low (10.15%, 43.93% and 63.62% for 800 bp Sanger reads, respectively) (Figure 5C). Though, in the taxonomic tree at genus level (Campylobacter, Lactococcus and Pseudomonas), the number of assigned reads virtually equals the number of sampled reads (Figure 7). This is due to the fact that closely related organisms from the same genus share a lot of genes. Again, the number of correct assignments is almost optimal.

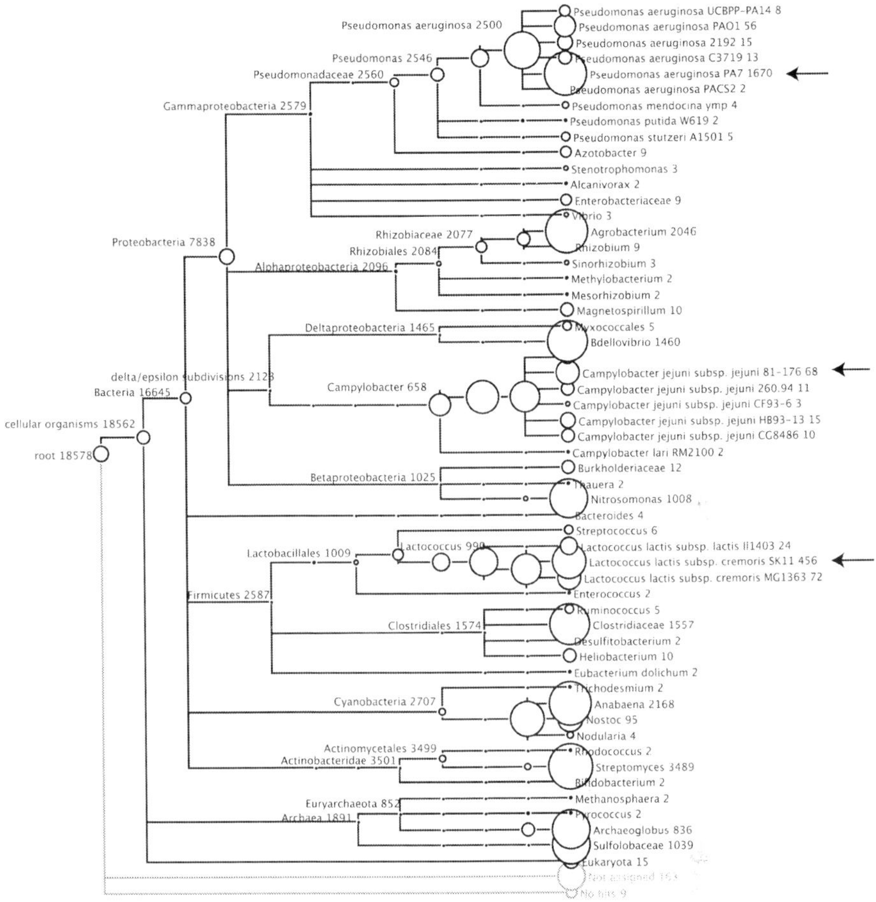

Figure 7: MEGAN visualization of the simHC data set (Sanger technology, read length ≈800 bp).

Arrows point out three of the 11 source genomes of the simulation run that show only few assigned reads at species level compared to the number of originally sampled reads. Due to the fact that C. jejuni subsp. jejuni 81-176, L. lactis subsp. cremoris SK11 and P. aeruginosa PA7 share genes with many closely related strains, most of the sampled reads were assigned by MEGAN to an high-order level in the tree (e.g. genus).

https://doi.org/10.1371/journal.pone.0003373.g007

Summary of MEGAN Results

The analysis of the nine artifical data sets help to reveal the pros and cons of taxonomical binning based on homology as done by MEGAN. All simulation runs indicated that MEGAN generally is capable of binning the majority of all generated reads correctly. Additionally, the number of unassigned reads i.e. reads that actually hit sequences in the database but could not be assigned due to MEGAN's parameter settings, is very low. However, reads having many homologous sequences in the database due to high conservation among microbial families and lineages, lead to a rather "diffuse", but still correct read assignment: due to gene sharing, MEGAN assigns these reads to high-order taxa closer to the root (e.g. at genus level), thus avoiding probable false-positive assignments (MEGAN uses an LCA-based algorithm [13].). This means that MEGAN classifies those reads rather "generally" but the assignment still is adequate for interpreting metagenomic data in sufficiency.

Another observation is that the amount of assigned reads correlates strongly with the read length. Obviously, with reads produced by Sanger sequencing (800 bp), taxonomical binning becomes easier compared to short reads (100–250 bp). This oberservation confirms the findings in [13]. Moreover, the positive effect of better assignment and classification of long reads comes with higher costs and workload in the sequencing phase.

The expense for sequencing with Sanger technology is about $500/Mbp (800 bp reads), whereas e.g. pyrosequencing with Roche's 454 technology is only about $100/Mbp (250 bp reads) yielding many more base pairs per run at the same time. On top of the positive cost factor, compared to the Sanger technology, next-generation sequencers do not suffer from cloning bias.

DISCUSSION

The appearance of next-generation sequencers on the market has boosted the number and scope of (meta-)genomic sequencing projects. A lot of data can be generated in less time demanding fast and precise analysis algorithms and software. However, especially in the field of metagenomics, the problem of producing individual, simulated test cases for benchmarking is open.

We try to fill this gap with MetaSim, a flexible tool for producing simulated read data sets, useful for designing metagenome projects and for testing and comparing metagenomic or assembly software. A lot of parameters can be adapted to generate user defined sequence sets that can serve as verified example data. Currently, the Sanger and Roche's 454 sequencing error model can be selected, as well as the Illumina error model which is based on empirical data. The empirical error model can easily be configured and adapted to other sequencing technologies or error probabilities.

Given this flexibility of the read simulator, it is possible to construct many kinds of individual fragments like, for example reads, contigs or expressed sequence tags (ESTs) derived from existing sequences. We plan to integrate further error models of upcoming sequencers. In addition, future version of MetaSim will extend the population sampler by using more sophisticated models for sequence evolution like HKY [36], as, e.g., already implemented in Seq-Gen [37].

ACKNOWLEDGMENTS

We would like to thank Korbinian Schneeberger and Stephan Ossowski for providing the empirical error models for the Illumina sequencing.

AUTHOR CONTRIBUTIONS

Conceived and designed the experiments: DCR AFA. Wrote the paper: DCR AFA DHH. Contributed to the implementation of the software: FO AFA RS DHH.

REFERENCES

1. Rusch DB, Halpern AL, Sutton G, Heidelberg KB, Williamson S, et al. (2007) The Sorcerer II Global Ocean Sampling expedition: northwest Atlantic through eastern tropical Pacific. PLoS Biol 5: e77. DB RuschAL HalpernG. SuttonKB HeidelbergS. Williamson2007The Sorcerer II Global Ocean Sampling expedition: northwest Atlantic through eastern tropical Pacific. PLoS Biol5e77
2. Tringe SG, von Mering C, Kobayashi A, Salamov AA, Chen K, et al. (2005) Comparative Metagenomics of Microbial Communities.

Science 308: 554–557. SG TringeC. von MeringA. KobayashiAA SalamovK. Chen2005Comparative Metagenomics of Microbial Communities. Science308554557

3. Tyson GW, Chapman J, Hugenholtz P, Allen EE, Ram RJ, et al. (2004) Community structure and metabolism through reconstruction of microbial genomes from the environment. Nature 428: 37–43. GW TysonJ. ChapmanP. HugenholtzEE AllenRJ Ram2004Community structure and metabolism through reconstruction of microbial genomes from the environment. Nature4283743

4. Gill SR, Pop M, Deboy RT, Eckburg PB, Turnbaugh PJ, et al. (2006) Metagenomic analysis of the human distal gut microbiome. Science 312: 1355–1359. SR GillM. PopRT DeboyPB EckburgPJ Turnbaugh2006Metagenomic analysis of the human distal gut microbiome. Science31213551359

5. Turnbaugh PJ, Ley RE, Mahowald MA, Magrini V, Mardis ER, et al. (2006) An obesity-associated gut microbiome with increased capacity for energy harvest. Nature 444: 1027–1031. PJ TurnbaughRE LeyMA MahowaldV. MagriniER Mardis2006An obesity-associated gut microbiome with increased capacity for energy harvest. Nature44410271031

6. Poinar HN, Schwarz C, Qi J, Shapiro B, Macphee RDE, et al. (2006) Metagenomics to paleogenomics: large-scale sequencing of mammoth DNA. Science 311: 392–394. HN PoinarC. SchwarzJ. QiB. ShapiroRDE Macphee2006Metagenomics to paleogenomics: large-scale sequencing of mammoth DNA. Science311392394

7. Margulies M, Egholm M, Altman WE, Attiya S, Bader JS, et al. (2005) Genome sequencing in microfabricated high-density picolitre reactors. Nature 437: 376–380. M. MarguliesM. EgholmWE AltmanS. AttiyaJS Bader2005Genome sequencing in microfabricated high-density picolitre reactors. Nature437376380

8. Pachter L (2007) Interpreting the unculturable majority. Nat Methods 4: 479–480. L. Pachter2007Interpreting the unculturable majority. Nat Methods4479480

9. Mavromatis K, Ivanova N, Barry K, Shapiro H, Goltsman E, et al. (2007) Use of simulated data sets to evaluate the fidelity of metagenomic processing methods. Nat Methods 4: 495–500. K.

MavromatisN. IvanovaK. BarryH. ShapiroE. Goltsman2007Use of simulated data sets to evaluate the fidelity of metagenomic processing methods. Nat Methods4495500

10. Bernal A, Ear U, Kyrpidesa N (2001) Genomes OnLine Database (GOLD): a monitor of genome projects world-wide. Nucleic Acids Res 29: 126–127. A. BernalU. EarN. Kyrpidesa2001Genomes OnLine Database (GOLD): a monitor of genome projects world-wide. Nucleic Acids Res29126127
11. Venter JC, Remington K, Heidelberg JF, Halpern AL, Rusch D, et al. (2004) Environmental Genome Shotgun Sequencing of the Sargasso Sea. Science 304: 66–74. JC VenterK. RemingtonJF HeidelbergAL HalpernD. Rusch2004Environmental Genome Shotgun Sequencing of the Sargasso Sea. Science3046674
12. Benson DA, Karsch-Mizrachi I, Lipman DJ, Ostell J, Wheeler DL (2005) GenBank. Nucleic Acids Res 1: D34–38. DA BensonI. Karsch-MizrachiDJ LipmanJ. OstellDL Wheeler2005GenBank. Nucleic Acids Res1D3438
13. Huson DH, Auch AF, Qi J, Schuster SC (2007) MEGAN analysis of metagenomic data. Genome Res 17: 377–386. DH HusonAF AuchJ. QiSC Schuster2007MEGAN analysis of metagenomic data. Genome Res17377386
14. Karlin S, Burge C (1995) Dinucleotide relative abundance extremes: a genomic signature. Trends Genet 11: 283–290. S. KarlinC. Burge1995Dinucleotide relative abundance extremes: a genomic signature. Trends Genet11283290
15. Deschavanne PJ, Giron A, Vilain J, Fagot G, Fertil B (1999) Genomic signature: characterization and classification of species assessed by chaos game representation of sequences. Mol Biol Evol 16: 1391–1399. PJ DeschavanneA. GironJ. VilainG. FagotB. Fertil1999Genomic signature: characterization and classification of species assessed by chaos game representation of sequences. Mol Biol Evol1613911399
16. Teeling H, Waldmann J, Lombardot T, Bauer M, Glšckner FO (2004) TETRA: a web-service and a stand-alone program for the analysis and comparison of tetranucleotide usage patterns in DNA sequences. BMC Bioinformatics 5: 163. H. TeelingJ. WaldmannT. LombardotM.

BauerFO Glšckner2004TETRA: a web-service and a stand-alone program for the analysis and comparison of tetranucleotide usage patterns in DNA sequences. BMC Bioinformatics5163

17. McHardy AC, Mart'n HGa, Tsirigos A, Hugenholtz P, Rigoutsos I (2007) Accurate phylogenetic classification of variable-length DNA fragments. Nat Methods 4: 63–72. AC McHardyHGa Mart'nA. TsirigosP. HugenholtzI. Rigoutsos2007Accurate phylogenetic classification of variable-length DNA fragments. Nat Methods46372
18. Chatterji S, Yamazaki I, Bai Z, Eisen JA Vingron M, Wong L, editors. (2008) CompostBin: A DNA composition-based algorithm for binning environmental shotgun reads. Lecture Notes in Computer Science Springer. 17–28. S. ChatterjiI. YamazakiZ. BaiJA Eisen2008CompostBin: A DNA composition-based algorithm for binning environmental shotgun reads. M. VingronL. WongLecture Notes in Computer ScienceSpringer1728
19. Tatusov RL, Koonin EV, Lipman DJ (1997) A genomic perspective on protein families. Science 278: 631–637. RL TatusovEV KooninDJ Lipman1997A genomic perspective on protein families. Science278631637
20. Kanehisa M, Goto S (2000) KEGG: kyoto encyclopedia of genes and genomes. Nucleic Acids Res 28: 27–30. M. KanehisaS. Goto2000KEGG: kyoto encyclopedia of genes and genomes. Nucleic Acids Res282730
21. Finn RD, Tate J, Mistry J, Coggill PC, Sammut SJ, et al. (2008) The Pfam protein families database. Nucleic Acids Res 36: D281–D288. RD FinnJ. TateJ. MistryPC CoggillSJ Sammut2008The Pfam protein families database. Nucleic Acids Res36D281D288
22. Overbeek R, Begley T, Butler RM, Choudhuri JV, Chuang H-Y, et al. (2005) The subsystems approach to genome annotation and its use in the project to annotate 1000 genomes. Nucleic Acids Res 33: 5691–5702. R. OverbeekT. BegleyRM ButlerJV ChoudhuriH-Y Chuang2005The subsystems approach to genome annotation and its use in the project to annotate 1000 genomes. Nucleic Acids Res3356915702
23. von Mering C, Jensen LJ, Snel B, Hooper SD, Krupp M, et al. (2005) STRING: known and predicted protein-protein associations,

integrated and transferred across organisms. Nucleic Acids Res 33: D433–D437. C. von MeringLJ JensenB. SnelSD HooperM. Krupp2005STRING: known and predicted protein-protein associations, integrated and transferred across organisms. Nucleic Acids Res33D433D437

24. Haft DH, Selengut JD, White O (2003) The TIGRFAMs database of protein families. Nucleic Acids Res 31: 371–373. DH HaftJD SelengutO. White2003The TIGRFAMs database of protein families. Nucleic Acids Res31371373
25. Rodriguez-Brito B, Rohwer F, Edwards RA (2006) An application of statistics to comparative metagenomics. BMC Bioinformatics 7: 162. B. Rodriguez-BritoF. RohwerRA Edwards2006An application of statistics to comparative metagenomics. BMC Bioinformatics7162
26. Krause L, Diaz NN, Goesmann A, Kelley S, Nattkemper TW, et al. (2008) Phylogenetic classification of short environmental DNA fragments. Nucleic Acids Res 36(7): 2230–2239. L. KrauseNN DiazA. GoesmannS. KelleyTW Nattkemper2008Phylogenetic classification of short environmental DNA fragments. Nucleic Acids Res36(7)22302239
27. Wheeler DL, Barrett T, Benson DA, Bryant SH, Canese K, et al. (2008) Database resources of the National Center for Biotechnology Information. Nucleic Acids Res 36: D13–D21. DL WheelerT. BarrettDA BensonSH BryantK. Canese2008Database resources of the National Center for Biotechnology Information. Nucleic Acids Res36D13D21
28. Meldrum D (2000) Automation for Genomics, Part one: Preparation for Sequencing. Genome Research 10: 1081–1092. D. Meldrum2000Automation for Genomics, Part one: Preparation for Sequencing. Genome Research1010811092
29. Meldrum D (2000) Automation for Genomics, Part two: Sequencers, Microarrays, and Future trends. Genome Research 10: 1288–1303. D. Meldrum2000Automation for Genomics, Part two: Sequencers, Microarrays, and Future trends. Genome Research1012881303
30. Bentley DR (2006) Whole-genome re-sequencing. Current Opinion in Genetics & Development 16: 545–552. DR Bentley2006Whole-genome re-sequencing. Current Opinion in Genetics &

Development16545552

31. Yule GU (1925) A mathematical theory of evolution, based on the conclusions of Dr. J. C. Willis. Philosophical Transactions of the Royal Society of London Ser B, Biol Sci 213: 21–87. GU Yule1925A mathematical theory of evolution, based on the conclusions of Dr. J. C. Willis. Philosophical Transactions of the Royal Society of London Ser B, Biol Sci2132187
32. Harding EF (1971) The Probabilities of Rooted Tree-Shapes Generated by Random Bifurcation. Advances in Applied Probability 3: 44–77. EF Harding1971The Probabilities of Rooted Tree-Shapes Generated by Random Bifurcation. Advances in Applied Probability34477
33. Jukes TH, Cantor CR (1969) Evolution of Protein Molecules. In: Munro HN, editor. Mammalian Protein Metabolism. New York, NY: Academic Press. pp. 21–132. TH JukesCR Cantor1969Evolution of Protein Molecules. HN MunroMammalian Protein MetabolismNew York, NYAcademic Press21132
34. Myers G (1999) A dataset generator for whole genome shotgun sequencing. pp. 202–210. G. Myers1999A dataset generator for whole genome shotgun sequencing202210
35. Engle ML, Burks C (1994) GenFrag 2. 1: new features for more robust fragment assembly benchmarks. Comput Appl Biosci 10: 567–568. ML EngleC. Burks1994GenFrag 2. 1: new features for more robust fragment assembly benchmarks. Comput Appl Biosci10567568
36. Hasegawa M, Kishino H, Yano T (1985) Dating of the human-ape splitting by a molecular clock of mitochondrial DNA. J Mol Evol 22: 160–174. M. HasegawaH. KishinoT. Yano1985Dating of the human-ape splitting by a molecular clock of mitochondrial DNA. J Mol Evol22160174
37. Rambaut A, Grassly NC (1997) Seq-Gen: an application for the Monte Carlo simulation of DNA sequence evolution along phylogenetic trees. Comput Appl Biosci 13: 235–238. A. RambautNC Grassly1997Seq-Gen: an application for the Monte Carlo simulation of DNA sequence evolution along phylogenetic trees. Comput Appl Biosci13235238

CHAPTER 8

CALIBRATION AND ANALYSIS OF GENOME-BASED MODELS FOR MICROBIAL ECOLOGY

Stilianos Louca[1], Michael Doebeli[2]

[1] Institute of Applied Mathematics, University of British Columbia, Vancouver, Canada

[2] Department of Zoology, University of British Columbia, Vancouver, Canada

ABSTRACT

Microbial ecosystem modeling is complicated by the large number of unknown parameters and the lack of appropriate calibration tools. Here we present a novel computational framework for modeling microbial ecosystems, which combines genome-based model construction with statistical analysis and calibration to experimental data. Using this framework, we examined the dynamics of a community of *Escherichia* *coli*strains that emerged in laboratory evolution experiments, during which an ancestral strain diversified into two coexisting ecotypes. We constructed a microbial community model comprising the ancestral and the evolved strains, which we calibrated using separate monoculture experiments. Simulations reproduced the successional dynamics in the evolution experiments, and pathway activation patterns observed in microarray transcript profiles. Our approach yielded detailed insights into the metabolic processes that drove bacterial diversification, involving acetate cross-feeding and competition for organic carbon and oxygen. Our framework provides a missing link towards a data-driven mechanistic microbial ecology.

eLife DIGEST

Microbes like bacteria and yeast play important roles in the environment, human health and even some industrial processes. However, it is difficult to understand the roles of microbes in these situations because many different types of microbes often live together in complex communities. Some of the microbes may compete with each other for resources like oxygen or sugar. Others may rely on one another for survival. For example, one microbe may feed on molecules that are released as waste from another microbe.

To better understand these microbial communities, we first need to understand the processes by which each microbe uses nutrients and releases waste molecules that influence other microbes. Researchers have used a technique called 'genome sequencing' to reconstruct the networks of genes and chemical reactions that are involved in these processes, and to build computer models of microbial communities in different environments.

However, the existing models can be labor intensive and do not allow researchers to easily use statistics to analyse them. To address this problem, Louca and Doebeli created a new computer model with built-in statistical tools that accurately predicts the interactions in communities that contain multiple strains of a bacterium called *Escherichia coli.* First, Louca and Doebeli grew a single strain of *E. coli* in the laboratory for many generations, which led to the evolution of the bacteria so that two new strains emerged. One of the new strains was more efficient at using sugar as a food source than the other and sometimes released a molecule called acetate. The other new strain became more efficient at using this acetate.

Next, Louca and Doebeli used data that had been collected for each individual strain, to test whether the model could recreate the way that the new strains had evolved together. The model accurately predicted that the two new strains would gradually replace the original strain. The strain that was more efficient at using sugar emerged first, which led to extra acetate being available for the other new strain that became more efficient at using acetate.

Louca and Doebeli›s findings demonstrate for the first time that data collected for individual microbes can be used to explain the dynamics and evolution of small communities of microbes using mathematical models. The next step is to test this approach on larger communities in the environment.

INTRODUCTION

Metabolic interactions are an emergent property of microbial communities (Morris et al., 2013; Chiu et al., 2014). Even the simplest life forms can only be understood in terms of biological consortia characterized by shared metabolic pathways and distributed biosynthetic capacities (Klitgord and Segrè, 2010; McCutcheon and Moran, 2012; Husnik et al., 2013). For example, glucose catabolism to carbon dioxide or methane is a multi-step process often involving several organisms that indirectly exchange intermediate products through their environment (Stams, 1994). Microbial communities are thus complex systems comprising several interacting components that cannot be fully understood in isolation. In fact, metabolic interdependencies between organisms are at least partially responsible for our current inability to culture the great majority of prokaryotes (Schink and Stams, 2006). Understanding the emergent dynamics of microbial communities is crucial to harnessing these multicomponent assemblages and using synthetic ecology for medical, environmental and industrial purposes (Brenner et al., 2008).

Genome sequencing has enabled the reconstruction of full-scale cell-metabolic networks (Henry et al., 2010), which have provided a firm basis for understanding individual cell metabolism (Varma and

Palsson, 1994; Duarte et al., 2004; Klitgord and Segrè, 2010). Recent work indicates that multiple cell models can be combined to understand microbial community metabolism and population dynamics (Stolyar et al., 2007; Klitgord and Segrè, 2010; Zengler and Palsson, 2012; Chiu et al., 2014; Harcombe et al., 2014). These approaches assume knowledge of all model parameters such as stoichiometric coefficients, maintenance energy requirements or extracellular transport kinetics, a requirement that is rarely met in practice (Feist et al., 2008; Harcombe et al., 2014). Experiments and monitoring of environmental samples could provide valuable data to calibrate microbial community models, for example, via statistical parameter estimation, but appropriate tools are lacking. So far, the standard approach has been to obtain each parameter through laborious specific measurements or from the available literature, or to manually adjust parameters to match observations (Mahadevan et al., 2002; Chiu et al., 2014; Harcombe et al., 2014). Furthermore, statistical model evaluation and sensitivity analysis is typically performed using ad hoc code, thus increasing the effort required for the construction of any new model. Consequently, the experimental validation of genome-based microbial community models and their application to biological questions are rare (Meadows et al., 2010; Harcombe et al., 2014).

We have developed MCM (Microbial Community Modeler), a mathematical framework and computational tool that unifies model construction with statistical evaluation, sensitivity analysis and parameter calibration. MCM is designed for modeling multi-species microbial communities, in which the metabolism and growth of individual cell species is predicted using genome-based metabolic models. Cells in the community interact in a dynamical environment in which metabolite concentrations and other environmental variables influence, and are influenced by, microbial metabolism. Unknown model parameters can be automatically calibrated (fitted) using experimental data such as cell densities, nutrient concentrations or rate measurements. To demonstrate the potential of MCM, we modeled a bacterial community that has emerged from in vitro evolution experiments, during which an ancestral strain repeatedly diversified into two distinct ecotypes. Experiments with microbes have an established tradition as model systems for understanding ecological and evolutionary processes (Elena and Lenski, 2003; Kassen and Rainey, 2004). We show that the predictions derived from MCM are in very good agreement with the outcomes of several monoculture and coculture experiments. While the experimental results described below have been found over the course of several years (Friesen et al., 2004; Spencer et al., 2007; Le Gac et al., 2008; Herron and Doebeli, 2013), it is only now that a mechanistic model has managed to unify them in a clear, unambiguous and synergistic manner. The analysis presented here thus provides strong credence to a large body of experimental work that was done in our lab over the course of roughly a decade.

Model

In MCM, a microbial community model is a set of differential equations for the population densities of the cell species comprising the community and of the ambient concentrations of utilized nutrients (metabolites), coupled to optimization problems for the cell-specific rates of reactions involving these metabolites. Each cell is characterized by its metabolic potential, that is, the genetically determined subset of reactions it can catalyze, as well as any available metabolite transport mechanisms. The reaction rates and metabolite exchange rates (i.e. the metabolism) of each cell are assumed to depend on its metabolic potential as well as on the current environmental conditions, such as metabolite concentrations. Through their metabolism, in turn, cells act as sinks and sources of metabolites in the environment. Additional metabolite fluxes, such as oxygen diffusion from the atmosphere into the growth medium of a modeled bacterial culture, can be included in the model.

At any point in time, individual cell metabolism is determined using flux balance analysis (FBA) (Orth et al., 2010), a widely used framework in cell-metabolic modeling (Varma and Palsson, 1994; Duarte et al., 2004; Klitgord and Segrè, 2010; Freilich et al., 2011; Chiu et al., 2014). In FBA, cell metabolism is assumed to be regulated in such a way that the rate of biosynthesis is maximized (Varma and Palsson, 1994; Feist and Palsson, 2010). The chemical state of cells is assumed to be steady, leading to stoichiometric constraints that need to be satisfied for any particular combination of intracellular reaction rates. Reaction rates, on the other hand, are limited due to finite enzyme capacities. Metabolite uptake/export rates can also be limited due to finite diffusion rates or limited transmembrane transporter efficiency. For example, uptake rates can be Monod-like functions of substrate concentrations (Mahadevan et al., 2002; Harcombe et al., 2014). Taken together, cell-metabolic potential, stoichiometric consistency, reaction rate limits and transport rate limits define the constraints of a linear optimization problem for each cell species at each point in time. The optimized biosynthesis rate is translated into a cell production rate by dividing by the cell's mass, thus defining the species' population growth (Figure 1).

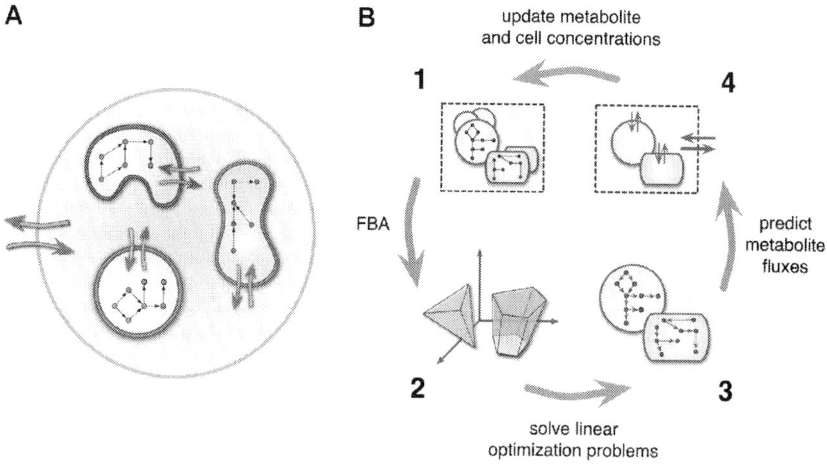

Figure 1: Framework used by MCM. (A) Conceptual framework used by MCM. Cells (colored shapes) optimize their metabolism for maximal growth and influence their environment via metabolite exchange (small colored arrows). Additional external fluxes can also affect the environment (large grey arrows). The environment, in turn, influences each cell's metabolism. (B) Computational framework used by MCM. Each iteration consists of four steps: flux balance analysis (FBA) is used to translate cell-metabolic potentials and environmental conditions (1) into a linear optimization problem for the growth rate of each cell species (2). The set of possible reaction rates corresponds to a polytope in high-dimensional space. Solving the optimization problems (3) yields predictions on microbial metabolite exchange rates (4). Metabolic fluxes and cell growth rates are used to predict metabolite and cell concentrations in the next iteration (1).

The central assumption of individual cells maximizing biosynthesis, subject to environmental and physiological constraints, is rooted in the idea that evolution has shaped regulatory mechanisms of unicellular organisms in such a way that they strive for maximum growth whenever possible. Biosynthesis has been experimentally verified as an objective for *Saccharomyces cerevisiae* and *E. coli* (Burgard and Maranas, 2003; Gianchandani et al., 2008; Harcombe et al., 2013). The assumption of maximized biosynthesis is less valid for genetically engineered organisms or those exposed to environments that are radically different from the environments that shaped their evolution (Segrè et al., 2002). Despite its limitations, FBA has greatly contributed to the understanding of several genome-scale metabolic networks and metabolic interactions between cells (Stolyar et al., 2007; Klitgord and Segrè, 2010; Orth et al., 2010; Freilich et al., 2011; Chiu et al., 2014; Harcombe et al., 2014). One advantage of FBA models over full biochemical cell models is their independence of intracellular kinetics and gene regulation, which limits the number of required parameters to stoichiometric coefficients and uptake kinetics.

The combination of FBA with a varying environmental metabolite pool, as implemented by MCM, is known as dynamic flux balance analysis (DFBA) (Mahadevan et al., 2002; Chiu et al., 2014; Harcombe et al., 2014). In contrast to conventional FBA, DFBA models are dynamical because cell densities and environmental metabolite concentrations both change with time, and the rate of change of each cell density and metabolite concentration depends on the current cell densities and metabolite concentrations (Mahadevan et al., 2002; Harcombe et al., 2014). Because metabolites can be depleted or produced by several cell species, the environmental metabolite pool mediates the metabolic interactions between cells (Schink and Stams, 2006). For example, oxygen uptake rates might depend on environmental oxygen concentrations, which in turn are reduced by cellular respiration. Similarly, cells might excrete acetate as a byproduct of glucose catabolism, which then becomes available to other cells. The metabolic optimization of individual cells striving for maximal growth, while modifying their environment, leads to non-trivial community dynamics that can include competition, cooperation and exploitation. The cell-centric nature of DFBA differs fundamentally from other flux balance analyses of microbial communities that assume an optimization of a community-wide objective such as total biomass synthesis (Stolyar et al., 2007; Klitgord and Segrè, 2011; Zomorrodi and Maranas, 2012). Such an assumption is at least questionable from an evolutionary perspective and likely not appropriate for communities comprising several species (Mitri and Foster, 2013).

Recent work suggests that DFBA is a promising approach to microbial ecological modeling (Meadows et al., 2010; Chiu et al., 2014; Harcombe et al., 2014). For example, Harcombe et al. (2014) designed a computational tool (COMETS) based on DFBA, which was able to accurately predict equilibrium compositions of mixed bacterial cultures grown on petri dishes. However, COMETS offers limited model versatility in terms of uptake and reaction kinetics and only has few environmental feedback mechanisms (namely, varying extracellular metabolite concentrations). Furthermore, it assumes complete knowledge of all required model parameters and provides no generic statistical model analysis. Hence, while COMETS sets an important precedent, considerable work is still needed to make DFBA a practical approach in microbial ecosystem modeling. MCM extends Harcombe et al.›s framework to more versatile microbial ecological models that include arbitrary reaction kinetics (e.g., subject to product-inhibition) as well as dynamical environmental variables (e.g., pH) that influence, and are influenced by, microbial metabolism. In addition, MCM supports cell models in which internal molecules act as dynamical constraints that further restrict the FBA solution space, for example to account for post-transcriptional regulation or delays in enzyme synthesis (Blazier and Papin, 2012). These so called regulatory FBA models have been shown to improve the fidelity of conventional FBA models for *E. coli* and *S. cerevisiae* (Covert et al., 2001; Covert and Palsson, 2002; Covert et al., 2004; Herrgård et al., 2006), however their application to microbial communities remains untested. MCM can statistically evaluate models against data, analyze their sensitivity to varying parameters (Cariboni et al., 2007), and estimate the uncertainty of model predictions in the face of stochasticity (Hammersley and Handscomb, 1964). Perhaps most importantly, MCM can automatically calibrate unknown model parameters to data, for example obtained from monoculture experiments (as demonstrated below), from bioreactor experiments involving multiple species (Louca and Doebeli, 2015) or from environmental samples of unculturable communities (Figure 2; see the ‘Materials and methods’). MCM can thus be used to understand the dynamics of realistic microbial ecosystems, ranging from the soil or groundwater to mixed laboratory cultures and bioreactors.

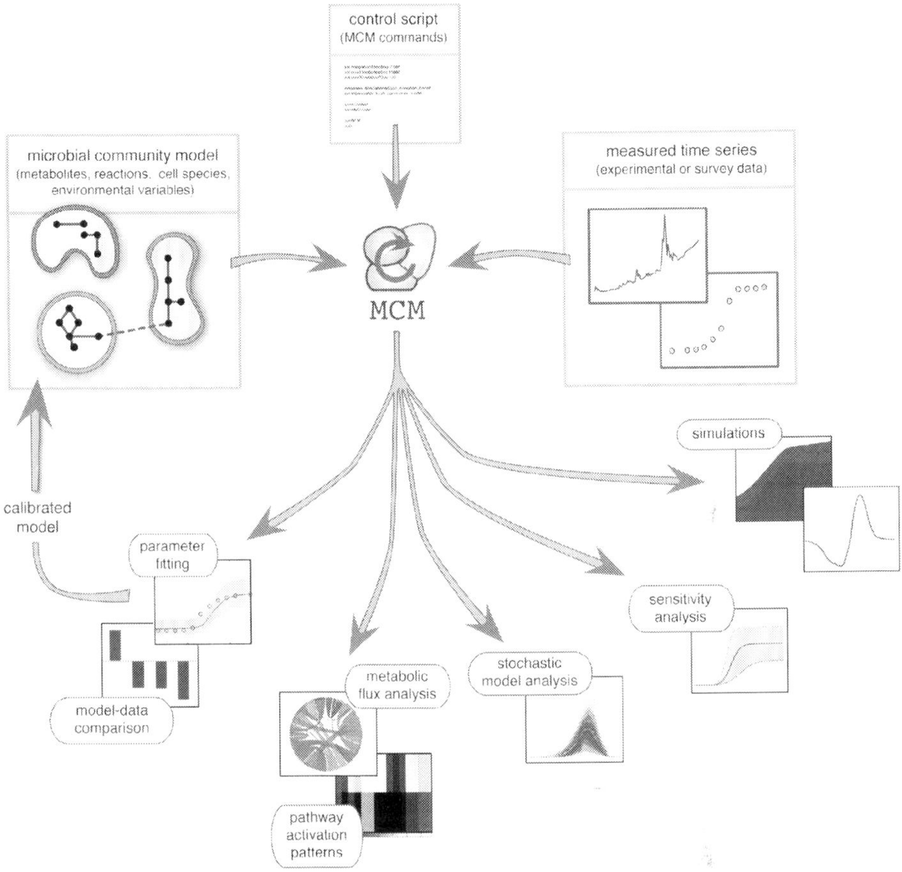

Figure 2: Overview of MCM's working principle and functionalities: A microbial community model is specified using human-readable configuration files in terms of metabolites, reactions, the metabolic potential of cell species and any additional environmental variables. Models with multiple ecosystem compartments are also possible. A script with MCM commands controls the analysis of the model and, if needed.

RESULTS AND DISCUSSION

Successional Dynamics of a Microbial Community

In a series of laboratory evolution experiments with *E. coli* (strain B REL606; Yoon et al., 2012) in glucose-acetate supplemented medium, two metabolically distinct strains consistently evolved from the ancestral

(A) strain (Le Gac et al., 2008; Spencer et al., 2008; Herron and Doebeli, 2013). When grown in monoculture with the same medium composition, all three strains exhibit diauxic growth curves with a fast glucose-driven growth phase followed by slower growth on acetate. However, the three strains differ in their efficiencies to catabolize glucose and acetate: Strain SS (slow switcher) is a better glucose utilizer when compared to strain A, and the depletion of glucose only leads to a slow switch to acetate consumption. On the other hand, the FS (fast switcher) strain has evolved to be a better acetate utilizer, initiating acetate consumption at higher remnant glucose concentrations than strains A and SS. This acetate specialization is based on a tradeoff in the citric acid cycle and comes at the cost of being a less competitive glucose consumer.

Replicated serial dilution experiments starting with strain A monocultures have shown a consistent phenotypic diversification, involving an initial invasion of the SS phenotype and a subsequent invasion of the FS phenotype, leading to the eventual extinction or near-extinction of the ancestor and the stable coexistence of the SS and FS phenotypes (Figure 3) (Le Gac et al., 2008; Spencer et al., 2008; Tyerman et al., 2008; Herron and Doebeli, 2013). Genome sequencing revealed that this metabolic diversification can be attributed to point-mutations in genes linked to glucose and acetate uptake kinetics and metabolism (Herron and Doebeli, 2013). The successional dynamics of the three phenotypes are thus likely driven by adaptations to a changing metabolic niche space, defined by fluctuating glucose, acetate and, potentially, oxygen availabilities (Le Gac et al., 2008; Tyerman et al., 2008; Herron and Doebeli, 2013). An understanding of the underlying ecological processes would shed light on the ecology and evolution of natural microbial communities with shared catabolic pathways.

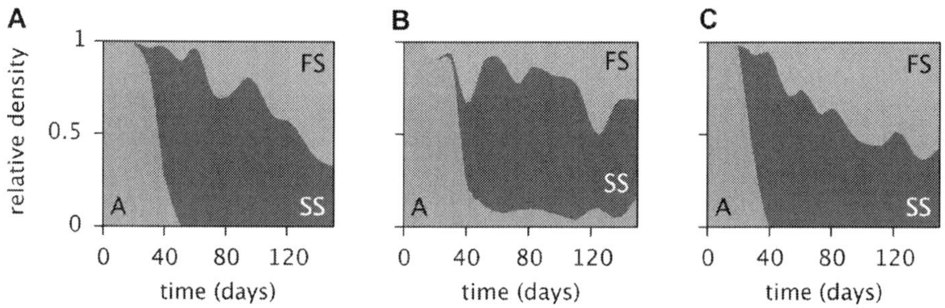

Figure 3: Estimated relative cell densities of the A, SS and FS types during

three replicated evolution experiments by Herron and Doebeli (2013, Figure 2), starting with the same ancestral E. coli strain. Within each of the three experiments (A–C), the illustrated SS or FS lineage comprises several strains with varyingly pronounced SS or FS phenotypes, respectively. Cell generations were translated to days by assuming an average of 6.7 generations per day (Herron and Doebeli, 2013).

To mechanistically explain the observed community dynamics, we used MCM to construct a model comprising the ancestral and the two evolved *E. coli* types. By keeping track of pathway activation, cell densities, metabolic fluxes and nutrient concentrations, we gained detailed insight into the processes driving the successional dynamics of metabolic diversification.

Experimental Calibration

Based on a published cell-metabolic template for the ancestral *E. coli* strain comprising over 2000 reactions (Yoon et al., 2012), we first constructed three separate cell models for the phenotypes A, SS and FS, respectively. In these preliminary models, cells grew on a substrate pool that resembled previous batch-fed monoculture experiments with glucose-acetate supplemented minimal medium (Le Gac et al., 2008). Cell-specific oxygen, acetate and glucose uptake rate limits were Monod-like functions of substrate concentrations (Emerson and Hedges, 2008; Millero, 2013). We calibrated several physiological parameters for each cell type to measured chemical concentration and cell density profiles, using least squares fitting (Figure 4). MCM automatically calibrates free parameters to data through an optimization algorithm that involves step-wise exploration of parameter space and repeated simulations (see 'Materials and methods').

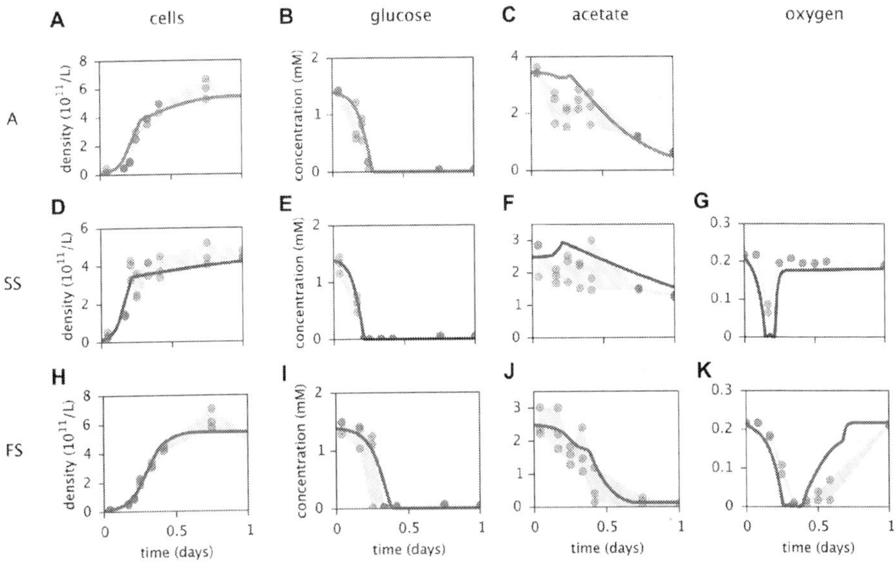

Figure 4: Calibration of E. coli cell models. Continuous curves: Time course of cell densities, glucose concentration, acetate concentration and oxygen concentration (columns 1–4, respectively) predicted by MCM for monocultures of strain A, SS and FS (rows 1–3, respectively) grown on glucose-acetate medium. Points are data used for model calibration, and were obtained from analogous monoculture growth experiments (Le Gac et al., 2008). Oxygen data were not available for strain A.

We then constructed the microbial community (MC) model by combining the three calibrated cell models into a community growing in a common substrate pool. The environmental context resembles Herron & Doebeli's evolution experiments (Herron and Doebeli, 2013). In particular, the model includes realistic oxygen depletion-repletion dynamics (Gupta and Rao, 2003), glucose and acetate depletion by microbial consumption, as well as daily dilutions into fresh glucose-acetate supplemented medium at a factor 1:100. The microbial community initially consists mostly of type A (10^{10}cells/l), while both SS as well as FS cells are assumed to be rare (1 cell/l). Because the model is deterministic, the invasion or extinction of each type only depends on its growth rate in a possibly changing environment, but not on random mutation events, nor on demographic stochastic fluctuations.

Predicting Microbial Community Dynamics

Simulations of the MC model reproduced the successional dynamics observed in Herron & Doebeli's experiments: An initial replacement of the ancestor by the SS type is followed by an invasion of the FS type, leading to the eventual coexistence of the SS and FS types and extinction of the ancestral strain (Figure 5A).

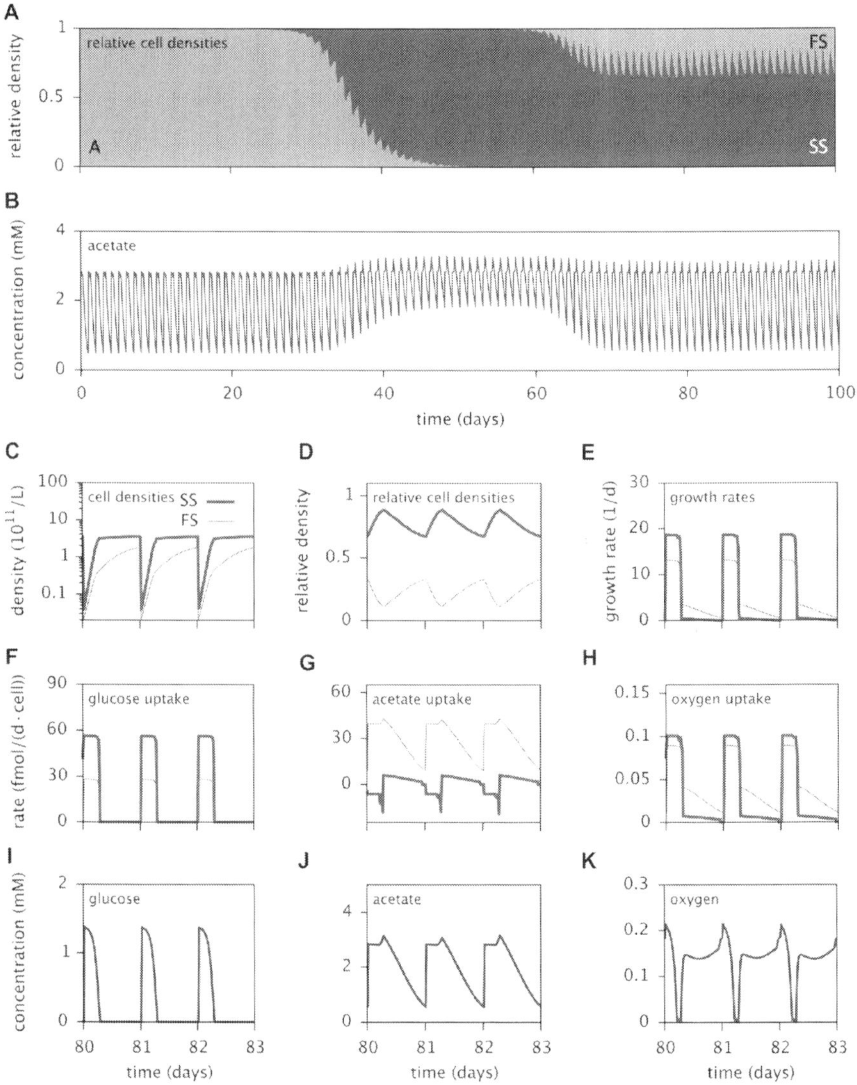

Figure 5: Dynamics of the E. coli microbial community model. (A) Relative cell densities of the A, SS and FS types over time. (B) Acetate

concentration over time. (C), (D) and (E): SS and FS cell densities, relative cell densities and growth rates over time, respectively, during stable coexistence. (F), (G) and (H): Cell-specific glucose, acetate and oxygen uptake rates over time, respectively. Negative values correspond to export. (I), (J) and (K): Glucose, acetate and oxygen concentrations over time, respectively. Diurnal fluctuations in all figures are due to daily dilutions into fresh medium. Tics on the time axes in (C–K) mark points of dilution.

Interestingly, FS can also invade in the absence of SS, however invasion occurs much slower and FS reaches lower densities than in the presence of SS (Figure 5). This is consistent with an early presence of the FS lineage at low densities in the evolution experiments (Figure 3), indicating that some of the first FS mutations already confer a slight advantage over the ancestor when FS is rare (Herron and Doebeli, 2013).

Time series of acetate concentrations (Figure 5B) link the observed successional dynamics of the three types to a gradually changing metabolic niche space: The replacement of type A by the more efficient glucose specialist SS leads to an accumulation of acetate and facilitates the invasion of the FS type. The specialization of the SS and FS types on glucose and acetate, respectively (Figure 6A), enables their long-term coexistence on glucose-acetate enriched medium through frequency dependent competition (Friesen et al., 2004; Le Gac et al., 2008; Herron and Doebeli, 2013). In fact, cell-specific acetate exchange rates reveal that the SS type temporarily excretes acetate during short intervals, which is concurrently and subsequently consumed by the FS type (Figure 5G). This periodic acetate cross-feeding is an evolutionarily emergent property of the microbial community (Treves et al., 1998). The temporary production of acetate by the SS type is consistent with previous SS-FS coculture experiments, which revealed slightly increased acetate concentrations towards the end of the SS exponential growth phase (Spencer et al., 2007). An evolved increase of acetate excretion by *E. coli* in glucose minimal medium has also been reported by Harcombe et al. (2013).

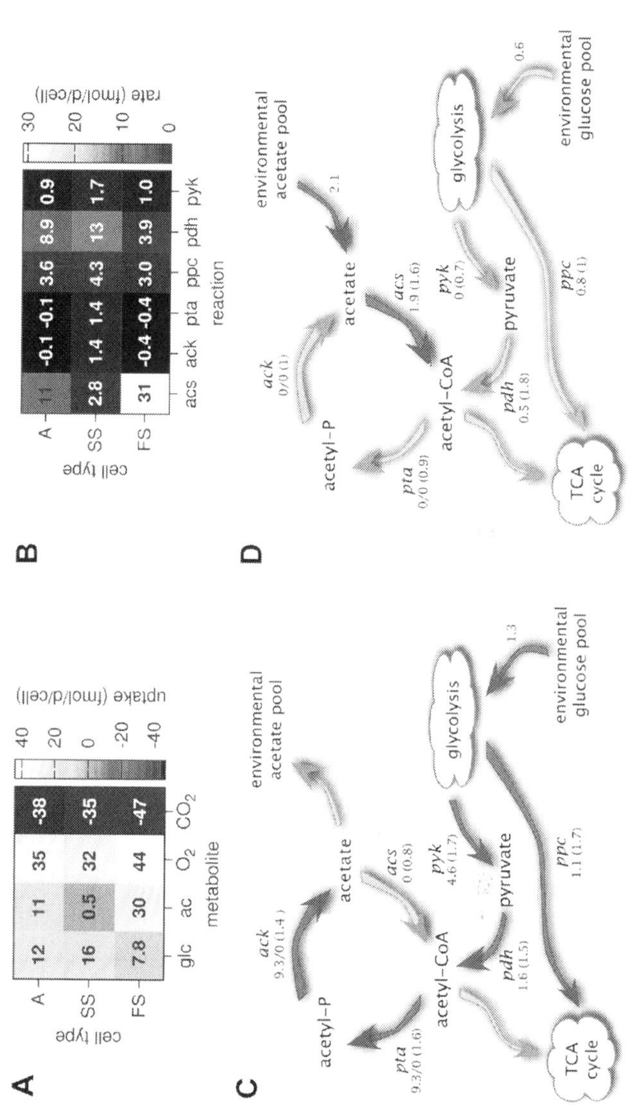

Figure 6: Metabolic differentiation of the A, SS and FS types. (A) Predicted cell-specific net metabolite uptake rates in coculture. (B) Predicted cell-specific reaction rates in coculture, for acs (acetyl-CoA synthesis), ack (acetate synthesis), pta (acetyl phosphate synthesis), ppc (oxaloacetate synthesis from phosphoenolpyruvate), pdh (decarboxylation of pyruvate to acetyl-CoA) and pyk (pyruvate synthesis from phosphoenolpyruvate). Rates in (A) and (B) are averaged over all time points within the first 100 days of evolution, with reversed reactions or net metabolite export represented by negative rates. (C) and (D): Simplified model subset of E. coli acetate and glucose metabolism, showing

pathway activations in type SS (C) and FS (D) relative to type A during exponential growth in monoculture. Non-bracketed numeric values are ratios of predicted fluxes in the evolved types over fluxes in type A. Bracketed values are ratios of mRNA harvested from monoculture experiments by Le Gac et al. (2008), for comparison. A ratio of 0/0 indicates zero flux in both the evolved and ancestral type, a ratio of 1 corresponds to an unchanged flux or mRNA, a ratio of 0 corresponds to complete deactivation in the evolved type. Darker arrows indicate increased predicted fluxes in the evolved type. Flux predictions correspond to the time points of mRNA measurements, that is, 3.5 hr after dilution for SS and 4 hr after dilution for A and SS (Le Gac et al., 2008).

It should be noted that cell metabolism depends on substrate concentrations and is subject to strong temporal variation. In particular, acetate excretion by SS cells correlates strongly with oxygen limitation (Figure 5G,K). The excretion of acetate by *E. coli* as a byproduct of oxygen-limited glucose catabolism has been observed experimentally and explained using flux balance analysis (Mahadevan et al., 2002). In the absence of oxygen limitation, complete aerobic glucose catabolism to carbon dioxide is preferred over incomplete glucose catabolism with acetate excretion. On the other hand, oxygen limitation leads to an energetic tradeoff between complete glucose catabolism and efficient oxygen utilization, resulting in the excretion of acetate.

Furthermore, the depletion of oxygen during cell growth makes oxygen a temporary limiting resource for all cells (Figure 5K). Shortly after dilution into fresh medium, the exponential growth of the SS type on glucose leads to a rapid drop of oxygen to nanomolar concentrations. Despite oxygen diffusion into the medium, oxygen remains at sub-saturation levels for several more hours because the slow-growing acetate-consuming FS cells still consume oxygen after the growth of SS cells has halted. Differences in SS and FS growth rates (Figure 5C,E) thus mitigate competition for oxygen through temporal niche separation. Hence, oxygen likely plays an important role in the metabolic diversification, as previously hypothesized by Le Gac et al. (2008). This shows that the splitting of metabolic pathways across specialists can be caused by the composite effects of competition for electron donors and electron acceptors.

Consistent with differential substrate usage, average cell-specific reaction rates (Figure 6B) reveal differences in pathway activation: The transformation of acetate into acetyl-CoA by acetyl-CoA synthetase (*acs*) is decreased in type SS and increased in type FS, when compared to the ancestral type. Furthermore, the conversion of phosphoenolpyruvate to oxaloacetate (*ppc*), the conversion of phosphoenolpyruvate to pyruvate (*pyk*) and the decarboxylation of pyruvate to acetyl-CoA (*pdh*), linking the glycolysis pathway to the citric acid cycle, are all upregulated in the SS type when compared to the FS type. Similar differences in pathway activation also exist during early exponential growth in monoculture (Figure 6C,D), because FS grows partly on acetate and SS excretes acetate (Figure 4F,J). Previous microarray profiles of mRNA concentrations during exponential growth in monocultures (Le Gac et al., 2008) found an upregulation of acetate consumption genes in FS and acetate excretion genes in SS compared to A, qualitatively confirming our predictions (Figure 6C,D). Interestingly, our simulations suggest a significant downregulation of glucose catabolism (*pyk*, *pdh* and *ppc*) in FS compared to A, which contradicts the transcript profiles (Figure 6D). However, mRNA was harvested from well-aerated flasks, while the monoculture experiments (Figure 4) and evolution experiments (Figure 3) were performed in test tubes where oxygen can become limiting (Andersen and von Meyenburg, 1980). Oxygen becomes particularly scarce in the FS tubes (Figure 4K) and temporarily limits glucose catabolism, which would explain the strong downregulation not reflected in the transcript profiles (Le Gac et al., 2008). Furthermore, while broad pathway activation patterns could be qualitatively reproduced in our system, this might be harder in other cases due to post-transcriptional regulation or post-translational modifications (Blazier and Papin, 2012).

The periodic (seasonal) changes in glucose and acetate concentrations in batch culture have previously been shown to promote coexistence of the SS and FS types, in analogy to the maintenance of phytoplankton diversity via fluctuations of resource availability (Sommer, 1984; Spencer et al., 2007). Experiments with SS-FS batch cocultures revealed that the SS type quickly dominates over the FS type, when restricted to the first glucose-rich season through frequent dilution into fresh growth medium. Reciprocally, when SS and FS are grown in solution resembling the second glucose-depleted acetate-rich season, the FS type quickly dominates over the SS type (Spencer et al., 2007). Accordingly, in a full batch cycle the relative SS cell density has been shown to culminate within 4–8 hr and to gradually decrease afterwards (Friesen et al., 2004, Figure 6B), in consistence with our simulations (Figure 5D). Simulations of the SS and FS batch coculture restricted to the first or second season, analogous to Spencer et al.'s experiments, reproduce these observations and verify the role of periodic variation of glucose and acetate concentrations in maintaining the coexistence of both types (Figure 7, see the 'Materials and methods' for details).

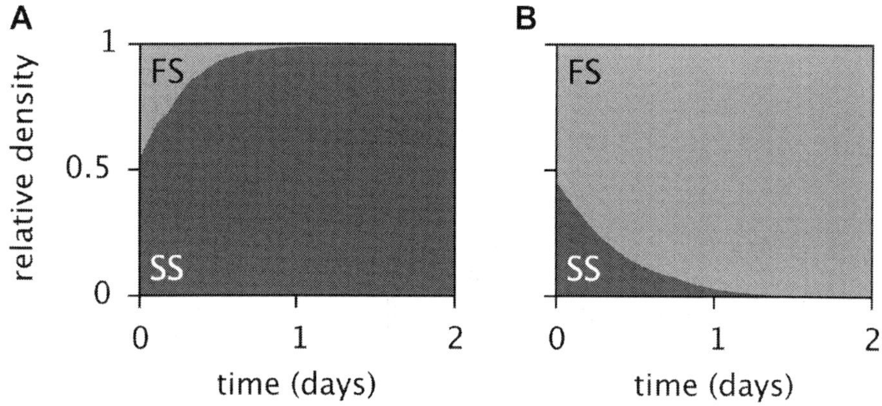

Figure 7: Predicted relative cell densities of the SS and FS types in batch coculture when restricted to either the first glucose-rich (A) or second glucose-depleted (B) season. In (A), restriction to the first season was achieved by shorter dilution periods which prevented the complete depletion of glucose. In (B), restriction to the second season was achieved by using the glucose-depleted acetate-rich solution, produced by the full-batch coculture, as growth medium (see the Methods for details). See Figure 7 for results from analogous experiments by Spencer et al. (2007).

CONCLUSIONS

The models presented here make detailed predictions about the microbial dynamics in the considered experiments. First, after calibration the cell models largely explain the data from the monoculture experiments (Figure 4). Second, the predictions for pathway activation in the three strains (Figure 6) are roughly consistent with transcription profiles. Third, simulations of the microbial community consisting of all three strains (Figure 5) reproduce the successional dynamics of diversification observed in the evolution experiments (Figure 3). Fourth, simulations of the SS-FS cocultures restricted to either the glucose-rich or glucose-depleted season reproduce the dominance of the SS or FS type (Figure 7), respectively, in consistence with previous co-culture experiments. It is important to note that only data from monoculture experiments were used to calibrate the cell models for the three strains (A, SS and FS). In particular, no information from co-culture experiments was used in the setup of the microbial community model, and thus there was no a priori knowledge about what the emergent community dynamics would be. Hence, our work conceptually produced non-trivial predictions that could be compared to experimental observations, although all experiments had already been performed.

Our work sheds light on the fundamental problem of metabolic diversification and the emergence of shared catabolic pathways. In particular, our model allowed quantitative predictions for the metabolic fluxes for each strain in coculture, revealing temporary cross-feeding as an emergent property of the evolved community (Treves et al., 1998). Cross-feeding, conventionally seen as a beneficial interaction (Morris et al., 2013), thus emerged as a form of niche segregation driven by competition for organic carbon and oxygen. Because both evolved types prefer glucose whenever available at high concentrations, but exchange acetate under oxygen limitation, the community constantly switches between competitive and beneficial interactions. Natural microbial populations might thus also oscillate between negative and positive interactions, for example depending on oxygen levels. Our findings also support previous suggestions that microbial evolution can be driven by deterministic ecological processes (Wood et al., 2005; Oxman et al., 2008; Herron and Doebeli, 2013). In this case, the observed diversification is due to competition for limiting resources whose use is constrained by basic metabolic tradeoffs. Other instances of ecological diversification in microbial evolution experiments, for example as reported by Plucain et al. (2014), might be explained using a similar approach.

More generally, we have demonstrated how MCM can be used to experimentally calibrate and combine genome-based cell models to predict the emergent dynamics of microbial communities. Our framework thus provides a starting point for designing microbial communities with particular metabolic properties, such as optimized catabolic performance. While MCM is designed for genome-based metabolic models, it can also accommodate conventional functional group models. In these models, different ecological functions such as photosynthesis, heterotrophy or nitrification are performed by distinct populations whose metabolic activity is determined, for example, by Michaelis–Menten kinetics and whose growth is described by simple substrate-biomass yield factors (Hood et al., 2006; Reed et al., 2014). Hence, natural microbial communities could be modeled even if annotated genomes are not available for each member species. While functional group models general require fewer parameters, their calibration remains a challenge (Panikov and Sizova, 1996). In MCM, model calibration becomes analogous to coefficient estimation in conventional multivariate regression, and can be used to estimate poorly known parameters such as stoichiometric coefficients, growth kinetics or extracellular transport coefficients (MCM user manual). To our knowledge, no existing comparable framework offers the flexibility combined with the statistical functionality of MCM. In view of the increasing availability of genome-scale metabolic models (Feist et al., 2008), our work provides a missing link to a predictive and synthetic microbial ecology.

MATERIALS AND METHODS

MCM Overview

MCM is a mathematical and computational framework for the construction, simulation, statistical analysis and calibration of microbial community models (Figure 2). Models are specified in special files that define all metabolites, reactions, cell species and environmental variables. MCM is controlled through custom scripts, that is, text files containing a sequence of special commands, such as for running simulations or fitting parameters. MCM includes tools for the conversion of conventional genome-scale FBA models, such as generated by the Model SEED pipeline (Henry et al., 2010) based on sequenced genomes, into a draft MC model.

MCM can accommodate microbial communities comprising genome-based cell models with arbitrary environmental variables, metabolite exchange kinetics and regulatory mechanisms. For example, environmental variables may be stochastic processes (e.g., representing climate), or specified using measured data (e.g., redox potential in bioreactor experiments), or depend on metabolite concentrations (e.g., pH determined by acetate concentration) or even be dynamical (e.g., temperature increasing at a rate proportional to biomass production rates). This versatility allows for the incorporation of complex environmental feedbacks, such as host immune responses in gut microbiota (Karlsson et al., 2011). Metabolite uptake and export rate limits can be arbitrary functions of metabolite concentrations or environmental variables. Similar interdependencies are possible for reaction rate limits, thus allowing the inclusion of inhibitory or regulatory mechanisms (Covert et al., 2008). Metabolite concentrations can be explicitly specified, for example, using measured time series, or depend dynamically on microbial export and other external fluxes. Effects of phage predation (Jensen et al., 2006), reaction energetics (Reed et al., 2014) or stochastic environments can also be incorporated.

MCM keeps track of a multitude of output variables such as cell densities, reaction rates, metabolite concentrations and metabolite exchange rates. Because each reaction can be formally associated with a particular enzyme, in turn encoded by a particular gene, MCM also makes predictions about gene densities as a product of cell densities and gene copy numbers per cell. Metabolic activity statistics (e.g., Figure 6A,B) facilitate the identification of metabolic interactions such as cross-feeding (Morris et al., 2013). The predicted time courses of output variables can be statistically evaluated against time series ranging from chemical concentrations, rate measurements to cell densities and metagenomics.

MC models can include arbitrary abstract (symbolic) numeric parameters with a predefined value range or probability distribution. Symbolic parameters can represent, for example, stoichiometric coefficients, gene copy numbers, cell life expectancies, half-saturation constants or environmental variables. The inclusion of symbolic parameters enables a high-level analysis of microbial communities: For example, MCM can automatically calibrate (fit) unknown symbolic parameters to time series using maximum–likelihood parameter estimation (Eliason, 1993). The likelihood of the data, given a particular parameter choice, is calculated by assuming a mixed deterministic-stochastic model in which the deterministic part is given by the model predictions, and the stochastic part is given by normally distributed errors. The likelihood is minimized using an iterative optimization algorithm involving step-wise parameter adjustments and repeated simulations. Other fitting algorithms are also available, such as maximization of the average coefficient of determination (R^2), which is equivalent to weighted least-squares fitting. Because MCM can calibrate unknown measurement units, raw uncalibrated data (e.g., optical cell densities with no calibration to colony forming units, Figure 4A) can also be used.

In this paper single-cell models were calibrated to monoculture experiments, however models can also be calibrated using data from experimental or natural communities that include unculturable species (MCM user manual, Louca and Doebeli, 2015). In general, fitted parameters need not be directly connected to the data used for calibration, as long as a change in the parameters influences the predictions that are being compared to the data. While this is a general principle of parameter estimation (Tarantola, 2005), in practice the uncertainty of calibrated parameters (e.g., in terms of confidence intervals) increases when their influence on the 'goodness of fit' is weaker. Moreover, alternative parameter combinations can sometimes yield a comparable match to the data, especially if multiple parameters influence the same variables (inverse problem degeneracy). Local fitting optima can be detected through repeated randomly seeded calibrations (see next section), and overfitting can be partially avoided by keeping the number of free parameters at a bare minimum. Nevertheless, in certain cases good knowledge of the system or previous literature may be required to identify the most plausible calibrations. Finally, we emphasize that MCM is, after all, merely a framework enabling the construction, calibration and analysis of microbial community models. MCM models are thus limited by the same caveats and assumptions as other constraint-based metabolic models (Blazier and Papin, 2012; Antoniewicz, 2013) and any predictions made by MCM should be subject to similar scrutiny.

Calibration of *E. coli* Cell Models

E. coli strains were obtained from an evolution experiment performed in a batch culture environment with daily dilutions into glucose-acetate supplemented Davis minimal medium (Spencer et al., 2008; Tyerman et al., 2008). For each phenotype, three clones were isolated from population 20 after 150 days and used for three independent monoculture growth experiments. Optical densities, as well as glucose, acetate and oxygen concentration data from these experiments were used to calibrate the individual cell-metabolic models for the A, SS and FS phenotypes. Oxygen measurements were not available for type A. Experimental details and results are described by Le Gac et al. (2008).

In the models, the limiting nutrients are assumed to be oxygen, glucose and acetate; all other nutrients can be taken up at an arbitrary rate. Oxygen, glucose and acetate uptake rate limits were described by Monod-like kinetics. The maximum cell-specific oxygen uptake rate was set to 1.008×10−13mol/(d·cell)1.008×10−13 mol/(d·cell), according to Varma and Palsson (1994). The oxygen half-saturation constant was set to 1.21×10^{-7} M according to Stolper et al. (2010). Oxygen was assumed to be initially at atmospheric saturation levels (0.217 mM at 37° C) and repleted at a rate proportional to its deviation from saturation (Gupta and Rao, 2003).

The fitted parameters for each cell type were the maximum cell-specific uptake rates and half-saturation constants for glucose and acetate, as well as initial cell densities and non-growth associated ATP maintenance energy requirements. The initial glucose and acetate concentrations were set to the average value measured at the earliest sampling point (1 hr after incubation) for each type. The oxygen mass transfer coefficient (M/dayM/day per M deviation) was initially fitted individually for each type together with all other parameters, and then fixed to the average of all three initial fits. All other parameters were then again fitted individually for each type. Parameter fitting was done by maximizing the average coefficient of determination (R^2) using the MCM command *fitMCM*. A total of 237 data points were used to fit 19 parameters. To reduce the possibility of only reaching a local maximum, fitting was repeated 100 times for each strain starting at random initial parameter values and the best fit among all 100 runs was used. While some fitting runs reached alternative local maxima, the best overall fit was reached in most cases.

Cell densities were directly compared to measurements. The appropriate calibrations were estimated by MCM and ranged within $8.2\times10^{11}-1.3\times10^{12}$cells/(L·OD)$8.2\times10^{11}-1.3\times10^{12}$ cells/(L·OD). These estimates are consistent with previous experimental calibrations (Lawrence and Maier, 1977) yielding 0.26gdryweight/(L·OD),0.26 g dry weight/(L·OD), which corresponds to 1.4×10^{12}cells/

(L·OD)1.4×10^{12} cells/(L·OD) (assuming a cell dry weight of 1.8×10^{-13} g in the stationary phase; Fagerbakke et al., 1996).

Simulation of the Microbial Community Model

The microbial community model was simulated using the MCM command *runMCM*. Initial glucose and acetate concentrations were set to the average of all values measured at the earliest sampling point of the monoculture incubations. Cell death was not explicitly included, because of lack of appropriate data for calibration and because daily dilutions by far exceeded cell death as a factor of cell population reduction.

Robustness of the SS-FS Coexistence

To verify the robustness of the stable SS-FS coexistence in coculture, we randomly varied each fitted model parameter uniformly within an interval spanning 10% above and 10% below its calibrated value. Both types coexisted in 50 out of 50 random simulations (Figure 5).

Seasonal Restriction of the SS-FS Cocultures

Simulations of the SS-FS cocultures restricted to the first glucose-rich or second glucose-depleted season, as opposed to the full batch cycle, were performed in analogy to the experiments by Spencer et al. (2007). More precisely, to model the first season experiment we changed the dilution rate to 1/321/32 every 5 hr, so that at the end of each batch cycle glucose was not yet completely depleted. Similarly, for the second season experiment we changed the dilution rate to 1/321/32 every 19 hr, and adjusted the growth medium to resemble the glucose-depleted acetate-rich solution reported by Spencer et al. (no glucose, 3.59 mM acetate). Initial cell densities were set to 1010cells/l1010 cells/l for both types. All other model parameters were kept unchanged. The original experiments by Spencer et al. (2007) were performed at higher dilution rates (4 and 15 hr for the first and second season experiment, respectively), however in our simulations neither the FS nor SS type could persist at these high dilution rates. We note that the strains used in our work (Le Gac et al., 2008) had evolved in separate evolution experiments using a different growth medium than those by Spencer et al. (2007).

REFERENCES

1. Andersen KB, von Meyenburg K. 1980. Are growth rates of Escherichia coli in batch cultures limited by respiration? Journal of Bacteriology 144:114–123.
2. Antoniewicz MR. 2013. Dynamic metabolic flux analysis – tools for probing transient states of metabolic networks. Current Opinion in Biotechnology 24:973–978. doi: 10.1016/j.copbio.2013.03.018.
3. Blazier AS, Papin JA. 2012. Integration of expression data in genome-scale metabolic network reconstructions. Frontiers in Physiology 3. doi: 10.3389/fphys.2012.00299.
4. Brenner K, You L, Arnold FH. 2008. Engineering microbial consortia: a new frontier in synthetic biology. Trends in Biotechnology 26:483–489. doi: 10.1016/j.tibtech.2008.05.004.
5. Burgard AP, Maranas CD. 2003. Optimization-based framework for inferring and testing hypothesized metabolic objective functions. Biotechnology and Bioengineering 82:670–677. doi: 10.1002/bit.10617.
6. Cariboni J, Gatelli D, Liska R, Saltelli A. 2007. The role of sensitivity analysis in ecological modelling. Ecological Modelling 203:167–182. doi: 10.1016/j.ecolmodel.2005.10.045.
7. Chiu HC, Levy R, Borenstein E. 2014. Emergent biosynthetic capacity in simple microbial communities. PLOS Computational Biology 10:e1003695. doi: 10.1371/journal.pcbi.1003695.
8. Covert MW, Knight EM, Reed JL, Herrgard MJ, Palsson BO. 2004. Integrating high-throughput and computational data elucidates bacterial networks. Nature 429:92–96. doi: 10.1038/nature02456.
9. Covert MW, Palsson BØ. 2002. Transcriptional regulation in constraints-based metabolic models of Escherichia coli. Journal of Biological Chemistry 277:28058–28064. doi: 10.1074/jbc.M201691200.
10. Covert MW, Schilling CH, Palsson B. 2001. Regulation of gene expression in flux balance models of metabolism. Journal of Theoretical Biology 213:73–88. doi: 10.1006/jtbi.2001.2405.
11. Covert MW, Xiao N, Chen TJ, Karr JR. 2008. Integrating metabolic, transcriptional regulatory and signal transduction models in

Escherichia coli. Bioinformatics 24:2044–2050.
12. Duarte NC, Herrgard MJ, Palsson BØ. 2004. Reconstruction and validation of ° Saccharomyces cerevisiae iND750, a fully compartmentalized genome-scale metabolic model. Genome Research 14:1298–1309. doi: 10.1101/gr.2250904.
13. Elena SF, Lenski RE. 2003. Evolution experiments with microorganisms: the dynamics and genetic bases of adaptation. Nature Reviews Genetics 4:457–469. doi: 10.1038/nrg1088.
14. Eliason SR. 1993. Maximum likelihood estimation: logic and practice. Newbury Park, CA: SAGE Publications.
15. Emerson S, Hedges J. 2008. Chemical oceanography and the marine carbon cycle. Cambridge, UK: Cambridge University Press.
16. Fagerbakke K, Heldal M, Norland S. 1996. Content of carbon, nitrogen, oxygen, sulfur and phosphorus in native aquatic and cultured bacteria. Aquatic Microbial Ecology 10:15–27. doi: 10.3354/ame010015.
17. Feist AM, Herrgard MJ, Thiele I, Reed JL, Palsson BØ. 2008. Reconstruction of biochemical networks in ° microorganisms. Nature Reviews Microbiology 7:129–143. doi: 10.1038/nrmicro1949.
18. Feist AM, Palsson BO. 2010. The biomass objective function. Current Opinion in Microbiology 13:344–349. doi: 10. 1016/j. mib.2010.03.003.
19. Freilich S, Zarecki R, Eilam O, Segal ES, Henry CS, Kupiec M, Gophna U, Sharan R, Ruppin E. 2011. Competitive and cooperative metabolic interactions in bacterial communities. Nature Communications 2:589. doi: 10.1038/ ncomms1597.
20. Friesen ML, Saxer G, Travisano M, Doebeli M. 2004. Experimental evidence for sympatric ecological diversification due to frequency-dependent competition in Escherichia coli. Evolution 58:245–260.
21. Gianchandani EP, Oberhardt MA, Burgard AP, Maranas CD, Papin JA. 2008. Predicting biological system objectives de novo from internal state measurements. BMC Bioinformatics 9:43. doi: 10.1186/1471–2105-9–43.
22. Gupta A, Rao G. 2003. A study of oxygen transfer in shake flasks using a non-invasive oxygen sensor. Biotechnology and

Bioengineering 84:351–358. doi: 10.1002/bit.10740.

23. Hammersley JM, Handscomb DC. 1964. Monte carlo methods. Methuen.
24. Harcombe WR, Delaney NF, Leiby N, Klitgord N, Marx CJ. 2013. The ability of flux balance analysis to predict evolution of central metabolism scales with the initial distance to the optimum. PLOS Computational Biology 9:e1003091. doi: 10.1371/journal.pcbi.1003091.
25. Harcombe WR, Riehl WJ, Dukovski I, Granger BR, Betts A, Lang AH, Bonilla G, Kar A, Leiby N, Mehta P, Marx CJ, Segre D. 2014. Metabolic resource allocation in individual microbes determines ecosystem interactions and ` spatial dynamics. Cell Reports 7:1104–1115. doi: 10.1016/j.celrep.2014.03.070.
26. Henry CS, DeJongh M, Best AA, Frybarger PM, Linsay B, Stevens RL. 2010. High-throughput generation, optimization and analysis of genome-scale metabolic models. Nature Biotechnology 28:977–982. doi: 10.1038/nbt.1672.
27. Herrgard MJ ° , Lee BS, Portnoy V, Palsson BØ. 2006. Integrated analysis of regulatory and metabolic networks reveals novel regulatory mechanisms in Saccharomyces cerevisiae. Genome Research 16:627–635.
28. Herron MD, Doebeli M. 2013. Parallel evolutionary dynamics of adaptive diversification in Escherichia coli. PLOS Biology 11:e1001490. doi: 10.1371/journal.pbio.1001490.
29. Hood RR, Laws EA, Armstrong RA, Bates NR, Brown CW, Carlson CA, Chai F, Doney SC, Falkowski PG, Feely RA, Friedrichs MA, Landry MR, Keith Moore J, Nelson DM, Richardson TL, Salihoglu B, Schartau M, Toole DA, Wiggert JD. 2006. Pelagic functional group modeling: Progress, challenges and prospects. Deep Sea Research II 53:459–512. doi: 10.1016/j.dsr2.2006.01.025.
30. Husnik F, Nikoh N, Koga R, Ross L, Duncan RP, Fujie M, Tanaka M, Satoh N, Bachtrog D, Wilson ACC, von Dohlen CD, Fukatsu T, McCutcheon JP. 2013. Horizontal gene transfer from diverse bacteria to an insect genome enables a tripartite nested mealybug symbiosis. Cell 153:1567–1578. doi: 10.1016/j.cell.2013.05.040.
31. Jensen MA, Faruque SM, Mekalanos JJ, Levin BR. 2006. Modeling

the role of bacteriophage in the control of cholera outbreaks. Proceedings of the National Academy of Sciences of USA 103:4652–4657. doi: 10.1073/pnas. 0600166103.

32. Karlsson FH, Nookaew I, Petranovic D, Nielsen J. 2011. Prospects for systems biology and modeling of the gut microbiome. Trends in Biotechnology 29:251–258. doi: 10.1016/j.tibtech.2011.01.009.
33. Kassen R, Rainey PB. 2004. The ecology and genetics of microbial diversity. Annual Review of Microbiology 58:207–231. doi: 10.1146/annurev.micro.58.030603.123654.
34. Klitgord N, Segre D. 2010. Environments that induce synthetic microbial ecosystems. ` PLOS Computational Biology 6:e1001002. doi: 10.1371/journal.pcbi.1001002.
35. Klitgord N, Segre D. 2011. Ecosystems biology of microbial metabolism. ` Current Opinion in Biotechnology 22:541–546. doi: 10.1016/j.copbio.2011.04.018.
36. Lawrence J, Maier S. 1977. Correction for the inherent error in optical density readings. Applied and Environmental Microbiology 33:482–484.
37. Le Gac M, Brazas MD, Bertrand M, Tyerman JG, Spencer CC, Hancock REW, Doebeli M. 2008. Metabolic changes associated with adaptive diversification in Escherichia coli. Genetics 178:1049–1060.
38. Louca S, Doebeli M. 2015. Transient dynamics of competitive exclusion in microbial communities. Environmental Microbiology. doi: 10.1111/1462-2920.13058.
39. Mahadevan R, Edwards JS, Doyle FJ III. 2002. Dynamic flux balance analysis of diauxic growth in Escherichia coli. Biophysical Journal 83:1331–1340.
40. McCutcheon JP, Moran NA. 2012. Extreme genome reduction in symbiotic bacteria. Nature Reviews Microbiology 10:13–26. doi: 10.1038/nrmicro2670.
41. Meadows AL, Karnik R, Lam H, Forestell S, Snedecor B. 2010. Application of dynamic flux balance analysis to an industrial Escherichia coli fermentation. Metabolic Engineering 12:150–160. doi: 10.1016/j.ymben.2009. 07.006.

42. Millero FJ. 2013. Chemical oceanography. Marine science series. Bosa Roca: Taylor & Francis. 4th edn.
43. Mitri S, Foster KR. 2013. The genotypic view of social interactions in microbial communities. Annual Review of Genetics 47:247–273. doi: 10.1146/annurev-genet-111212-133307.
44. Morris BE, Henneberger R, Huber H, Moissl-Eichinger C. 2013. Microbial syntrophy: interaction for the common good. FEMS Microbiology Reviews 37:384–406. doi: 10.1111/1574-6976.12019.
45. Orth JD, Thiele I, Palsson BO. 2010. What is flux balance analysis? Nature Biotechnology 28:245–248. doi: 10.1038/ nbt.1614.
46. Oxman E, Alon U, Dekel E. 2008. Defined order of evolutionary adaptations: experimental evidence. Evolution 62:1547–1554. doi: 10.1111/j.1558-5646.2008.00397.x.
47. Panikov NS, Sizova MV. 1996. A kinetic method for estimating the biomass of microbial functional groups in soil. Journal of Microbiological Methods 24:219–230. doi: 10.1016/0167-7012(95)00074-7.
48. Plucain J, Hindre T, Le Gac M, Tenaillon O, Cruveiller S, M ′ edigue C, Leiby N, Harcombe WR, Marx CJ, Lenski RE, ′ Schneider D. 2014. Epistasis and allele specificity in the emergence of a stable polymorphism in Escherichia coli. Science 343:1366–1369. doi: 10.1126/science.1248688.
49. Reed DC, Algar CK, Huber JA, Dick GJ. 2014. Gene-centric approach to integrating environmental genomics and biogeochemical models. Proceedings of the National Academy of Sciences of USA 111:1879–1884. doi: 10.1073/ pnas.1313713111.
50. Schink B, Stams A. 2006. Syntrophism among prokaryotes. In: Dworkin M, Falkow S, Rosenberg E, Schleifer KH, Stackebrandt E, editors. The prokaryotes. New York: Springer. p. 309–335.
51. Segre D` , Vitkup D, Church GM. 2002. Analysis of optimality in natural and perturbed metabolic networks. Proceedings of the National Academy of Sciences of USA 99:15112–15117.
52. Sommer U. 1984. The paradox of the plankton: fluctuations of phosphorus availability maintain diversity of phytoplankton in flow-through cultures. Limnology and Oceanography 29:633–636. doi: 10.4319/lo.1984.29.3.0633.

53. Spencer CC, Saxer G, Travisano M, Doebeli M. 2007. Seasonal resource oscillations maintain diversity in bacterial microcosms. Evolutionary Ecology Research 9:775.
54. Spencer CC, Tyerman J, Bertrand M, Doebeli M. 2008. Adaptation increases the likelihood of diversification in an experimental bacterial lineage. Proceedings of the National Academy of Sciences of USA 105:1585–1589. doi: 10.1073/pnas.0708504105.
55. Stams AJ. 1994. Metabolic interactions between anaerobic bacteria in methanogenic environments. Antonie van Leeuwenhoek 66:271–294. doi: 10.1007/BF00871644.
56. Stolper DA, Revsbech NP, Canfield DE. 2010. Aerobic growth at nanomolar oxygen concentrations. Proceedings of the National Academy of Sciences of USA 107:18755–18760. doi: 10.1073/pnas.1013435107.
57. Stolyar S, Van Dien S, Hillesland KL, Pinel N, Lie TJ, Leigh JA, Stahl DA. 2007. Metabolic modeling of a mutualistic microbial community. Molecular Systems Biology 3:92. doi: 10.1038/msb4100131.
58. Tarantola A. 2005. Inverse problem theory and methods for model parameter estimation. Philadelphia: Society for Industrial and Applied Mathematics.
59. Treves DS, Manning S, Adams J. 1998. Repeated evolution of an acetate-crossfeeding polymorphism in long-term populations of Escherichia coli. Molecular Biology and Evolution 15:789–797. doi: 10.1093/oxfordjournals. molbev.a025984.
60. Tyerman JG, Bertrand M, Spencer CC, Doebeli M. 2008. Experimental demonstration of ecological character displacement. BMC Evolutionary Biology 8:34. doi: 10.1186/1471–2148-8–34.
61. Varma A, Palsson BO. 1994. Stoichiometric flux balance models quantitatively predict growth and metabolic byproduct secretion in wild-type Escherichia coli W3110. Applied and Environmental Microbiology 60:3724–3731.
62. Wood T, Burke J, Rieseberg L. 2005. Parallel genotypic adaptation: when evolution repeats itself. Genetica 123:157–170. doi: 10.1007/s10709-003-2738-9.
63. Yoon SH, Han MJ, Jeong H, Lee CH, Xia XX, Lee DH, Shim JH,

Lee SY, Oh TK, Kim JF. 2012. Comparative multi-omics systems analysis of Escherichia coli strains B and K-12. Genome Biology 13:R37. doi: 10.1186/ gb-2012-13-5-r37.

64. Zengler K, Palsson BO. 2012. A road map for the development of community systems (CoSy) biology. Nature Reviews Microbiology 10:366–372. doi: 10.1038/nrmicro2763.

65. Zomorrodi AR, Maranas CD. 2012. OptCom: a multi-level optimization framework for the metabolic modeling and analysis of microbial communities. PLOS Computational Biology 8:e1002363. doi: 10.1371/journal.pcbi.1002363.

CHAPTER 9

ESTIMATION OF BACTERIAL DIVERSITY USING NEXT GENERATION SEQUENCING OF 16S RDNA: A COMPARISON OF DIFFERENT WORKFLOWS

Jorge Barriuso, Jose R Valverde and Rafael P Mellado

Centro Nacional de Biotecnología (CSIC), c/Darwin 3, 28049 Madrid, Spain

ABSTRACT

Background

Next generation sequencing (NGS) enables a more comprehensive analysis of bacterial diversity from complex environmental samples. NGS data can be analysed using a variety of workflows. We test several simple and complex workflows, including frequently used as well as recently published tools, and report on their respective accuracy and efficiency under various conditions covering different sequence lengths, number of sequences and real world experimental data from rhizobacterial populations of glyphosate-tolerant maize treated or untreated with two different herbicides representative of differential diversity studies.

Results

Alignment and distance calculations affect OTU estimations, and multiple sequence alignment exerts a major impact on the computational time needed. Generally speaking, most of the analyses produced consistent results that may be used to assess differential diversity changes, however, dataset characteristics dictate which workflow should be preferred in each case.

Conclusions

When estimating bacterial diversity, ESPRIT as well as the web-based workflow, RDP pyrosequencing pipeline, produced good results in all circumstances, however, its computational requirements can make method-combination workflows more attractive, depending on sequence variability, number and length.

BACKGROUND

The application of nucleic acid-based techniques is a useful tool for diversity studies in natural habitats [1] and a number of culture-independent nucleic acid-based methods have been used to characterise microbial communities. Next Generation Sequencing (NGS) of hypervariable regions from small-subunit ribosomal RNA genes is a conventional tool to analyse the composition and diversity of microbial communities in several habitats [2, 3, 4]. NGS allows gene sequencing from complex environmental samples [2, 5, 6] favouring the analysis of bacterial diversity in a comprehensive manner [7].

Taxonomy-independent studies are used to analyse diversity at different similarity levels [8, 9, 10, 11, 12]. Several analytical methods included in different software packages are available for these processes [9, 12, 13, 14, 15, 16, 17, 18, 19, 20, 21, 22].

Typical diversity data analysis workflows start by assessing data quality and removing primers and noise. This is usually followed by a multiple sequence alignment (MSA) used for distance calculation, which is the basis for clustering sequences into Operational Taxonomic Units (OTUs) at the desired dissimilarity, usually 3% for species and 5% for genera [2, 20]. Additional filtering steps may be inserted to remove redundant gaps, even sequence ends, and detect repeated or closely related sequences to reduce the amount of data to be processed [20, 21, 22, 23, 24, 25, 26]. Filtering processes are also used to improve sequence quality [24, 25, 26]. Each step can be carried out using a variety of tools, and different tool combinations are commonly used to tailor the analysis to the original data [e. g. 26]. Some approaches avoid MSA by using pairwise alignments to compute distances [20, 22, 23].

Observed OTU counts and relative abundances are representative of actual diversity, yet we cannot be sure that total diversity has been identified unless an appropriate sample size has been employed, which depends on diversity and hence is difficult to predict. For this reason estimates of species richness must be considered, such as rarefaction curves and ACE or Chao1 estimators, among others.

Comparative studies of diversity in environmental samples are usually carried out either by comparing the above estimators or using phylogenetic information, as implemented in UniFrac [27]. Lately, approaches to derive OTU numbers from taxonomic classifications produced by the RDP classifier [28] have been proposed, however, this approach is limited by the existing data in the databases [29].

Recent reports have compared some of the methods available and their potential advantages [11, 25, 30] computing OTU counts, however, there is still little knowledge on how these combinations affect workflow performance under different conditions, and the specific suitability for differential diversity studies. In order to acquire useful rules and advice on the choice of workflow, we employed the most commonly used tool combinations to generate the corresponding workflows.

Application of these techniques has been greatly facilitated by the availability of tool collections packaged for easy setup and use, such as QIIME [31], which includes many of the tools analysed here, allowing scientists to select and combine specific tools to suit their needs, highlighting the requirement for studies to compare the relative merits of each method combination workflow. We tested three different alignment strategies: *ab initio* alignments using the progressive alignments of MUSCLE [14], the MAFFT partition tree method [16], and reference guided alignments as implemented in Mothur [19]. The effectiveness of filtering and pre-clustering was tested on Mothur alignments. We checked the effect of alternative distance calculation approaches using the Jukes-Cantor correction for multiple nucleotide substitution, as implemented in DNADIST [2, 12, 32, 33, 34], the uncorrected distance with the gap count method from Mothur [19] (hereinafter referred to as «Mothur distance»), and the k-mer based distance method from MAFFT [16] (referred to as «MAFFT distance»). Finally, all combinations of MSA and distance matrices were clustered using Mothur [19].

In addition to these combinations, we also considered other popular streamlined OTU identification workflows: Mothur used as the only tool for the whole pipeline with and without pre-clustering, ESPRIT [20], which uses pairwise alignments to compute distances, the recently published CROP [23], which uses an unsupervised Bayesian clustering method, Otupipe, which is based on UCLAST/USEARCH [22], and the RDP pyrosequencing pipeline [21], which uses Infernal [35, 36] alignments and its own complete linkage clustering method.

To compare these workflows, we needed to resort to well-defined datasets representative of typical experimental set ups: we used reference 16S sequence data of various lengths using synthetic data developed by us, as well as datasets proposed by Youssef [37], Quince [26], Hao [23, 38] and Huse [24]. These sets test accuracy by using a single sample with a limited number of species. To analyse workflow suitability for differential diversity studies, we used real-world field data of hypervariable V6 16S rDNA sequences [32] retrieved at different times under different environmental conditions. The V6 region is a common choice for this kind of analysis as it has been proven to yield results similar than longer sequences [11, 37], and is a sensible choice for short-length sequencing approaches. In order to compare clustering results one needs to account for potential method-dependent variability due to randomness in the workflow (e.g. Mothur randomly selects the cluster to group in case of a tie and CROP uses a Markov Chain Monte Carlo -MCMC- approach). We have collected statistics on the variability results obtained with Mothur, ESPRIT and CROP.

RESULTS

Reference set of Short-Length Hypervariable V6 16S rDNA

The results obtained with short V6 region sequences using our synthetic data sets derived from 60 species are shown in Table 1, which indicates the number of OTUs obtained at 3%, 5% and 10% dissimilarity levels with each workflow. Manual checking of the selected sequences revealed that two of them were identical, (EU930421 and DQ989484), two pairs of sequences had a dissimilarity level below 5% (X90760 and AM743175, AF328681 and EU551675), and five pairs of sequences had a dissimilarity level below 10% (AM935529 and EU631792, AF114805 and AY894325, AY234615 and EF173341, AF114805 and AY894325, U855746 and EU360125). Therefore, the expected number of OTUs should be 59 at 3% dissimilarity, 57 at 5% dissimilarity, and 52 at 10% dissimilarity, which are precisely the results obtained with ESPRIT, Otupipe and the RDP pipeline. CROP correctly identified the number of OTUs at 3% when only 60 sequences were used, yet overestimated them at 5% and when 50 repeated copies of them were used. Mothur, on the other hand, underestimated the number at all dissimilarity levels, both with and without the pre-clustering step, due to the removal of sequence fragments during the reference-guided alignment step, converting divergent sequences into identical or more closely-related fragments (http://www.mothur.org/wiki/Align.seqs).

Table 1: Accuracy test

	60 seqs			**50 × 60 seqs**		
Reference expected values	59	57	52	59	57	52
CROP	59	59	59	60	60	60
ESPRIT	59	57	52	59	57	52
MAFFT+JC	58	56	50	58	57	51
MAFFT+MAFFT	59	59	59	59	59	59
MAFFT+Mothur	59	56	51	59	57	54
Mothur+JC	55	55	55	2992	2992	2992
Mothur+MAFFT	41	40	36	252	251	243
Mothur+Mothur	48	48	48	48	48	48

Mothur+PreC+JC	44	44	44	43	43	43
Mothur+PreC+MAFFT	47	47	47	47	47	47
Mothur+PreC+Mothur	48	48	48	48	48	48
MUSCLE+JC	58	56	50	63	59	59
MUSCLE+MAFFT	59	59	59	59	59	59
MUSCLE+Mothur	59	56	55	66	62	56
Otupipe	59	57	52	59	57	52
RDP	59	57	52	59	57	52

OTUs observed with each of the workflows analysed at distances of 3, 5 and 10% for the datasets containing 60 test sequences and 50 replicas of the same (50 × 60). JC stands for Jukes-Cantor and PreC for the pre-clustering step applied after Mothur MSA. Combined workflows are indicated stating first the method used for alignment (MAFFT, Mucle, Mothur or Mothur with pre-clustering), and then the distance method used (Jukes-Cantor, MAFFT or Mothur). Clustering was performed with Mothur for all combined workflows.

Regarding combined workflows, it is noticeable that both MAFFT and MUSCLE alignments produce good results, and that the combinations of Mothur alignments with either Jukes-Cantor corrected or MAFFT distance calculations give abnormally high counts with the 50 replica data set. The MAFFT distance calculation had trouble to discriminate between 3%, 5% or 10% dissimilarity levels (59 OTUs), in contrast to Mothur or Jukes-Cantor corrected distances.

Reference sets of Short Length Mutated V6 rDNA

To highlight the effect of sequence alignment on clustering and diversity estimation we mutated the 60 reference sequences and arranged them in two different groups: stacked or interspersed. Since random mutations have been added, we now expect diversity to be higher at 3% distance and conserved at 5% or more. Results are summarized in Table 2: CROP gives abnormally large counts, ESPRIT overestimates at 3% dissimilarity but returns to acceptable estimations at 5% and 10% distances. Mothur alignments exceedingly overestimate diversity on account of the mentioned abnormal alignments, with the preclustering step reducing OTU counts, which is likely because these discard many sequences as if they were sequencing errors. Once again MAFFT and

MUSCLE alignments give better results when combined with Jukes-Cantor corrected or Mothur distances, but not with MAFFT distances.

Table 2: Alignment test

	50 × 60 mutated (interleaved)			50 × 60 mutated (stacked)		
	3%	**5%**	**10%**	**3%**	**5%**	**10%**
Reference expected values	59	57	52	59	57	52
CROP	1959	1954	1955	1865	1850	1850
ESPRIT	193	59	56	205	59	56
MAFFT+JC	141	101	85	132	113	113
MAFFT+MAFFT	2289	1708	80	2289	1777	1270
MAFFT+Mothur	261	119	96	279	121	121
Mothur+JC	2947	2947	2948	2985	2985	2985
Mothur+MAFFT	899	685	477	2999	2999	2999
Mothur+Mothur	1087	923	736	1100	930	888
Mothur+PreC+ JC	1198	1198	1198	1333	1333	1333
Mothur+PreC+MAFFT	1328	1328	1328	1346	1346	1346
Mothur+PreC+Mothur	1080	938	938	1087	940	915
MUSCLE+JC	70	59	44	2999	2999	2999
MUSCLE+MAFFT	2287	1707	80	2288	1785	1269
MUSCLE+Mothur	264	64	57	571	466	466
Otupipe	139	91	59	144	95	61
RDP	59	58	53	59	57	52

OTUs observed with each of the workflows analysed at distances of 3, 5 and 10% for the datasets containing 50 different mutated replicas of the 60 test sequences stacked or interleaved.

There are obvious differences in the calculations depending on sequence ordering. A strong influence arises from the alignment step, verified by looking at the MSA. We know that each sequence contains only one change and therefore, deviations are easily identifiable. In both cases (stacked or interspersed) MAFFT and MUSCLE produced shorter alignments than Mothur, with or without pre-clustering (400 MAFFT, 686 MUSCLE sites vs. 2641 Mothur sites for stacked mutants, and 445 MAFFT, 372 MUSCLE sites vs. 2553 and 2436 Mothur sites for interspersed mutants). Evening sequence ends in Mothur produced useless alignments. Strong differences can also be detected between both datasets, with the combination of MUSCLE and Jukes-Cantor yielding very high values with stacked data. ESPRIT is more resilient to sequence order, as it avoids the MSA step, showing only minor differences in results between datasets and consistently producing good Chao1 estimates. Otupipe, which also avoids the MSA step, displays a similar behaviour. Only RDP produced results consistent with the expected values and was able to avoid the influence of sequence order in the input data.

Reference Sets Simulating 454 NGS data of 16S rDNA

Table 3 summarizes the results obtained when analysing the V5 region datasets from Quince et al. [26]. CROP, ESPRIT, Otupipe and the RDP pipeline results are in line with published data [23, 26], with CROP and Otupipe giving low estimates. When Mothur was used to compute the alignment with only vertical filtering to remove gaps, extremely large OTU counts were obtained, independently of the distance algorithm used (data not shown). However, Mothur behaviour could be corrected by removing uneven ends in the filtering step («trump»), producing estimates closer to those previously published but lower than ESPRIT. MUSCLE alignments were able to produce acceptable results with the Artificial but not the Priest Pot data. MAFFT alignments produced very large OTU counts. In all cases, the worst results were obtained when MAFFT k-mer distances were used.

Table 3: Quince's 454 data

	Artificial			**Priest Pot**		
	3%	**5%**	**10%**	**3%**	**5%**	**10%**
CROP	41	25	15	562	246	42
ESPRIT	248	77	38	1115	773	394
MAFFT+JC	686	686	686	3764	3764	3764
MAFFT+MAFFT	31933	31933	31933	15984	15984	15984
MAFFT+Mothur	1756	1756	1756	6672	6672	6672
Mothur+JC	49	36	33	640	537	537
Mothur+MAFFT	4276	4276	4276	2824	2824	2824
Mothur+Mothur	113	53	53	766	642	642
Mothur+PreC+ JC	61	40	36	662	482	482
Mothur+PreC+MAFFT	3864	3141	3141	1905	1625	1625
Mothur+PreC+Mothur	136	65	46	810	575	575
MUSCLE+JC	146	146	146	4059	4059	4059
MUSCLE+MAFFT	33491	33491	33491	15718	15718	15718
MUSCLE+Mothur	258	258	258	6433	6433	6433
Otupipe	66	39	24	793	570	302
RDP	250	94	43	1209	862	456

OTUs observed with each of the workflows analysed at distances of 3, 5 and 10% for Quince's Artificial and Priest Pot datasets. For the Priest Pot data, 855 OTUs at 3% and 699 at 5% were previously estimated by Quince et al. [26].

Reference Set using a Large Number of Short Length Reads

Recent advances in pyrosequencing have increased the size of reads that can be collected, making this technology an attractive tool for biodiversity studies. To analyse the likely behaviour of these new approaches we have resorted to a well- known and established dataset consisting of a large number of sequences (340150) previously reported by Huse *et al.*, [24]. Results are shown in Table 4: CROP failed to successfully deal with this large dataset. For the remaining approaches, the main problem arises from the size of the distance matrices generated, which can easily reach terabyte file sizes, becoming unmanageable for clustering programs such as Mothur. In these cases, the data can be processed either by including a preclustering step (e.g. in Mothur or combining MUSCLE with USEARCH), or analyzing unique sequences (albeit at the cost of losing information on abundance), or by obviating the need for an MSA (like ESPRIT or Otupipe). The RDP pipeline was able to build the alignment, but it could not be processed in the standard manner, and it did require specific RDP staff assistance to process this dataset.

Table 4: Huse short read data

	3%	**5%**	**10%**
CROP	NC	NC	NC
ESPRIT	6464	3308	1402
Unique:MAFFT+JC	23442	23442	23442
Unique: MAFFT+MAFFT	23445	23445	23445
Unique: MAFFT+Mothur	23445	23445	23445
Unique:Mothur+JC	23441	23441	23441
Unique: Mothur+MAFFT	23441	23441	23441
Unique: Mothur+Mothur	18210	18210	18210
Mothur+PreC+ JC	15594	15594	15594

Mothur+PreC+MAFFT	15601	15601	15601
Mothur+PreC+Mothur	14776	14776	14776
Unique:MUSCLE+JC	22816	22816	22816
Unique:MUSCLE+MAFFT	23444	23444	23444
Unique:MUSCLE+Mothur	21318	21318	21318
Otupipe	2149	1422	878
RDP	4228	2932	1777

OTUs observed with each of the workflows analysed at distances of 3, 5 and 10% for a large number of short length reads dataset [24]. NC = non computable.

None of the workflows involving an MSA followed by distance-matrix calculation and clustering could be completed due to the huge size of the distance-matrix. When the 23445 unique sequences in the dataset were used to reduce dataset size, with or without preclustering, all of them gave very high OTU counts and were unable to make distinctions at different similarity levels. Only ESPRIT, Otupipe and the RDP pipeline were able to produce different results at each dissimilarity level (Table 4).

Reference Sets using Near-Full Length 16S rDNA Sequences

We have analysed two near-full length sets with similar sequence counts: Hao's axillary skin reference data comprised low diversity, while Youssef's prairie soil samples contained higher diversity. The results are summarized in Table 5: CROP results show a strikingly high number of OTUs. In all cases fairly similar results were obtained with both datasets and Mothur gives acceptable figures either alone or in combined workflows, again with the exception of k-mer based distance calculation (MAFFT), which overestimates OTU numbers, albeit much less than the estimates obtained when shorter sequences were analysed.

Table 5: Near-full length sequences

	Skin axillary			**Prairie soil**		
	3%	**5%**	**10%**	**3%**	**5%**	**10%**
CROP	1009	1009	1009	1128	1128	1128

ESPRIT	59	43	26	504	340	162
MAFFT+JC	47	39	39	490	346	183
MAFFT+MAFFT	266	126	109	1007	895	714
MAFFT+Mothur	50	39	36	503	351	181
Mothur+JC	47	39	36	491	352	184
Mothur+MAFFT	269	126	109	977	872	689
Mothur+Mothur	54	40	38	535	396	204
Mothur+PreC+ JC	47	39	36	491	353	268
Mothur+PreC+MAFFT	287	127	109	977	877	836
Mothur+PreC+Mothur	55	40	38	536	397	305
MUSCLE+JC	50	39	37	480	337	175
MUSCLE+MAFFT	266	126	109	1007	895	714
MUSCLE+Mothur	58	40	38	487	339	173
Otupipe	49	37	26	490	336	160
RDP	60	43	27	504	348	175

OTUs observed with each of the workflows analysed at distances of 3, 5 and 10% for the skin axillary microbiome data described by Hao *et al.* [23], and tall grass prairie soil data [37].

Experimental V6 16S rDNA Sequences from Soil Samples

These datasets compare differential diversity estimations before and after herbicide treatment with a control: differences in diversity should be consistent across workflows. The results are shown in Table 6. CROP gives the lowest diversity estimates in all cases. In the remaining workflows, the untreated and glyphosate-treated soils in practically all cases are more diverse than the GTZ-treated soil when comparing sampling times. The actual figures varied considerably depending on the workflow used. Combined workflows showed again a tendency to produce higher OTU counts when MAFFT was used to calculate distances.

Table 6: Comparative diversity analysis

	Control		GTZ		**Glyphosate**	
	t1	**t2**	**t1**	**t2**	**t1**	**t2**
CROP	812	286	414	625	454	477
ESPRIT	1631	1053	1227	922	1951	1102

MAFFT+JC	1936	1087	1025	975	1977	1088
MAFFT+MAFFT	2792	1588	1971	1329	3296	1577
MAFFT+Mothur	2112	1215	1655	1163	2646	1326
Mothur+JC	3464	1622	3842	823	5024	1090
Mothur+MAFFT	2789	1589	1969	1328	3323	1574
Mothur+Mothur	1730	1128	1310	982	2122	1209
Mothur+PreC+JC	2479	1460	1958	826	3134	1498
Mothur+PreC+MAFFT	2471	1459	1961	1252	3135	1495
Mothur+PreC+Mothur	2063	1143	1890	1010	2516	1221
MUSCLE+JC	1762	1247	1257	950	2020	1238
MUSCLE+MAFFT	2790	1587	1970	1330	3296	1577
MUSCLE+Mothur	2278	1352	1800	1059	2772	1477
Otupipe	1314	881	938	795	1428	948
RDP	1762	1106	1236	901	1932	1094

OTUs observed with each of the workflows analysed. For simplicity, only results at 3% dissimilarity are shown from pooled samples collected at two different times from control, GTZ and glyphosate treated soils [32].

Computational Variability of Results

Most calculations were repeated to confirm the observed results. This showed, in several instances, the same computation producing different results due to intrinsic algorithmic randomness. These expected variations were measured repeating the calculations 20 times with three selected datasets under the same conditions. The results obtained are presented in Table 7. ESPRIT did not show any variability, Mothur showed a slight variability in some instances, which should not change OTU estimates, and CROP showed a slightly higher variability, which may induce minor estimate changes.

Table 7: Observed output variability

		50 × 60 seqs mut stacked				50 × 60 seqs mut interleaved				Prairie soil			
	N	μ	Err	Min	Max	μ	Err	Min	Max	μ	Err	Min	Max
CROP 3%	20	1852.2	2.16	1847.6	1856.7	1954.6	0.83	1952.8	1956.3	1128	0	1128	1128
CROP 5%	20	1853.5	2.43	1848.4	1858.5	1952.4	0.66	1951	1953.7	1128	0	1128	1128
CROP 10%	20	1850.9	1.65	1847.4	1854.4	1951.7	0.62	1950.4	1953	1128	0	1128	1128
ESPRIT 3%	20	205	0	205	205	193	0	193	193	565	0	565	565
ESPRIT 5%	20	59	0	59	59	59	0	59	59	381	0	381	381
ESPRIT 10%	20	56	0	56	56	56	0	56	56	180	0	180	180
Mothur 3%	20	1465.3	0.23	1464.8	1465.8	1173	0	1173	1173	541	0	541	541
Mothur 5%	20	1465.3	0.23	1464.8	1465.8	1173	0	1173	1173	400	0	400	400
Mothur 10%	20	1465.3	0.23	1464.8	1465.8	1173	0	1173	1173	347	0	347	347
Mothur+PreC 3%	20	1490.6	0.16	1490.3	1490.9	1197.3	0.12	1197	1197.6	541	0	541	541
Mothur+PreC 5%	20	1490.6	0.16	1490.3	1490.9	1197.3	0.12	1197	1197.6	400	0	400	400
Mothur+PreC 10%	20	1490.6	0.16	1490.3	1490.9	1197.3	0.12	1197	1197.6	348	0	348	348

Observed output variability with 20 equal runs of CROP, ESPRIT and Mothur (with (PreC) and without preclustering). N is the number of observations for all samples; μ is the observed mean of the sample. Standard error (Err) and confidence intervals (Min and Max) were calculated from these values, as described in Methods. Many of the CROP calculations failed to complete in this test, as reflected by N (Prairie soil), and statistics were adjusted accordingly.

Since CROP estimates are obtained via independent runs for each distance, this may result in a seemingly contradictory output, where OTU counts at 5% might be greater than at 3% (e.g. one of the runs for the stacked mutants produced 1865 OTUs at 3%, and another 1869 at 5%).

DISCUSSION AND CONCLUSIONS

We have analysed the most commonly used workflows applied for bacterial diversity studies including simple, one-tool, workflows and tool combinations, and compared them for accuracy, results variability and efficiency under a variety of conditions representative of Illumina-based, 454-based, near full-length and differential diversity studies.

While more complex workflows can be devised (e. g. using RDP web alignments or ESPRIT distance calculations or clustering as optional combinatorial steps), these are rarely reported except when evaluating new algorithms [23, 26] and are inconvenient for routine use. The most relevant patterns from the wealth of information generated by our analysis are highlighted below.

Sequence Processing

The relevance of initial data quality to the reliability of the results has been repeatedly noted, leading to the recommendation of filtering raw data according to quality, removal of chimeras and experimental noise, and ensuring all sequences correspond to the same region of 16S rDNA. This last step usually relies on the selection of experimental primers, but may sometimes benefit from the additional removal of non-overlapping sequences ("trump" filtering in Mothur). This step may arguably remove important elements in an alignment (e. g. if terminal gaps are significant), but it was the only way to obtain sensible results with the datasets from Quince *et al.* [26]. However, its use with our artificially mutated sequences resulted in useless alignments, showing that this option must be applied with caution.

When a large number of short length sequences are used [24], it is easy to find many individuals with the same sequence. Identifying and using unique sequences exerts a major impact on reducing computation time, especially when dealing with modern vast datasets (in the hundreds of thousands or millions of sequences).

Alignments

ESPRIT, which uses pairwise alignments, generally produces more accurate results than workflows relying on MSAs. Pairwise alignments; however, do not consider likely evolutionary relationships or structural properties, which is why MSA tools are often preferred. Alignments based on a reference alignment are shown to be very useful, while improving quality, as can already be seen from the RDP pipeline results; however, we observe that using MAFFT or MUSCLE to obtain *ab initio* MSAs gave better results than Mothur reference-guided alignments for short sequence lengths. This is most evident with our engineered mutated datasets, and is likely due to reference guided alignments selecting different reference sequences for related mutants, leading to an MSA where related sequences include more variability than actually exists. Hence, Mothur MSAs should always be filtered to remove redundant gaps if other tools are to be used for distance calculation [19]. MAFFT and MUSCLE frequently identify the mutation correctly and produce more accurate alignments, as can be seen by inspecting the aligned stacked dataset. This advantage is diluted as sequence length increases (datasets from Quince, Hao and Youssef), since the availability of more information increases the accuracy of reference identification.

Another major difference is related to the computational time required for constructing the alignments. Both MAFFT and MUSCLE have a well-defined behaviour: MAFFT time complexity is $O(N^2L)$ [39], N being the number of sequences and L their length, and MUSCLE uses multiple iterations with a cost of $O(NL+L^2)$ per iteration; however, ESPRIT, CROP and Mothur compute more expensive Needleman-Wunsch alignments (cost $O(L_1 \cdot L_2)$, or $O(L^2)$ for equal-length sequences) and rely on a previous filtering step to reduce N, the number of sequences to be compared. ESPRIT and CROP require $(N \cdot (N-1))$ comparisons to build the distance matrix; one may expect ESPRIT k-mer filtering or CROP unique identification step to become less efficient for longer, relatively distant sequences. In the case of Mothur only N comparisons -each sequence against its reference, selected using one of various filtering methods (k-mer, blastn and suffix trees)- are needed.

Summarizing, ESPRIT works better for short and closely related sequence reads and is usually very fast; however, it is much slower when tens of thousands of sequences are considered. MUSCLE is preferred for short sequence lengths, but its long running time puts it at a disadvantage when sequences are long, deficiently grouped, or too many are used. Mothur generally achieves the fastest alignments when using a reference database, although there are concerns with the relative merits of this approach [19, 35]. In the case of long sequences, MAFFT offers a very good balance between speed and quality.

Special mention must be made for MSAs of huge datasets in the order of hundreds of thousands of sequences: in these cases, MUSCLE is unable to directly align the data and requires a previous "clumping" and filtering step with USEARCH to guide the final alignment, or selection of unique sequences prior to alignment. In any case, subsequent processing makes it advisable to favour working with unique sequences or a pre-clustering step when available, both to reduce computation time and produce manageable datasets. Regarding computation time, this is the most expensive step, but only becomes a real issue with modern computer architectures when dealing with very large sequences (which is not usually the case in diversity studies) or huge numbers of sequences. In the last case, the selection of unique sequences and/or preclustering helps to reduce computation to acceptable times (from days or even months down to hours). In the case of MUSCLE it may additionally be required to reduce the number of iterations and increase the maximum memory used (--maxiter 2 --maxmb 10000).

Distance Calculation

It has been argued that the direct calculation of distances from pairwise comparisons yields better distance matrices than distances calculated from MSAs, and indeed one expects less gaps in a pairwise than a multiple alignment, as reflected in the ESPRIT package, which tends to produce the lowest number of OTUs, in agreement with earlier findings

[20, 25]. However, the computational cost increases with the number of sequences, making it more attractive to use MSA-based methods for large datasets. We see that both Jukes-Cantor corrected distances and Mothur distances give consistent and sensible results. Mothur distances are inclined to produce lower diversity and MAFFT distances tend to give 30-35% overestimations in the number of OTUs. The behaviour of MAFFT distance calculation is explained by its k-mer-based approach, where few changes in short sequences will have a bigger impact on overall distance than similar changes in longer ones.

In conclusion, distances derived from pairwise alignments produce better results, but may become inconvenient for long or large numbers of sequences. In these cases, Mothur distances produce lower OTU counts than Jukes-Cantor. K-mer-based distances are not currently recommended.

It is worth noting here that distance matrices inherently grow in size with the number of sequences squared. This may become a problem with modern datasets consisting of hundreds of thousands or millions of sequences, where terabyte-sized matrices may be generated. Thus, methods that avoid generating distance matrices may be more advisable for this data.

New versions of other popular alignment methods tuned for NGS analysis are expected soon (Higgins, D, personal communication), and it will be interesting to test them as soon as they become available.

Clustering Analysis

We have compared the results obtained with different clustering strategies: CROP, ESPRIT, Mothur, USEARCH and RDP. To simplify the study, we did not consider other clustering tools such as CD-HIT-454, which uses heuristics that are too stringent for richness analysis [40]. It is difficult to compare the different clustering strategies as ESPRIT, CROP, USEARCH and RDP have only been considered in their own workflow, except for the fact that Mothur is the easiest to integrate in complex workflows and offers a wider range of clustering options (average, furthest and nearest neighbour). However, Mothur failed to process the huge, terabyte-sized, distance matrices derived from full MSAs of the extensive dataset [24], forcing us to reduce information content by including only unique sequences in the MSA. Regarding the results obtained with CROP, while proposing a new, promising and rigorous approach, it was unable to complete the analyses or produced anomalous results in several cases. This may be due to CROP being a new method that is still actively refined.

We have also considered variability in the results of the clustering algorithms. There are two sources of variability. One is due to the sub sampling required to build rarefaction curves and estimate actual diversity. The other source depends on algorithmic randomness, and after Studentized correction, the results from 20 repeated analyses show that ESPRIT consistently produces the same results, Mothur produces a slight variability, depending on the analysed dataset, and will not normally affect the diversity results, and CROP, which uses an MCMC step, has a higher but still low variability that should have a minimal impact on the results, although this may rarely lead to counter-intuitive diversity estimates as separate runs are used for each similarity level.

Suitability for Differential Diversity Studies

The results obtained from the analysis of the short-length hypervariable V6 16S rDNA sequences from soil were in all cases (with the exception of CROP for GTZ and glyphosate data) in good agreement with those previously reported for the same set of sequences, confirming that the reduction in species richness was more evident in the GTZ-treated soil at the first sampling time, and that the relative recovery from this herbicide treatment appeared to be poorer than that from the glyphosate treatment [32]. All of the workflows used, except CROP, were able to show differences in species richness among the different herbicide treatments used, and the untreated soil was the one that contained the highest bacterial diversity. This is in accordance with the results we have obtained from other field studies (data not shown), yet a general conclusion may not be reached just from our soil data.

Workflow Accuracy

Use of our reference set of 60 short-length hypervariable V6 16S rDNA sequences shows that MSAs obtained using MUSCLE and MAFFT produce the same results, regardless of the distance calculation method (with the exception of MAFFT), while the Mothur guided alignment underestimates diversity when dealing with a small number of OTUs. The method used for distance calculation also affects the number of OTUs, as MAFFT is unable to discriminate between different dissimilarity levels (even at 18%, data not shown), considering the small number of OTUs found. Jukes-Cantor or Mothur distances slightly underestimate the number of OTUs. No aberrant combinations were found when nearly full-length 16S rDNA sequences were analysed, thus indicating that when dealing with long sequences, Mothur reference-guided alignments may be preferred. When dealing with short length sequences, the ESPRIT package and the RDP pipeline are the most accurate on the three dissimilarity levels. When analysing a large number of long sequences, computational time requirements adversely affect the use of the ESPRIT package due to the longer time needed, while the use of Mothur alone produces acceptable estimates in a reasonable time.

For even larger numbers of short length sequences in the range of hundreds of thousands [24], a filtering step that reduces sequence numbers is needed. ESPRIT and Otupipe automatically apply an initial filtering step and hence can directly work with these huge datasets, producing acceptable results. The RDP web-based pipeline may experience troubles or long delays with some datasets, but does generally give good results. In the case of MAFFT, Mothur and MUSCLE, the initial selection of unique sequences and optional preclustering steps helps by significantly reducing the data to manageable sizes. In conclusion, the RDP web-based pipeline is most convenient for general use, but when dealing with very large or many datasets, or when timely results are needed, ESPRIT and Otupipe are very efficient and can be used locally to produce acceptable results in every circumstance. In all the cases tested, RDP and ESPRIT always rendered more accurate and similar results, although perhaps in specific circumstances, and depending on the number of sequences, length and variability, combination workflows may still be an attractive option.

METHODS

Test Datasets of 16S rDNA Sequences

To check for accuracy, we constructed a reference data set with sequences retrieved from the NCBI and trimmed to obtain the V6 region (from positions 963 to 1063 in *E. coli*) using 60 sequences of different bacterial species from 59 different genera, according to NCBI taxonomy (accession numbers: AF538931, AF363135, DQ310706, AY167839, X90760, AM743175, AY367026, AF530131, U855746, EU360125, AM747393, FJ486138, AY691545, DQ366688, EF076758, AY234615, AM936268, AJ233945, FJ418118, AM935473, EF606819, AJ298940, EF173341, EF019646, CU466738, AM935145, AM935820, EF212893, AM59107, FM209153, EU375221, EF054879, EF466123, EF466120, EU722519, AF328681, EU58528, EU593733, AM935529, EU631792, AJ519368, AM935078, AY730501, EU630729, EU634621, DQ263467, X99390, DQ676361, AY852181, EU930421, DQ989484, AF448044, AJ224039, AB360346, AF114805, AY894325, EU551675, EU427317, AJ314848, AB037012). The accuracy of non-parametric estimators (ACE, Chao1) is expected to depend on changes in the ratio of OTUs to sequence numbers and OTU cluster size. To test their accuracy, the base dataset was copied 50 times to build a second set of 3000 sequences, and with the same number of OTUs.

To test the MSA step and richness estimators (ACE, Chao1), we simulated natural variability by generating two new 3000 sequence datasets mutating 50 copies of each sequence using EMBOSS [41] to introduce a single random mutation (indel or point mutation): one set had the 50 different mutants of each sequence stacked and the second one had them interspersed. Alignments were inspected using Seaview [42].

Recent studies tend to consider regions longer than V6 obtained from 454 sequencing. We obtained two datasets described by Quince [26], an artificial community pyrosequenced GSFLX dataset of 34308 V5 sequences with an average length of 266 nt and a maximum length of 289 nt, from 90 different clones and the Priest Pot pyrosequenced environmental DNA GSFLX data with 16222 sequences (both after initial noise removal) with an average length of 257 nt and a maximum length of 303 nt. These datasets were processed as above but in addition, we also calculated Mothur alignments with "trump" filtering to remove uneven ends.

To test the effect of nearly-full length sequences, two additional experimental data sets were used: one consisting of 1132 16S rDNA sequences from tall grass prairie soil samples with high diversity [37], with an average length of 1487 nt and a maximum length of 1542 nt, and a second one consisting of 1130 sequences from the axillary vault of patient HV5 of the human skin microbiome data with low diversity [23, 38], with an average length of 1344 nt and a maximum length of 1387 nt, by retrieving the sequences from the EBI and assembling each dataset in a single file.

Recent advances in pyrosequencing technology have enabled Illumina reads to reach a length that is suitable for hypervariable region sequencing, yielding huge numbers of sequence reads. We have simulated this situation using the well-known dataset of Huse *et al.*, [24] consisting of 340150 sequences (23445 of which are unique), with an average length of 96 nt and a maximum length of 169 nt.

To evaluate suitability for the study of differential environmental effects we have used a real-world experimental dataset consisting of short length sequences from rhizobacterial V6 16S rDNA obtained as previously described [32]. In synthesis, we collected soil samples from glyphosate-tolerant maize, event NK603 cultivars that had been treated in pre-emergence with the herbicide Harness®GTZ (GTZ henceforth) or in post-emergence with glyphosate (Roundup®Plus), and from untreated soil. Samples were pooled seven days after glyphosate application (first sampling time) and just before crop harvesting (final sampling time). Sequence counts obtained from the soils at the first sampling time were 3467 (untreated) with an average length of 103 nt and a maximum length of 243 nt, 5025 (glyphosate-treated) with an average length of 102 nt and a maximum length of 277 nt, and 3843 (GTZ-treated) with an average length of 99 nt and a maximum length of 269 nt, and at the final sampling time were 1814 with an average length of 101 nt and a maximum length of 157 nt, 1796 with an average length of 100 nt and a maximum length of 128 nt, and 1526 with an average length of 99 nt and a maximum length of 153 nt, respectively.

Sequence Processing

Differential soil data experimental sequences were filtered by 454 software, and additionally cleaned by eliminating those containing ambiguous base calls and sequences shorter than 50 nt after removing the primer sequence, as these account for 50% of all NGS errors [2]. We have omitted analysing the effect of different pre-processing strategies, as this was not within the scope of this study.

Sequence Analysis Workflows

For the combined workflows the first step consists of obtaining the MSA. We test three different alignment approaches: a) Iterative progressive alignments by Log-Expectation comparison as implemented in MUSCLE version 3.7 [14] with parameters -diags -maxiters 4 -stable. b) A fast group-to-group alignment based on partition trees implemented in MAFFT version 6.712 [16] with parameters -ep 0.123 -retree 1 -nofft -parttree, and c) an initial 8-mer search followed by Needleman-Wunsch pairwise alignments against a reference alignment implemented in Mothur version 1.14.0, using the SILVA-98 [43] and Greengenes [44] reference alignment, although only SILVA guided results are reported because it has been proven more effective in aligning the 16S rRNA hypervariable regions [11]. For each dataset we generated two Mothur alignments: one with and one without a pre-clustering step, filtering the result to remove redundant gaps using the standard approach (default parameters) described in the Mothur documentation, and fixing gap characters for compatibility with other programs.

Using these four initial MSA, we calculate the distance matrices using three different approaches for each alignment: a) evolutionary distance estimation using Jukes-Cantor correction [34] for multiple substitution as implemented in DNADIST from the PHYLIP software package version 3.67 [33]; b) simple distance estimation counting gaps only once, penalizing terminal gap, with a cut-off value of 0.11 and increasing precision to 1000 using Mothur [19]; and c) an approximate 6-mer distance calculation method implemented in MAFFT version 6.712 [16]. The MAFFT output file was converted to a lower triangular matrix to produce the clustering input.

The cluster analysis comprises the generation of OTUs and calculation of species richness with the Chao1 and ACE estimators at three dissimilarity levels (3%, 5% and 10%). Clustering was performed using the three methods (average, nearest and farthest neighbour) available from Mothur [19], although we report only average neighbour calculations, as they are generally considered more accurate [11, 25]. Beside these combined workflow procedures, we have included other methods that do not require an initial MSA: the ESPRIT software package [20], the recently published CROP package [23], the Otupipe pipeline (http://drive5.com/otupipe) and the RDP pyrosequencing pipeline [21]. ESPRIT uses an efficient k-tuple based distance filter before aligning sequences using the Needleman-Wunsch method [18] and computes pairwise distances using the "quickdist" algorithm [20] to perform complete-link hierarchical clustering; and CROP filters sequences first to reduce their number, then uses a modified Needleman-Wunsch alignment, performs a "quickdist" distance calculation, and finally an unsupervised Bayesian clustering method with MCMC sampling. Otupipe relies on UCHIME [45] to remove chimeras and on USEARCH to perform sequence comparisons and clustering [22]. RDP builds an initial MSA using the Infernal [36] tool and then directly proceeds to perform a complete linkage clustering using its own implementation.

To compare diversity estimation, we have compared rarefaction curves as well as ACE and Chao1 values obtained in all workflows involving Mothur and ESPRIT clustering. As CROP only produces OTU counts and composition, only these were considered in diversity estimate comparisons.

The results collected from all the different sources were processed to a common format (Mothur summary file format) and converted to tab-delimited data for further analysis and display using gnuplot.

Assessment of Clustering Consistency

Clustering methods often resort to random decisions, which may lead to different results in different runs using the same initial data and conditions. We have analysed the consistency of the results obtained with different datasets: analysis of short sequences using our synthetic mutated datasets (60 species mutated 50 times, with consecutive or interspersed sequences) and long sequences using a 16S rRNA dataset from Youssef et al. These datasets were analysed using CROP, ESPRIT and Mothur (with and without pre-clustering, using the average neighbour distance clustering method), repeating the analysis 20 times, collecting the number of OTUs reported, and calculating mean, standard deviation and standard error of the mean to derive corrected 95% confidence intervals using Student's T value for N-1 degrees of freedom to account for the relatively small sample size.

ACKNOWLEDGEMENTS

The Spanish Ministry of the Environment and Rural and Marine affairs (Grant No. EGO22008) has commissioned and supported this work. The Spanish Research Council has also supported this research (PIE No. 200420E397) in part. Additional support from CYTED 510RT0391 and SEQAHEAD COST networks is also acknowledged. We express our gratitude to the Informatics Services of the Spanish Research Council (CSIC) for the use of their cluster computing facilities. We want to thank the authors of the different tools for their advice and cooperation.

AUTHORS' CONTRIBUTIONS

JB and JRV carried out the data collection and analysis, and drafted the manuscript. RPM conceived the study, and participated in its design and coordination and helped to draft the manuscript. All authors read and approved the final manuscript.

REFERENCES

1. Amann RI, Ludwig W, Schleifer KH: Phylogenetic identification and in situ detection of individual microbial cells without cultivation. Microbiol Rev 1995, 59: 143–169.
2. Huber JA, Welch DBM, Morrison GH, Huse SM, Neal PR, Butterfield DA, Sogin ML: Microbial population structures in the deep marine biosphere. Science 2007, 318: 98–100.
3. Huse SM, Dethlefsen L, Huber JA, Welch D, Relman DA, Sogin ML: Exploring microbial diversity and taxonomy using SSU rRNA hypervariable tag sequencing. PLoS Genet 2008, 4: 1–10. 10.1371/journal.pgen.0040001
4. Margulies M, Egholm M, Altman WE, Attiya S, Bader JS, Bemben LA, Berka J, Braverman MS, Chen YJ, Chen ZT, Dewell SB, de Winter A, Drake J, Du L, Fierro JM, Forte R, Gomes XV, Godwin BC, He W, Helgesen S, Ho CH, Hutchison SK, Irzyk GP, Jando SC, Alenquer MLI, Jarvie TP, Jirage KB, Kim JB, Knight JR, Lanza JR, et al.: Genome sequencing in microfabricated high-density picolitre reactors. Nature 2005, 437: 376–380.
5. Lia CRS, Teixeira RS, Peixoto JC, Cury W, Jun S, Vivian H, Pellizari

JT, Alexandre SR: Bacterial diversity in rhizosphere soil from Antarctic vascular plants of Admiralty Bay, maritime Antarctica. ISME J 2010, 4: 989–1001. 10.1038/ismej.2010.35

6. López-Bueno A, Tamames J, Velázquez D, Moya A, Quesada A, Alcamí A: High diversity of the viral community from an Antarctic Lake. Science 2009, 326: 858–861. 10.1126/science.1179287
7. Schloss PD, Handelsman J: Toward a census of bacteria in soil. PLoS Comput Biol 2006, 2: 786–793.
8. Li W, Jaroszewski L, Godzik A: Clustering of highly homologous sequences to reduce the size of large protein database. Bioinformatics 2001, 17: 282–283. 10.1093/bioinformatics/17.3.282
9. Pandey RV, Nolte V, Schlötterer C: CANGS: a user-friendly utility for processing and analyzing 454 GS-FLX data in biodiversity studies. BMC Res Notes 2010, 3: 3. 10.1186/1756-0500-3-3
10. Russell DJ, Way SF, Benson AK, Sayood K: A grammar-based distance metric enables fast and accurate clustering of large sets of 16S sequences. BMC Bioinformatics 2010, 11: 601. 10.1186/1471-2105-11-601
11. Schloss PD: The Effects of alignment quality, distance calculation method, sequence filtering, and region on the analysis of 16S rRNA gene-based studies. PLoS Comput Biol 2010, 6: e1000844. 10.1371/journal.pcbi.1000844
12. Schloss PD, Handelsman J: Introducing DOTUR, a computer program for defining operational taxonomic units and estimating species richness. Appl Environ Microbiol 2005, 71: 501–1506. 10.1128/AEM.71.1.501-506.2005
13. Ahola V, Aittokallio T, Vihinen M, Uusipaikka E: Model-based prediction of sequence alignment quality. Bioinformatics 2008, 24: 2165–2171. 10.1093/bioinformatics/btn414
14. Edgar RC: MUSCLE: multiple sequence alignment with high accuracy and high throughput. Nucleic Acids Res 2004, 32: 1792–1797. 10.1093/nar/gkh340
15. Essoussi N, Boujenfa K, Limam M: A comparison of MSA tools. Bioinformation 2008, 2: 452–455.
16. Katoh K, Misawa K, Kuma K, Miyata T: MAFFT: a novel method for

rapid multiple sequence alignment based on fast Fourier transform (describes the FFT-NS-1, FFT-NS-2 and FFT-NS-i strategies). Nucleic Acids Res 2002, 30: 3059–3066. 10.1093/nar/gkf436

17. Mori H, Maruyama F, Kurokawa K: VITCOMIC: visualization tool for taxonomic compositions of microbial communities based on 16S rRNA gene sequences. BMC Bioinformatics 2010, 11: 332. 10.1186/1471-2105-11-332
18. Needleman SB, Wunsch CD: A general method applicable to the search for similarities in the amino acid sequence of two proteins. J Mol Biol 1970, 48: 443–453. 10.1016/0022-2836(70)90057-4
19. Schloss PD, Westcott SL, Ryabin T, Hall JR, Hartmann M, Hollister EB, Lesniewski RA, Oakley BB, Parks DH, Robinson CJ, Sahl JW, Stres B, Thallinger GG, Van Horn DJ, Weber CF: Introducing Mothur: Open-source, platform-Independent, community-supported software for describing and comparing microbial communities. Appl Environ Microbiol 2009, 75: 7537–7541. 10.1128/AEM.01541-09
20. Sun Y, Cai Y, Liu L, Yu F, Farrell ML, McKendree W, Farmerie W: ESPRIT: estimating species richness using large collections of 16S rRNA pyrosequences. Nucleic Acids Res 2009, 37: e76. 10.1093/nar/gkp285
21. Cole JR, Wang Q, Cardenas E, Fish J, Chai B, Farris RJ, Kulam-Syed-Mohideen AS, McGarrell DM, Marsh T, Garrity GM, Tiedje JM: The Ribosomal Database Project: improved alignments and new tools for rRNA analysis. Nucleic Acids Res 2009, 37: D141. 10.1093/nar/gkn879
22. Edgar RC: Search and clustering orders of magnitude faster than BLAST. Bioinformatics 2010, 26: 2460–2461. 10.1093/bioinformatics/btq461
23. Hao X, Jiang R, Chen T: Clustering 16S rRNA for OTU prediction: a method of unsupervised Bayesian clustering. Bioinformatics 2011, 27: 611–618. 10.1093/bioinformatics/btq725
24. Huse SM, Huber JA, Morrison HG, Sogin ML, Welch DM: Accuracy and quality of massively parallel DNA pyrosequencing. Genome Biol 2007, 8: R143. 10.1186/gb-2007-8-7-r143
25. Huse SM, Welch DM, Morrison HG, Sogin ML: Ironing out the wrinkles in the rare biosphere through improved OUT clustering.

Environ Microbiol 2010, 12: 1889–1898. 10.1111/j.1462-2920.2010.02193.x

26. Quince C, Lanzen A, Curtis TP, Davenport RJ, Hall N, Head IM, Read LF, Sloan WT: Accurate determination of microbial diversity from 454 pyrosequencing data. Nat Methods 2009, 6: 639-U27. 10.1038/nmeth.1361
27. Lozupone CA, Hamady M, Kelley ST, Knight R: Quantitative and qualitative beta diversity measures lead to different insights into factors that structure microbial communities. Appl Environ Microbiol 2007, 73: 1576–1585. 10.1128/AEM.01996-06
28. Wang Q, Garrity GM, Tiedje JM, Cole JR: Naive Bayesian classifier for rapad assignment of rRNA sequences into the new bacterial taxonomy. Appl Environ Microbiol 2007, 73: 5261–7. 10.1128/AEM.00062-07
29. Sul WJ, Cole JR, Jesus Eda C, Wang Q, Farris RJ, Fish JA, Tiedje JM: Bacterial community comparisons by taxonomy-supervised analysis independent of sequence alignment and clustering. Proc Natl Acad Sci USA 2011, 30(108):14637–14642.
30. White JR, Navlakha S, Nagarajan N, Ghodsi MR, Kingsford C, Pop M: Alignment and clustering of phylogenetic markers implications for microbial diversity studies. BMC Bioinformatics 2010, 11: 152. 10.1186/1471-2105-11-152
31. Caporaso JG, Kuczynski J, Stombaugh J, Bittinger K, Bushman FD, Costello EK, Fierer N, Peña AG, Goodrich JK, Gordon JI, Huttley GA, Kelley ST, Knights D, Koenig JE, Ley RE, Lozupone CA, McDonald D, Muegge BD, Pirrung M, Reeder J, Sevinsky JR, Turnbaugh PJ, Walters WA, Widmann J, Yatsunenko T, Zaneveld J, Knight R: QIIME allows analysis of high-throughput community sequencing data. Nat Methods 2010, 7: 335–336. 10.1038/nmeth.f.303
32. Barriuso J, Marín S, Mellado RP: Effect of the herbicide glyphosate on glyphosate-tolerant maize rhizobacterial communities: a comparison with pre-emergency applied herbicide consisting of a combination of acetochlor and terbuthylazine. Environ Microbiol 2010, 12: 1021–1030. 10.1111/j.1462-2920.2009.02146.x
33. Felsenstein J: PHYLIP - Phylogeny Inference Package. Cladistics

1989, 5: 164–166.**Google Scholar**

34. Jukes TH, Cantor CR: Evolution of protein molecules. (Munro H N ed.) Mammalian protein Metabolism, III. New York: Academic Press; 1969:21–132.
35. Letsch HO, Kueck P, Stocsits RR, Misof B: The Impact of rRNA Secondary Structure Consideration in Alignment and Tree Reconstruction: Simulated Data and a Case Study on the Phylogeny of Hexapods. Mol Biol Evol 2010, 27: 2507–2521. 10.1093/molbev/msq140
36. Nawrocki EP, Kolbe DL, Eddy SR: Infernal 1.0: inference of RNA alignments. Bioinformatics 2009, 25: 1713–1713. 10.1093/bioinformatics/btp326
37. Youssef N, Sheik CS, Krumholz LR, Najar FZ, Roe BA, Elshahed MS: Comparison of Species Richness Estimates Obtained Using Nearly Complete Fragments and Simulated Pyrosequencing-Generated Fragments in 16S rRNA Gene-Based Environmental Surveys. Appl Environ Microbiol 2009, 75: 5227–36. 10.1128/AEM.00592-09
38. Grice EA, Kong HH, Conlan S, Deming CB, Davis J, Young AC, Bouffard GG, Blakesley RW, Murray PR, Green ED, Turner ML, Segre JA: Topographical and temporal diversity of the human skin microbiome. Science 2009, 324: 1190–1192. 10.1126/science.1171700
39. Katoh K, Toh H: Recent developments in the MAFFT multiple sequence alignment program. Brief Bioinformatics 2008, 9: 286–298. 10.1093/bib/bbn013
40. Niu B, Fu L, Sun S, Li W: Artificial and Natural. Duplicates in Pyrosequencing Reads of Metagenomic data. BMC Bioinformatics 2010, 11: 187. 10.1186/1471-2105-11-187
41. Rice P, Longden I, Bleasby A: EMBOSS: the European Molecular Biology Open Software Suite. Trends Genet 2010, 16: 276–277.
42. Gouy M, Guindon S, Gascuel O: SeaView version 4: a multiplatform graphical user interface for sequence alignment and phylogenetic tree building. Mol Biol Evol 2010, 27: 221–224. 10.1093/molbev/msp259
43. Pruesse E, Quast C, Knittel K, Fuchs BM, Ludwig W, Peplies J,

Glöckner FO: SILVA: a comprehensive online resource for quality checked and aligned ribosomal RNA sequence data compatible with ARB. Nucleic Acids Res 2007, 35: 7188–7196. 10.1093/nar/gkm864

44. DeSantis TZ, Hugenholtz P, Keller K, Brodie EL, Larsen N, Piceno YM, Phan R, Andersen GL: NAST: a multiple sequence alignment server for comparative analysis of 16S rRNA genes. Nucleic Acids Res 2006, 34: 394–399.
45. Edgar RC, Haas BJ, Clemente JC, Quince C, Knight R: UCHIME improves sensitivity and speed of chimera detection. Bioinformatics 2011, 15(27):2194–200.

CHAPTER 10

RAPID GENETIC DIVERGENCE IN RESPONSE TO 15 YEARS OF SIMULATED CLIMATE CHANGE

Catherine H. Ravenscroft[1], Raj Whitlock[2,3], and Jason D. Fridley[1]

[1] Department of Biology, Syracuse University, 107 College Place, Syracuse, NY 13244, USA

[2] Institute of Integrative Biology, University of Liverpool, The Biosciences Building, Crown Street, Liverpool L69 7ZB, UK

[3] Department of Animal and Plant Sciences, University of Sheffield, Western Bank, Sheffield S10 2TN, UK

ABSTRACT

Genetic diversity may play an important role in allowing individual species to resist climate change, by permitting evolutionary responses.

Our understanding of the potential for such responses to climate change remains limited, and very few experimental tests have been carried out within intact ecosystems. Here, we use amplified fragment length polymorphism (AFLP) data to assess genetic divergence and test for signatures of evolutionary change driven by long-term simulated climate change applied to natural grassland at Buxton Climate Change Impacts Laboratory (BCCIL). Experimental climate treatments were applied to grassland plots for 15 years using a replicated and spatially blocked design and included warming, drought and precipitation treatments. We detected significant genetic differentiation between climate change treatments and control plots in two coexisting perennial plant study species (Festuca ovinaand Plantago lanceolata). Outlier analyses revealed a consistent signature of selection associated with experimental climate treatments at individual AFLP loci in P. lanceolata, but not in F. ovina. Average background differentiation at putatively neutral AFLP loci was close to zero, and genomewide genetic structure was associated neither with species abundance changes (demography) nor with plant community-level rcsponses to long-term climate treatments. Our results demonstrate genetic divergence in response to a suite of climatic environments in reproductively mature populations of two perennial plant species and are consistent with an evolutionary response to climatic selection in P. lanceolata. These genetic changes have occurred in parallel with impacts on plant community structure and may have contributed to the persistence of individual species through 15 years of simulated climate change at BCCIL.

Keywords: adaptation, climate change, evolution, genetic differentiation, grassland

INTRODUCTION

Climate change is expected to impose strong directional selection pressures on plant populations (Davis & Woods, 1986; Davis & Shaw, 2001; Jump & Peñuelas, 2005; Bradshaw & Holzapfel, 2006; Reusch & Wood, 2007; Anderson et al., 2012). In response, plant populations may adapt in situ via selection on standing genetic variation (Hoffmann &

Willi, 2008; Jump et al., 2008a). This adaptive response could be an important component of species' resistance to climate change, because it would provide an in situ'option' for persistence in spite of environmental change (Jump et al., 2008a).

Plant populations are often adapted to their local environments (Leimu & Fischer, 2008). These evolutionary responses can occur at fine spatial scales, despite gene flow. For example, populations of the grass Anthoxanthum odoratum show adaptive responses to heavy metal contamination at the boundaries of metal ore mines (Antonovics & Bradshaw, 1970) and have evolved in response to sharp boundaries separating grassland management regimes in the Park Grass Experiment (Snaydon & Davies, 1976; Gould et al., 2014). Genetic differentiation in response to gradients of climate (e.g. temperature) and climate-driven abiotic factors (e.g. water availability) has been observed in plants at both fine spatial scales (Kelly et al., 2003; Parisod & Christin, 2008; Jump et al., 2009; Manel et al., 2010; Franks, 2011) and over landscapes (Hamrick & Allard, 1972; Hamrick & Holden, 1979; Li et al., 1999; Owuor et al., 2003; Jump et al., 2006). In some cases, associations between the climate and genetic structure are repeated at different spatial scales or with temporal climate changes (e.g. Hamrick & Allard, 1972; Hamrick & Holden, 1979; Jump et al., 2006). These patterns of genetic structuring are highly indicative of adaptive differentiation in response to environmental selection. More recently, genetic responses to the climate have been identified using genomic and quantitative genetic approaches (e.g. Jump et al., 2008b; Franks, 2011). Together, these studies indicate that climatic factors can act as potent forces of selection, driving adaptive differentiation between populations, and within populations at fine spatial scales despite potentially high levels of gene flow.

Although previous studies suggest adaptive responses to environmental selection, there have been very few direct experimental tests of genetic responses to climate change in plant species within intact ecosystems (Jump et al., 2008b; Avolio et al., 2013). Such tests are important, because they tell us about the capacity for neutral or adaptive genetic change in ecologically realistic settings that may incorporate competition from coexisting species, or fine-scale abiotic heterogeneity that interferes with climatic selection. Avolio et al. (2013) used AFLP markers to demonstrate changes in clonal structure of the grass Andropogon gerardii in response to manipulation of precipitation regimes in a prairie ecosystem, but did not identify a clear adaptive component to these changes. Jump et al. (2008b) used an outlier analysis to identify adaptive divergence in response to simulated drought and warming in Fumana thymifolia within a Mediterranean shrub community, focussing on the establishing (seedling) phase of the plant life cycle. Thus, we know very little regarding either the extent to which climatic selection can drive adaptive divergence within plant populations occupying intact ecosystems, or about adaptive responses in the established (reproductively mature) phase of perennial plant species.

At the Buxton Climate Change Impacts Laboratory (BCCIL) in northern England, intact species-rich limestone grassland has been subjected to experimentally manipulated climate treatments since 1993 (involving increased temperature, modified precipitation and factorial combinations of these; Fig. 1). BCCIL maintains some of the longest-running multifactorial climate manipulations in the world. Community composition has remained relatively stable in all experimental treatments (Grime et al., 2000, 2008). Such stability is rare; the majority of experimental climate manipulation studies report rapid community and ecosystem responses (e.g. Harte & Shaw, 1995; Grime et al., 2000; Zavaleta et al., 2003; Evans et al., 2011). Genetic adaptation to climatic selection has been hypothesized as one potential mechanism supporting the apparent resistance of this grassland community to long-term climate manipulations (Grime et al., 2008). Genetic variation in phenotype has been documented at fine spatial scales (< 10 m) in several of the species that are common at BCCIL, and this diversity can influence plant community structure (Booth & Grime, 2003; Fridley et al., 2007; Whitlock et al., 2007, 2010; Bilton et al., 2010). Recently, Ravenscroft et al. (2014) have shown that subpopulations of Plantago lanceolatasubjected to long-term experimental climate treatments at BCCIL differed significantly in phenotype when grown under common environment conditions. This finding suggests that plant phenotypes have evolved in response to simulated climate change treatments and may be consistent with an adaptive response to the climatic environment.

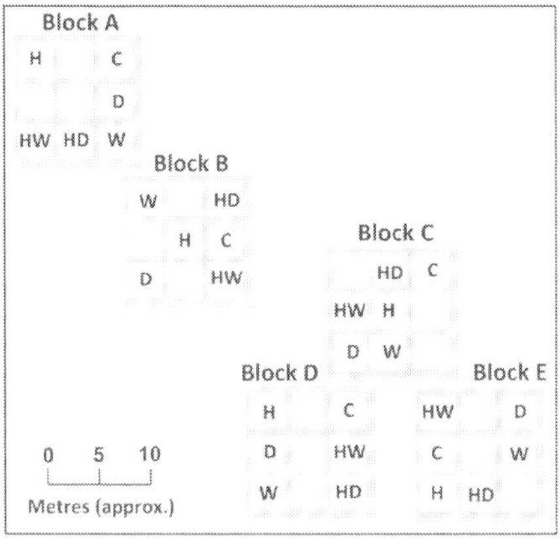

Figure 1: Layout of experimental plots at BCCIL. Each climate treatment is replicated five times in a randomized block design. C: Control, D: Drought, H: Heated, W: Watered, HD: Heated-Drought, HW: Heated-Watered.

We used amplified fragment length polymorphism (AFLP) markers to test for genetic differentiation in coexisting populations of established individuals of P. lanceolata and Festuca ovina exposed to 15 years of experimental climate manipulations at BCCIL. Simulated climate change could lead to genetic differentiation in this system through two processes: genetic drift and ecologically adaptive evolution. Genetic drift may be induced if a novel climate leads to altered demography (specifically, changes in subpopulation size or modified gene flow between treatments and individual plots; Ellstrand & Elam, 1993). The effects of drift are expected to be genomewide, leading to a consistent pattern of genetic differentiation across loci. Adaptive responses, on the other hand, are expected to accrue through selection on variation at individual gene loci that control adaptive phenotypes. In this case, we would expect to see a pattern of excess (outlier) differentiation at a limited number of selected loci against a background of genomewide differentiation imposed by drift. Crucially, neutral spatial genetic structure is not expected to lead to genetic structuring among climate treatments at BCCIL, as treatment plots are evenly replicated over spatial blocks in this experiment (Fig. 1). We combined multivariate and outlier analyses of genetic data with information on species-level demography and abundance to assess first whether 15 years of simulated climate change had resulted in genetic differentiation and, second, whether genetic change was consistent with an evolutionary (adaptive) response to climatic selection.

MATERIALS AND METHODS

Study System and Focal Species

Prior to the onset of experimentation in 1993, the vegetation at the BCCIL site comprised an ancient, unimproved sheep pasture located on the side of a dale (valley) near Harpur Hill, Buxton, UK. The experiment now running at BCCIL was set up as a series of 3 × 3 m plots within this pasture, each containing intact species-rich limestone grassland vegetation. These plots have been exposed to annual climatic treatments since 1993 using a replicated, spatially blocked design. Treatments include (i) winter warming, where soil surface is maintained at 3 °C above ambient (nonmanipulated) temperature from November to April; (ii) drought treatments, where rainfall in July and August is intercepted by automated rain shelters; (iii) water addition (watered), where water is added to experimental plots from June to September at a rate of 20% above the long-term average; (iv) warming and drought (heated/drought); (v) warming and increased precipitation (heated/watered); and (vi) control. Experimental plots (3 × 3 m) are replicated five times in a randomized block design (Fig. 1). Grime et al. (2000) provide further details on the experimental design at BCCIL. Soil depth is a major source of fine-scale environmental heterogeneity within plots, varying from 0 cm (bare rock) to > 30 cm, and influences plant community structure at local scales (Fridley et al., 2011).

Festuca ovina L. (sheep's fescue, a grass) and Plantago lanceolata L. (ribwort plantain, a forb) are perennial (polycarpic) plants with wide geographical distributions; both species are self-incompatible (obligate outcrossers) and wind pollinated. Both study species are common in all experimental plots at BCCIL, and both have a limited capacity for clonal growth: P. lanceolata via lateral rosette formation and F. ovina through tillering, typically leading to the formation of loose tussocks. Festuca ovina was both the most frequent of all the species (occurring in 223 of 240 permanent 10 × 10 cm quadrats distributed throughout the grassland at BCCIL) and the species with greatest cover in the plots at BCCIL (Fridley et al., 2011). This species was significantly more abundant in heated, drought-treated and heated/drought plots and less abundant in watered plots after 15 years of climatic manipulation (Fridley et al., 2011; Fig. 2a). Soil depth was not a significant predictor of F. ovina abundance

(Fridley et al., 2011). The abundance of P. lanceolata varied with climate treatment, soil depth and a treatment × soil depth interaction (Fridley et al., 2011). The abundance of this species was highest in warming treatments, was lowest in treatments that included drought and varied considerably with soil depth in all experimental treatments (Fridley et al., 2011; Fig. 2b).

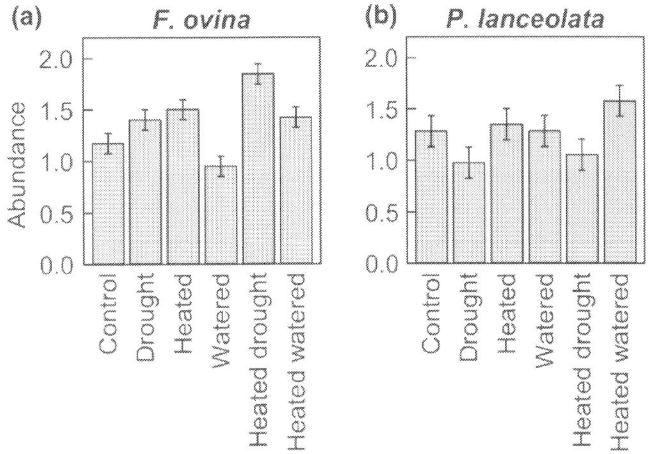

Figure 2: Species-level abundance of F. ovina and P. lanceolata within simulated climate change treatments at BCCIL (data from Fridley et al., 2011). Abundance was measured as ordinal vegetation cover class (0 = 0–4%; 1 = 5–24%; 2 = 25–49%; 3 = 50–74%; 4 = 75% +) within 10 × 10 cm permanent quadrats located within each 3 × 3 m grassland plot (eight quadrats per plot, five plots per treatment). Error bars show one standard error of the mean.

Sample Collection

In the following, we use 'total population' or 'population' to refer to the individuals of a species present across the BCCIL site. Subsets of individuals within climate treatments, blocks or plots are referred to as 'subpopulations'. We collected leaf tissue samples of each species from control, drought-treated, warmed, watered, heated-drought and heated-watered plots at BCCIL in June 2009 (30 plots in total), after 15 years of experimental treatments. We sampled 12 individuals of each species from each plot, excluding a 0.3-m boundary zone at the edge of each plot.

We used eight previously established permanent 10 × 10 cm quadrats (Fridley et al., 2011) and four newly established quadrats within each plot to guide sampling. Leaf tissue was recovered from the individual that was rooted most closely to the centre of each quadrat (12 samples per plot in total). The spatial locations of both the centre of each quadrat and of the sampled plant relative to this point were recorded to the nearest 5 mm. Soil depth at the 240 existing quadrats was known (Fridley et al., 2011). Local soil depths were determined for our newly established quadrats and for any plants rooted outside quadrats (i.e. for plants that were located > 5 cm from the quadrat centre). Our sampling scheme led to a total of 60 samples per climate treatment subpopulation. Samples from each treatment spanned the entire range of soil depths occupied by our study species at the site. Each leaf tissue sample was stored in self-indicating silica gel (1–3 mm grain size; Merck-Millipore, Darmstadt, Germany) for genetic analysis. Replicate tissue samples were collected from 40 randomly selected individuals to estimate genotyping error.

Molecular Methods

Laboratory protocols for DNA extraction from leaf tissue samples and subsequent AFLP analysis followed Whitlock et al. (2008a). Modifications and further details, including methods for quality control. AFLP peak-height data for each individual were generated using genemapper version 3.0 (Applied Biosystems), and AFLP phenotypes were scored using aflpscore (Whitlock et al., 2008b). AFLP band presence–absence data for our study species comprised 270 loci and 222 individuals for P. lanceolata, and 999 loci and 303 individuals for F. ovina.

Population Density and Abundance Estimates

We estimated population density in order to understand whether the climatic environments at BCCIL had led to demographic changes that could have driven changes in genetic structure. Closest individual density estimation was used to assess population density for each study species, exploiting the known distance between the centres of the 100 cm^2 quadrats and the nearest individual of each of our study species (Cottam et al., 1953; Cottam & Curtis, 1956). Population density at BCCIL can be used as a proxy for population size on the basis that each of the plots and each of the treatments have identical areas. Existing species abundance data, collected for 240 of the quadrats, and measured as ordinal vegetation cover class (0 = 0–4%; 1 = 5–24%; 2 = 25–49%; 3 = 50–74%; 4 = 75% +; Fridley et al., 2011) were used to create plot-level and treatment-level summaries of species abundance. This cover-based measure of species abundance is a function of both local plant density and plant morphology.

Genetic Diversity and Spatial Structure

Allele frequencies were calculated using Zhivotovsky's Bayesian method (1999) assuming Hardy–Weinberg equilibrium. Allele frequency estimates for globally polymorphic loci ($0.05 \leq q \leq 0.95$) were used to calculate gene diversity (an unbiased estimate of the expected heterozygosity) for climatic and soil-depth subpopulations at BCCIL for each study species (Lynch & Milligan, 1994).

We tested for the presence of spatial genetic structure (SGS) within each species by regressing pairwise kinship coefficients on the distance separating plant individuals (implemented using the software spagediversion 1.4; Hardy & Vekemans, 2002). Any loci identified as showing outlier patterns of differentiation among climate treatments under the analyses described below (i.e. those loci putatively responding to climatic selection) were removed prior to the assessment of SGS (F. ovina, 48 loci removed; P. lanceolata, 25 loci removed). Significance of the regression slope was tested using 2000 permutations of AFLP genotypes among sampling locations. The extent of SGS was estimated by quantifying Wright's neighbourhood size, which measures the effective population size of locally panmictic units within the total population, and σ (the gene dispersal parameter), which defines the physical size of these panmictic units (equal to half the expected parent–offspring dispersal distance). These analyses assume that populations have reached drift–dispersal equilibrium and that dispersal is isotropic and also require an estimate of effective population density. Populations were assumed to be in Hardy–Weinberg equilibrium (i.e. $F_{IS} = 0$) on the basis of previous isozyme studies carried out on our study species (Bos et al., 1986; Weibull et al., 1991).

Climate-induced Genetic Differentiation

Genomewide (multilocus) differentiation between climatic environments was assessed using Cockerham & Weir's (1993) β estimator of F_{ST} (R-functions given as Supporting Information). The significance of F_{ST} was determined by creating a null distribution for F_{ST} via 5000 permutations of individuals among climatic subpopulations (incorporating the observed F_{ST} as a datum). We used permutational analysis of variance to determine whether genetic distances among individuals could be explained by climate treatment, soil depth, treatment × soil depth interaction or block, using function 'adonis' in r package vegan 1.17-4 (Legendre & Anderson, 1999; McArdle & Anderson, 2001). Permutational anova is an alternative to amova (Excoffier et al., 1992) that provides a flexible approach to partitioning variation in a distance matrix, allowing for continuously distributed explanatory variables, interaction terms and nonindependence due to grouping factors (incorporated by restricting permutations within different 'strata'). Permutational anova analyses were conducted on Euclidian genetic distance matrices computed from the AFLP presence–absence data. First we carried out a global permutational anova analysis that included all treatment and control climates, in which a treatment × soil depth interaction and block main effects were fitted (P-values for treatment effects were identical whether or not blocks were fitted as strata). Second, we performed additional analyses that paired each climate treatment with control. In these permutational anova, we fitted a treatment × soil depth interaction and incorporated blocks as strata. All of these analyses used 1999 permutations of the genetic distance matrix to compute significance tests on the parameter estimates.

To determine whether genetic structure was associated with local demographic history, we carried out a regression of plot-level gene diversity (expected heterozygosity) against plot-level harmonic mean subpopulation abundance over the 15 years of monitoring at BCCIL and against current subpopulation density. These former abundance data are point quadrat records drawn from Grime et al. (2008); population abundance and population density are correlated in this system (see results). Genetic drift is expected to lead to a loss of genetic diversity that is related to local subpopulation size (Crow & Kimura, 1970). Thus, a positive relationship between local subpopulation size and genetic diversity would provide evidence in support of a role for local demographic isolation and genetic drift (neutral genetic change) in driving population structure at BCCIL. Harmonic mean abundance was used as a predictor for genetic diversity because the effects of drift on diversity in a temporally fluctuating population are expected to be dictated largely by periods of low effective population size (Wright, 1931; Crow, 1964). These analyses assume that census population sizes or abundance estimates are proportional to effective population sizes.

We tested for the genomic signature of adaptive differentiation in response to simulated climate change and soil-depth heterogeneity using two outlier analyses. First, we conducted analyses using bayescan v. 2.1 (Foll & Gaggiotti, 2008). This software identifies individual marker loci showing patterns of differentiation consistent with departures from neutrality by modelling a parameter for neutral differentiation shared by all loci, and locus-specific parameters for differentiation representing the effects of selection. For each locus, the weight of evidence in favour of a response to selection is calculated as an odds ratio of posterior probabilities for models containing either both the parameters or containing only the parameter for neutral differentiation. We ran bayescan models with a burn-in period of 50 000 iterations and then extracted samples of size 1000 from MCMC chains, with a thin interval of 1000 iterations (1.05×10^6 iterations in total). The prior odds for the neutral model were set to 10. We set a beta prior for F_{IS} with mean 0.05 and standard deviation 0.1 to incorporate information regarding the mating system of our study species (Bos et al., 1986; Weibull et al., 1991). This prior incorporates previous evidence for a value of F_{IS} close to zero, but allows for uncertainty in the precise value and for nonzero F_{IS} in our study populations. Loci responding to selection were identified when the posterior odds for a response to selection exceeded three (equivalent to 'substantial' support for selection as defined by Jeffrey's scale of evidence for Bayes factors; Jeffreys, 1961). We tested for responses to climatic selection by undertaking a global analysis including all climatic environments and by constructing pairwise outlier analyses for each combination of climate treatment with control, pooling individuals across plots within treatments (five analyses, each with treatment and control subpopulations). Plot-level sample sizes (n = 12) were insufficient to support separate, robust outlier analyses for each climate treatment within each block. We also carried out a bayescan analysis that used experimental blocks as subpopulations, to test for signatures of selection resulting from spatial variation in the environment at BCCIL. The second outlier analysis used logistic regression to investigate associations between the presence–absence phenotypes at individual AFLP loci and climate treatment and soil depth (soil depth could not be modelled in the bayescan analysis). Logistic regressions were fitted separately for each locus, using the R function bayesglm, with a logit link (package arm; Gelman et al., 2009). For each locus, we fitted a treatment × soil depth interaction and block main effect. This model was reduced sequentially, and likelihood ratio tests were used to assess the treatment × soil depth interaction, and the treatment, soil depth and block main effects. To identify loci putatively responding to environmental selection, we estimated the tail area-based false discovery rate for each locus within each model effect, using the total set of P-values across loci within model effects (implemented in R package fdrtool; Strimmer, 2008a,b). Model effects for particular loci were considered to show an outlier pattern of genetic differentiation when their corresponding Q-value was below 10%.

Genetic vs. Community-Level Responses to Climate Change

We carried out two analyses in order to determine whether responses to simulated climate change at BCCIL at the community level were coupled with multilocus genetic responses within species. First, we performed correlations of plot-level species diversity (species richness and Shannon diversity index) with multilocus gene diversity [species abundance data were taken from Fridley et al. (2011)]. Second, we calculated the ecological distance between all pairs of plots using the Bray–Curtis index and compared the resulting dissimilarity matrix with a matrix of multilocus pairwise genetic distances (F_{ST}), using a simple mantel test.

RESULTS

Population Structure and Diversity

Population density varied significantly with climate change treatment for both study species

(F. ovina: $F_{5,295}=4.3, P=0.046$; P. lanceolata: $F_{5,213}=18.8, P<0.001$; Fig. 3a, b). Both study species responded positively to the warming treatment, but showed opposing responses to the drought treatment (increase in rooted plant density in F. ovina, decrease in P. lanceolata; Fig. 3a, b). Both species also exhibited a pattern of increasing plant density with soil depth ($P<0.05$ in each case). Plot-level estimates of population density were associated positively with plot-level abundance (cover) within the permanent quadrats (F. ovina: $r^2 = 0.27$, $P = 0.007$; P. lanceolata: $r^2 = 0.69$, $P < 0.001$).

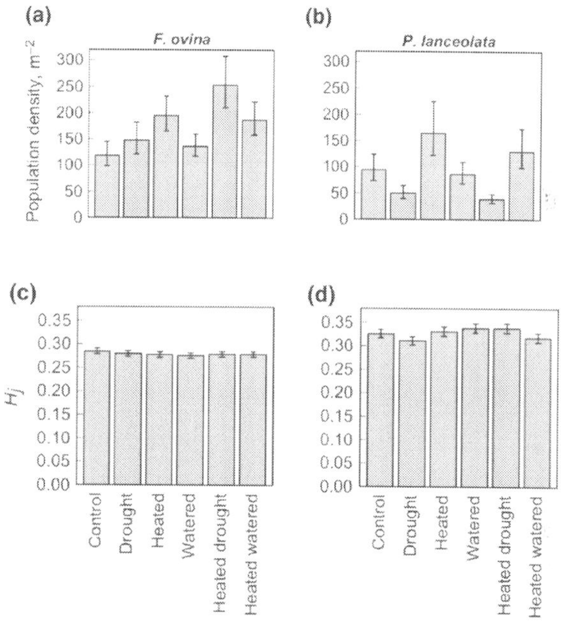

Figure 3: Population density and gene diversity (expected heterozygosity) of subpopulations of F. ovina and P. lanceolata growing in different climatic environments at BCCIL. (a, b) Population density estimated from the distance between the centre of each 10 9 10 cm permanent quadrat and the nearest (sampled) individual of F. ovina or P. lanceolata. (c, d) Gene diversity estimates (H_j) based on polymorphic AFLP loci (613 loci, F. ovina; 221 loci, P. lanceolata). Estimates of H_j were based on identical sample sizes for each climate environment subpopulation. Error bars show one standard error of the mean.

Festuca ovina and P. lanceolata populations occupying intact limestone grassland at BCCIL were genetically diverse; gene diversity estimates (H_j) based on polymorphic AFLP loci were 0.281 and 0.332 for F. ovina and P. lanceolata, respectively. All individuals possessed a unique multilocus AFLP genotype, indicating that all sampled plants represented distinct genetic individuals. Subpopulations of F. ovina that had been exposed for 15 years to different simulated climatic environments possessed closely similar gene diversities (Fig. 3c). Standard error estimates for gene diversity in P. lanceolata indicated weak variation in diversity among climate treatment subpopulations (Fig. 3d). Drought-treated subpopulations of P. lanceolata had the lowest genetic diversity (H_j = 0.311), and watered and heated-drought subpopulations were the most genetically diverse (H_j = 0.338; Fig. 3d). Levels of genetic diversity within these species did not vary consistently along the soil-depth gradient present at the study site (between shallow and deep soils).

Both study species showed evidence of spatial genetic structure (regression slopes of pairwise kinship coefficients on distance were significantly negative $P < 0.001$). Neighbourhood sizes (the effective size of locally panmictic units within the population) were 320 and 140, and neighbourhood areas were 9.29 and 9.97 m^2 for F. ovina and P. lanceolata, respectively (assuming an N_E/N ratio of 50%). These neighbourhood areas are approximately the size of individual plots at BCCIL (9 m^2). Neighbourhood sizes and areas were substantially greater for F. ovina when we assumed an N_E/N ratio of 10% (636 individuals and 70.68 m^2, respectively). The procedure used to estimate dispersal parameters did not converge using an assumed N_E/N ratio of 10% when applied to P. lanceolata.

Genetic Differentiation

Subpopulations of both study species showed weak but significant genetic differentiation among the climatic environments at BCCIL (F. ovina: F_{ST} = 0.006, permutation test $P < 0.001$; P. lanceolata: F_{ST} = 0.012, $P < 0.001$). Permutational anova analysis supported the presence of climate-associated differentiation (partial $R^2 = 0.02$ and 0.03 for F. ovina and P. lanceolata, respectively; $P < 0.001$; Table 1) and also indicated significant genetic differentiation among experimental blocks (partial R^2 = 0.02 and

0.03; $P < 0.001$; Table 1). The majority of genetic variance in these analyses was unstructured with respect to both experimental treatments and experimental blocks (partial R^2 for residual variance = 0.94 and 0.91; Table 1). There were weak associations between the genetic distance between individuals and soil depth, for both F. ovina ($R^2 = 0.004$, $P = 0.018$) and P. lanceolata ($R^2 = 0.005$, $P = 0.057$; Table 1). Pairwise treatment–control comparisons for F. ovina revealed significant climate-induced genetic differentiation for all climatic environments relative to control (results for drought-treated and watered plots remained significant following Bonferroni correction; Table 1). Plantago lanceolatasubpopulations showed significant climate-induced genetic differentiation in all cases except the heated–control contrast. Only the watered–control pairwise contrast remained significant following Bonferroni correction (Table 1). The greatest genetic distances (F_{ST}) between individual climate treatments and control plots occurred for the watered treatment. There was no clear correspondence between the occurrence of control–climate treatment genetic differentiation (Table 1) and differences in subpopulation density (Fig. 3a, b). For example, in P. lanceolata, we observed no genetic differentiation of the heated (winter warming) from control subpopulations (Table 1), yet winter warming had modified this species' density substantially (73.6% increase in density; Fig. 3b). Both drought-treated and watered subpopulations of F. ovina were significantly differentiated from the control subpopulation (Table 1), but the drought and watering treatments elicited only modest changes in population density (increase in density < 25%; Fig. 3a).

Table 1: Permutational ANOVA analyses show that subpopulations of Festuca ovina and Plantago lanceolata exposed to 15 years of simulated climate change have become genetically differentiated. Permutational ANOVA analyses were carried out on Euclidian distance matrices derived from multilocus AFLP genotypes for sampled individuals. Results are given for models that included all treatments and for models that used individual pairwise treatment–control comparisons. All analyses following the global analysis were stratified by experimental block, and so block terms are not shown for these analyses

	Festuca ovina					*Plantago lanceolata*				
	df	SS	F	R^2	§Pr (> F)	df	SS	F	R^2	§Pr (> F)
All treatments										
Treatment	5	507	1.23	0.020	0.0005***	5	210	1.32	0.030	0.0005***
Depth	1	96	1.17	0.004	0.0175*	1	38	1.20	0.005	0.0565†
Block	4	453	1.37	0.018	0.0005***	4	226	1.78	0.032	0.0005***
Treatment × Depth	5	425	1.03	0.017	0.1735	5	171	1.08	0.024	0.0865†
Residuals	287	23680		0.941		204	6472		0.909	
Total	302	25162		1.000		219	7117		1.000	
Drought–Control										
Treatment	1	108	1.28	0.018	0.0005**	1	40	1.29	0.019	0.0090
Depth	1	95	1.14	0.016	0.1330	1	30	0.96	0.014	0.5710
Treatment × Depth	1	75	0.90	0.013	0.9450	1	34	1.07	0.016	0.2775
Residuals	68	5696		0.953		64	2004		0.951	
Total	71	5974		1.000		67	2108		1.000	
Heated–Control										
Treatment	1	99	1.19	0.013	0.0270	1	37	1.14	0.016	0.1010
Depth	1	93	1.12	0.012	0.3255	1	36	1.13	0.016	0.2740
Treatment × Depth	1	91	1.09	0.012	0.1815	1	31	0.96	0.014	0.7505
Residuals	91	7563		0.964		68	2185		0.955	
Total	94	7845		1.000		71	2289		1.000	
Watered–Control										
Treatment	1	110	1.32	0.014	0.001*	1	44	1.33	0.017	0.0035†
Depth	1	99	1.19	0.012	0.0470	1	36	1.10	0.014	0.4080
Treatment × Depth	1	86	1.03	0.011	0.3655	1	27	0.81	0.010	0.9745
Residuals	92	7625		0.963		75	2473		0.958	
Total	95	7919		1.000		78	2580		1.000	
Heated-drought–Control										
Treatment	1	103	1.23	0.014	0.0185	1	43	1.34	0.019	0.0055†
Depth	1	92	1.11	0.012	0.4560	1	32	1.00	0.014	0.5265
Treatment × Depth	1	93	1.11	0.013	0.1225	1	38	1.18	0.017	0.0970
Residuals	85	7094		0.961		67	2176		0.950	
Total	88	7381		1.000		70	2290		1.000	
Heated-watered–Control										
Treatment	1	100	1.19	0.013	0.0245	1	43	1.37	0.019	0.002*
Depth	1	94	1.12	0.012	0.3055	1	37	1.16	0.016	0.1015
Treatment × Depth	1	80	0.96	0.010	0.7295	1	31	0.97	0.013	0.5365
Residuals	91	7602		0.965		70	2207		0.952	
Total	94	7876		1.000		73	2317		1.000	

§P-values are given as raw (unadjusted) values.

***,**,*,†Indicate that Bonferroni-adjusted P-values were < 0.001, < 0.01, < 0.05 or < 0.10, respectively. No Bonferroni correction was applied to the models that included all treatments.

To determine whether the data supported a role for drift in shaping genetic structure, we regressed plot-level gene diversity on 'long-term' (15-year) population abundance estimates and on current population density. Gene diversity was associated with neither population abundance (F. ovina: $R^2 = 0.010$, $P = 0.625$; P. lanceolata: $R^2 = 0.039$, $P = 0.293$) nor current population density (F. ovina: $R^2 = 0.005$, $P = 0.725$; P. lanceolata: $R^2 = 0.019$, $P = 0.468$).

Response to Climatic Selection

Outlier analyses using bayescan failed to detect any outlier loci associated with experimental climatic environment for F. ovina (Q-values were all > 0.5; posterior probabilities for a response to selection were all < 0.5). Two of the AFLP loci typed in P. lanceolata showed evidence of genetic divergence in response to climatic selection (pc1_ned_60: Q-value = 0.124, posterior probability = 0.876, F_{ST} (Cockerham & Weir, 1993) = 0.201; pc1_fam_55: Q-value = 0.126, posterior probability = 0.873, F_{ST} = 0.218). These loci were identified in the analysis involving all climate treatment subpopulations.

Logistic regression identified two F. ovina AFLP loci (X306.3_TCT_CGG_Fo and X327_TCC_CGC_Fo) with significant variation in fragment frequency between climate change treatments (F_{ST} = 0.036 and 0.030, respectively). Similar analyses identified eighteen P. lanceolata loci as showing outlier patterns of genetic differentiation among climate change treatments, which included the two previously identified outlier loci for this species (range in F_{ST} = 0.021–0.218). Soil depth and the soil depth × treatment interaction were not significant predictors of AFLP fragment frequency for any locus in either study species. Genetic differentiation among climate treatments at putatively neutral (nonoutlier) loci was low for both species (F. ovina, median F_{ST} = 0.003; P. lanceolata, median F_{ST} = 0.003).

Outlier Loci Associated with Experimental Block

Outlier analysis using bayescan identified eight P. lanceolata loci that showed significant divergence among experimental blocks (i.e. with naturally occurring spatial variation in the block environment; all Q-values < 0.1, posterior probability for a response to selection = 0.744–0.999). The same eight loci were present in the sets of candidate outliers identified by analysis of the block environment using logistic regression. In contrast, analysis of the block environment using bayescan did not detect any F. ovinaAFLP locus with an outlier pattern of genetic differentiation (all Q-values ≥ 0.350).

Community-level vs. Genetic Structure

Correlations between species-level and genetic diversity were all small and nonsignificant for both study species ($|r_P| < 0.1$; $P > 0.05$). Similarly, associations between species-level ecological dissimilarity and genetic differentiation within species were weak and nonsignificant (F. ovina: $r_M = 0.174$, $P = 0.071$; P. lanceolata: $r_M = 0.086$, $P = 0.225$).

DISCUSSION

We investigated species' genetic responses to 15 years of simulated climate change applied to intact species-rich limestone grassland, focusing on two of the most abundant species within the community. Our aims were to understand the extent to which genetic responses to the climate are possible within grassland plants, and whether these responses are consistent with a process of evolution in response to climatic selection. We detected genetic differentiation among long-term climate treatments at BCCIL using both genomewide multilocus analyses (F_{ST}, permutational anova) and single locus-oriented outlier analyses. This differentiation has occurred within contiguous grassland over a very small spatial extent (thirty 9-m^2plots spread over approximately 0.2 ha grassland), and happened rapidly, during 15 years of climate change manipulations. Similarly, rapid genetic change occurring in response to a natural drought has been documented in the annual species Brassica rapa (Franks et al., 2007). Treatment-level genetic structure at BCCIL is unlikely to be dominated by neutral spatial genetic structuring because each treatment is replicated in each spatial experimental block. Furthermore, we detected two outlier AFLP loci associated with climate treatment, which possessed patterns of differentiation that were statistically distinct from the background of putatively neutral genomic differentiation. This signature of climatic selection may reflect climate-driven adaptive evolutionary change within populations at BCCIL.

Our survey data suggested that 15 years of experimental climate change had altered population density within some of the experimental treatments (in particular, the drought, heated and heated-drought treatments; Fig. 3a, b), indicating modified patterns of survival and recruitment within these environments. These changes are known to have altered community structure at fine spatial scales (Fridley et al., 2011). Such demographic responses may be sufficient to induce neutral genetic change through drift and the loss of genetic diversity at low population size, assuming that subpopulations at BCCIL are isolated demographically. Our data, however, did not provide evidence in support of this effect. Levels of gene diversity within individual plots were not associated with either (long-term) estimates of population abundance or population density, for either species. Furthermore, although we found evidence in both species for genetic differentiation in all treatments except the winter warming treatment (permutational anova analysis), there was no obvious link between the extent or presence of genetic differentiation and treatment-mediated changes in population density. Hence, neither genetic diversity nor genetic differentiation was associated consistently with sustained changes in population size at BCCIL.

Effective gene flow would be expected to homogenize the distribution of genetic variation at selectively neutral loci, limiting the effects of drift within subpopulations at BCCIL (Wright, 1931). Such gene flow may be especially likely in our study species, which produce wind-dispersed pollen and are highly outcrossing (Morjan & Rieseberg, 2004). Analyses of fine-scale spatial genetic structure were consistent with panmixia at putatively neutral AFLP loci at, or exceeding, the scale of the experimental blocks at BCCIL (~9 m^2; each climate treatment is represented in every block). Cross-fertilization between plant populations can occur at appreciable rates at scales of 100–1000 m (Ellstrand, 2014), and thus, the spatial scale of the experiment at BCCIL is unlikely to pose a problem for gamete (pollen) movement among climate change treatments. Moreover, the low levels of genetic differentiation observed among the climate treatments at putatively neutral (nonoutlier) AFLP loci (median F_{ST} = 0.003) are also consistent with a scenario of effective gene flow. F_{ST} can be used to calculate a crude measure of the effective number of migrants per generation ($N_e m$) exchanged between populations (Wright, 1931). Observed values of F_{ST} between climate change treatment pairs at BCCIL imply $N_e m > 10$ for all treatment pairs in either species (see Whitlock & McCauley, 1999 for the limitations of this approach). Additional analyses using the software structure (Pritchard et al., 2000) provide further support for the prevalence of gene flow, indicating only very weak genetic structure that was not associated with simulated climate change treatments. Finally, recent studies of floral phenology suggest substantial overlaps in flowering time between subpopulations occupying different climate treatments at BCCIL (S. Buckland, unpublished data). Thus, we have no reason to believe that any barrier to gene flow has arisen through asynchrony in floral timing, which could otherwise have facilitated genetic drift (Fox, 2003; Hendry & Day, 2005). We conclude that it is extremely unlikely that climatic subpopulations at BCCIL have become demographically isolated from each other. Hence, it also seems unlikely that genetic drift has had a marked impact on treatment-level genetic structure.

Outlier analyses detected a consistent signature of selection in response to simulated climate change at individual AFLP loci in P. lanceolata. Two outlier loci, identified in both outlier analyses, showed genetic differentiation ($F_{ST} \approx 0.2$) greatly exceeding the average background genetic differentiation at putatively neutral loci in this species (median $F_{ST} = 0.003$). Genome regions associated with the two outlier loci identified consistently in P. lanceolata represent candidate targets for climatic selection and may control phenotypes that are adaptive under the modified climates at BCCIL. The observed genetic differentiation at the putatively selected AFLP loci occurred against a background of low genomewide differentiation. As we argue above, low levels of background differentiation are likely to have been maintained by gene exchange among grassland plots. Thus, implying that climatic selection has had to work against gene flow to establish divergence at the genome regions putatively responding to selection.

Evidence for a response to climatic selection in F. ovina was equivocal; outlier loci associated with climate change treatment were only detected using logistic regression and were weakly differentiated ($F_{ST} \approx 0.03$; neutral background differentiation = 0.003). Failure to detect a stronger response to climatic selection in F. ovina may simply represent the absence of a response, but it may also reflect a genomic marker density insufficient to detect such an effect. In other words, linkage disequilibrium (LD), which may create associations between loci responding to climatic selection and neutral loci, may decay too rapidly for our genotyping strategy to be useful. For example, in maize,

LD can decay within 10 kb (Flint-Garcia et al., 2003; Yan et al., 2009), and coalescence simulations suggest that, in outbreeding species, LD can decay completely within 4 kb (Nordborg, 2000). The 2C DNA values for F. ovina (Grime et al., 1985) indicate that the AFLP markers will be spaced approximately every 4.7 Mb on average (assuming 978 Mb per picogram DNA).

The argument for a climatically adaptive response to selection is supported by recent common environment experiments using plants collected from BCCIL documenting climate-associated evolutionary changes in phenotype in both of our study species (Ravenscroft et al., 2014; R. Whitlock, unpublished data). For example, when grown in a common environment, P. lanceolata individuals from the drought plots at BCCIL make a greater investment in reproductive output, whereas those from control plots allocate more resources to vegetative growth (Ravenscroft et al., 2014). The expression of differentiated phenotypes under a common environment demonstrates a genetic basis to phenotypic differentiation. Thus, these studies show that populations of grassland plants at BCCIL are capable of evolutionary responses to climatic selection.

How do the community-level responses to simulated climate change documented previously at BCCIL contrast with impacts below the species-level documented in this study, and are these outcomes correlated? Species abundance and community composition at BCCIL have been shown to be mediated through the effects of both climate treatment and soil depth; these factors had comparable effect sizes on plant community structure (Fridley et al., 2011). In contrast, multilocus (genomewide) genetic responses to simulated climate change were dominated by treatment-level effects; the effects of soil depth were at least fivefold smaller in magnitude than those attributable to treatment (Table 1). Furthermore, there was no clear relationship between the extent of community-level compositional shifts in different climate treatments and the strength of genetic differentiation within individual species. For example, the highest multilocus genetic differentiation (F_{ST}) was observed for the watered treatment, yet this was the treatment with the lowest community-level compositional change relative to control plots (Fridley et al., 2011). Finally, we found no evidence to support relationships between species diversity and multilocus genetic diversity or between ecological distance (community dissimilarity) and multilocus genetic distance within species. Thus, genomewide genetic responses to simulated climate change within individual species were likely to have been decoupled from, or occurring independently of community-level compositional responses. We predict, however, that the abundance of individual species at BCCIL will be, in part, a function of genetic structure at a subset of genomic loci (including outlier loci) that mediate adaptation to particular climatic environments.

Our results are novel in demonstrating rapid genetic divergence in the established (reproductively mature) phase of perennial plant species in an intact ecosystem in response to a suite of simulated climate change environments. As the climate treatments are replicated in a spatially blocked experimental design (Grime et al., 2000), it seems likely that this divergence has arisen specifically in response to the climate treatments themselves, in parallel with community-level compositional change (Fridley et al., 2011). Furthermore, genetic divergence has occurred over a relatively small period of evolutionary time (15 years of experimentation is ≤ 15 generations for our study species). Our findings are consistent with a scenario in which climate-induced selection imposed at BCCIL has driven climatically adaptive genetic divergence, against a background of weak differentiation maintained by gene flow. However, it is important to note that, based on the available data, we cannot exclude the possibility that genetic drift has contributed to differentiation among climate change treatments at BCCIL. Future work at BCCIL will focus on verifying and functionally characterizing the genomic signature of climatic selection documented in this study and will seek to determine the relationship between climatically selected genotypes, climatically adaptive phenotypes and species abundance.

DATA ACCESSIBILITY

AFLP genotypes, spatial data and sample information (including experimental treatment, block and soil depth at sampling site): DRYAD entries.

ACKNOWLEDGEMENTS

The Buxton Climate Change Impacts Laboratory is supported by US National Science Foundation grant DEB-0716360 to JDF and was established with funds from the UK Natural Environmental Research Council (NERC) with space and logistical support provided by the Health and Safety Laboratory, Harpur Hill, Derbyshire, UK. RW was supported by the NERC UK Population Biology Network (UKPopNet grant to T. Burke) and a Fellowship from the Natural Environment Research Council (grant NE/H015647/1). CR was supported by a Graduate Fellowship from Syracuse University. We thank J. P. Grime and A. P. Askew for maintaining experimental treatments at BCCIL, T. Burke for assistance in devising the sampling strategy, S. Allen and J. Crossmore for assistance with molecular work, and T. Burke and two referees for constructive comments on the manuscript. The authors have no conflict of interests to declare.

REFERENCES

1. Anderson JT, Panetta AM, Mitchell-Olds T (2012) Evolutionary and ecological responses to anthropogenic climate change. Plant Physiology, 160, 1728–1740.
2. Antonovics J, Bradshaw AD (1970) Evolution in closely adjacent plant populations VIII. Clinal patterns at a mine boundary. Heredity, 25, 349–362.
3. Avolio ML, Beaulieu JM, Smith MD (2013) Genetic diversity of a dominant C(4) grass is altered with increased precipitation variability. Oecologia, 171, 571–581.
4. Bilton MC, Whitlock R, Grime JP, Pakeman R, Marion G (2010) Intraspecific trait variation in grassland plant species reveals trade-offs and ecological strategy variation that underpins performance in ecological communities. Botany-Botanique, 88, 939–952.
5. Booth RE, Grime JP (2003) Effects of genetic impoverishment on plant community diversity. Journal of Ecology, 91, 721–730.
6. Bos M, Harmens H, Vrieling K (1986) Gene flow in Plantago I. Gene flow and neighbourhood size in Plantago lanceolata . Heredity, 56, 43–54.
7. Bradshaw WE, Holzapfel CM (2006) Evolutionary response to rapid climate change. Science, 312, 1477–1478.

8. Cockerham CC, Weir BS (1993) Estimation of gene flow from F-statistics. Evolution, 47, 855–863.
9. Cottam G, Curtis JT (1956) The use of distance measures in phytosociological sampling. Ecology, 37, 451–460.
10. Cottam G, Curtis JT, Wilde Hale B (1953) Some sampling characteristics of a population of randomly dispersed individuals. Ecology, 34, 741–757.
11. Crow JF (1964) Breeding structure of populations. II. Effective population size In: Statistics and Mathematics in Biology (eds Kempthorne O, editor; , Bancroft TA, editor; , Gowen JW, editor; , Lush JL, editor.), pp. 543–557. Hafner, New York, NY.
12. Crow JF, Kimura M (1970) An Introduction to Population Genetics Theory. Harper & Row, New York, NY.
13. Davis MB, Shaw RG (2001) Range shifts and adaptive responses to quaternary climate change. Science, 292, 673–679.
14. Davis MB, Woods KD (1986) Dispersal versus climate: expansion of Fagus and Tsuga into the upper Great Lakes region. Vegetatio, 67, 93–103.
15. Ellstrand NC (2014) Is gene flow the most important evolutionary force in plants? American Journal of Botany, 101, 737–753.
16. Ellstrand NC, Elam DR (1993) Population genetic consequences of small population size: implications for plant conservation. Annual Review of Ecology and Systematics, 24, 217–242.
17. Evans SE, Byrne KM, Lauenroth WK, Burke IC (2011) Defining the limit to resistance in a drought-tolerant grassland: long-term severe drought significantly reduces the dominant species and increases ruderals. Journal of Ecology, 99, 1500–1507.
18. Excoffier L, Smouse PE, Quattro JM (1992) Analysis of molecular variance inferred from metric distances among DNA haplotypes: application to human mitochondrial DNA restriction data. Genetics, 131, 479–491.
19. Flint-Garcia SA, Thornsbury JM, Buckler ES (2003) Structure of linkage disequilibrium in plants. Annual Review of Plant Biology, 54, 357–374.
20. Foll M, Gaggiotti O (2008) A genome scan method to identify

selected loci appropriate for both dominant and codominant markers: a bayesian perspective. Genetics, 180, 977–993.

21. Fox GA (2003) Assortative mating and plant phenology: evolutionary and practical consequences. Evolutionary Ecology Research, 5, 1–18.
22. Franks SJ (2011) Plasticity and evolution in drought avoidance and escape in the annual plant Brassica rapa . New Phytologist, 190, 249–257.
23. Franks SJ, Sim S, Weis AE (2007) Rapid evolution of flowering time by an annual plant in response to a climate fluctuation. Proceedings of the National Academy of Sciences of the United States of America, 104, 1278–1282.
24. Fridley JD, Grime JP, Bilton M (2007) Genetic identity of interspecific neighbours mediates plant responses to competition and environmental variation in a species-rich grassland. Journal of Ecology, 95, 908–915.
25. Fridley JD, Grime JP, Askew AP, Moser B, Stevens CJ (2011) Soil heterogeneity buffers community response to climate change in species-rich grassland. Global Change Biology, 17, 2002–2011.
26. Gelman A, Jakulin A, Pittau MG, Su Y-S (2009) A weakly informative default prior distribution for logistic and other regression models. The Annals of Applied Statistics, 2, 1360–1383.
27. Gould B, McCouch S, Geber M (2014) Variation in soil aluminium tolerance genes is associated with local adaptation to soils at the Park Grass Experiment. Molecular Ecology, 23, 6058–6072.
28. Grime JP, Shacklock JML, Band SR (1985) Nuclear DNA contents, shoot phenology and species co-existence in a limestone grassland community. New Phytologist, 100, 435–445.
29. Grime JP, Brown VK, Thompson K et al (2000) The response of two contrasting limestone grasslands to simulated climate change. Science, 289, 762–765.
30. Grime JP, Fridley JD, Askew AP, Thompson K, Hodgson JG, Bennett CR (2008) Long-term resistance to simulated climate change in an infertile grassland. Proceedings of the National Academy of Sciences of the United States of America, 105, 10028–10032.

31. Hamrick JL, Allard RW (1972) Microgeographical variation in allozyme frequencies in Avena barbata . Proceedings of the National Academy of Sciences of the United States of America, 69, 2100–2104.
32. Hamrick JL, Holden LR (1979) Influence of microhabitat heterogeneity on gene frequency distribution and gametic phase disequilibrium in Avena barbata . Evolution, 33, 521–533.
33. Hardy OJ, Vekemans X (2002) SPAGeDi: a versatile computer program to analyse spatial genetic structure at the individual or population levels. Molecular Ecology Notes, 2, 618–620.
34. Harte J, Shaw R (1995) Shifting dominance within a montane vegetation community: results of a climate warming experiment. Science, 267, 876–880.
35. Hendry AP, Day T (2005) Population structure attributable to reproductive time: isolation by time and adaptation by time. Molecular Ecology, 14, 901–916.
36. Hoffmann AA, Willi Y (2008) Detecting genetic responses to environmental change. Nature Reviews: Genetics, 9, 421–432.
37. Jeffreys H (1961) Theory of Probability. Oxford University Press, Oxford, UK.
38. Jump AS, Peñuelas J (2005) Running to stand still: adaptation and the response of plants to rapid climate change. Ecology Letters, 8, 1010–1020.
39. Jump AS, Hunt JM, Martinez-Izquierdo JA, Peñuelas J (2006) Natural selection and climate change: temperature-linked spatial and temporal trends in gene frequency in Fagus sylvatica . Molecular Ecology, 15, 3469–3480.
40. Jump AS, Marchant R, Peñuelas J (2008a) Environmental change and the option value of genetic diversity. Trends in Plant Science, 14, 51–58.
41. Jump AS, Peñuelas J, Rico L, Ramallo E, Estiarte M, Martinez-Izquierdo JA, Lloret F (2008b) Simulated climate change provokes rapid genetic change in the Mediterranean shrub Fumana thymifolia . Global Change Biology, 14, 637–643.
42. Jump AS, Rico L, Lloret F, Peñuelas J (2009) Microspatial

population genetic structure of the Mediterranean shrub Fumana thymifolia . Plant Biology, 11, 152–160.

43. Kelly CK, Chase MW, De Bruijn A, Fay MF, Woodward FI (2003) Temperature-based population segregation in birch. Ecology Letters, 6, 87–89.
44. Legendre P, Anderson MJ (1999) Distance-based redundancy analysis: testing multi-species responses in multifactorial ecological experiments. Ecological Monographs, 69, 1–24.
45. Leimu R, Fischer M (2008) A meta-analysis of local adaptation in plants. PLoS ONE, 3, e4010.
46. Li YC, Fahima T, Beiles A, Korol AB, Nevo E (1999) Microclimatic stress and adaptive DNA differentiation in wild emmer wheat, Triticum dicoccoides . Theoretical and Applied Genetics, 98, 873–883.
47. Lynch M, Milligan BG (1994) Analysis of population genetic structure with RAPD markers. Molecular Ecology, 3, 91–99.
48. Manel S, Poncet BN, Legendre P, Gugerli F, Holderegger R (2010) Common factors drive adaptive genetic variation at different spatial scales in Arabis alpina . Molecular Ecology, 19, 3824–3835.
49. McArdle BH, Anderson MJ (2001) Fitting multivariate models to community data: a comment on distance-based redundancy analysis. Ecology, 82, 290–297.
50. Morjan CL, Rieseberg LH (2004) How species evolve collectively: implications of gene flow and selection for the spread of advantageous alleles. Molecular Ecology, 13, 1341–1356.
51. Nordborg M (2000) Linkage disequilibrium, gene trees and selfing: an ancestral recombination graph with partial self-fertilization. Genetics, 154, 923–929.
52. Owuor ED, Fahima T, Beiles A, Korol A, Nevo E (2003) Population genetic response to microsite ecological stress in wild barley, Hordeum spontaneum . Molecular Ecology, 6, 1177–1187.
53. Parisod C, Christin PA (2008) Genome-wide association to fine-scale ecological heterogeneity within a continuous population of Biscutella laevigata . New Phytologist, 178, 436–447.
54. Pritchard JK, Stephens M, Donnelly P (2000) Inference of population

structure using multilocus genotype data. Genetics, 155, 945–959.
55. Ravenscroft CH, Fridley JD, Grime JP (2014) Intraspecific functional differentiation suggests local adaptation to long-term climate change in a calcareous grassland. Journal of Ecology, 102, 65–73.
56. Reusch TBH, Wood TE (2007) Molecular ecology of global change. Molecular Ecology, 16, 3973–3992.
57. Snaydon RW, Davies MS (1976) Rapid population differentiation in a mosaic environment. IV. Populations of Anthoxanthum odoratum at sharp boundaries. Heredity, 37, 9–25.
58. Strimmer K (2008a) fdrtool: a versatile R package for estimating local and tail area-based false discovery rates. Bioinformatics, 24, 1461–1462.
59. Strimmer K (2008b) A unified approach to false discovery rate estimation. BMC Bioinformatics, 9, 303.
60. Weibull P, Ghatnekar L, Bengtsson BO (1991) Genetic variation in commercial varieties and natural populations of sheep's fescue, Festuca ovina s. l . Plant Breeding, 107, 203–209.
61. Whitlock MC, McCauley DE (1999) Indirect measures of gene flow and migration: $F_{ST} \neq 1/(4Nm + 1)$. Heredity, 82, 117–125.
62. Whitlock R, Grime JP, Booth RE, Burke T (2007) The role of genotypic diversity in determining grassland community structure under constant environmental conditions. Journal of Ecology, 95, 895–907.
63. Whitlock R, Hipperson H, Mannarelli M, Burke T (2008a) A high-throughput protocol for extracting high-purity DNA from plants and animals. Molecular Ecology Resources, 8, 736–741.
64. Whitlock R, Hipperson H, Mannarelli M, Butlin RK, Burke T (2008b) An objective, rapid and reproducible method for scoring AFLP peak-height data that minimizes genotyping error. Molecular Ecology Resources, 8, 725–735.
65. Whitlock R, Grime JP, Burke T (2010) Genetic variation in plant morphology contributes to the species-level structure of grassland communities. Ecology, 91, 1344–1354.
66. Wright S (1931) Evolution in Mendelian populations. Genetics, 16,

97–159.

67. Yan J, Shah T, Warburton ML, Buckler ES, McMullen MD, Crouch J (2009) Genetic characterization and linkage disequilibrium estimation of a global maize collection using SNP markers. PLoS ONE, 4, e8451.
68. Zavaleta ES, Shaw MR, Chiariello NR, Mooney HA, Field CB (2003) Additive effects of simulated climate changes, elevated CO_2, and nitrogen deposition on grassland diversity. Proceedings of the National Academy of Sciences of the United States of America, 100, 7650–7654.
69. Zhivotovsky LA (1999) Estimating population structure in diploids with multilocus dominant DNA markers. Molecular Ecology, 8, 907–913.

CHAPTER 11

METAGENOMIC ANALYSIS EXPLORING TAXONOMIC AND FUNCTIONAL DIVERSITY OF SOIL MICROBIAL COMMUNITIES IN CHILEAN VINEYARDS AND SURROUNDING NATIVE FORESTS

Luis E. Castañeda[1] and Olga Barbosa[1,2]

[1] Instituto de Ciencias Ambientales y Evolutivas, Facultad de Ciencias, Universidad Austral de Chile, Valdivia, Chile

[2] Instituto de Ecologı´a y Biodiversidad (IEB-Chile), Santiago, Chile

ABSTRACT

Mediterranean biomes are biodiversity hotspots, and vineyards are important components of the Mediterranean landscape. Over the last few decades, the amount of land occupied by vineyards has augmented rapidly, thereby increasing threats to Mediterranean ecosystems. Land use change and agricultural management have important effects on soil biodiversity, because they change the physical and chemical properties of soil. These changes may also have consequences on wine production considering that soil is a key component of *terroir*. Here, we describe the taxonomic diversity and metabolic functions of bacterial and fungal communities present in forest and vineyard soils in Chile. To accomplish this goal, we collected soil samples from organic vineyards in central Chile and employed a shotgun metagenomic approach to sequence the microbial DNA. Additionally, we studied the surrounding native forest to obtain a baseline of the soil conditions in the area prior to the establishment of the vineyard. Our metagenomic analyses revealed that both habitats shared most of the soil microbial species. The most abundant genera in the two habitats were the bacteria *Candidatus Solibacter* and *Bradyrhizobium* and the fungus *Gibberella*. Our results suggest that the soil microbial communities are similar in these forests and vineyards. Therefore, we hypothesize that native forests surrounding the vineyards may be acting as a microbial reservoir buffering the effects of the land conversion. Regarding the metabolic diversity, we found that genes pertaining to the metabolism of amino acids, fatty acids, and nucleotides as well as genes involved in secondary metabolism were enriched in forest soils. On the other hand, genes related to miscellaneous functions were more abundant in vineyard soils. These results suggest that the metabolic function of microbes found in these habitats differs, though differences are not related to taxonomy. Finally, we propose that the implementation of environmentally friendly practices by the wine industry may help to maintain the microbial diversity and ecosystem functions associated with natural habitats.

Keywords: Conservation, Bacterial diversity, Ecosystem services, Wine, Shotgun sequencing, Pyrosequencing, Fungal diversity

INTRODUCTION

Being one of the main drivers of global change, land use change affects many important ecosystem properties and functions (Vitousek et al., 1997). Land use change (also referred to here as land conversion) has consequences at ecosystem scales because ecological functions can be lost during the conversion of native habitats (Griffiths & Philippot, 2013). Particularly in Mediterranean biomes, land conversion has occurred at very rapid rates over the last decades (Cincotta, Wisnewski & Engelman, 2000; Lauber et al., 2008; Underwood et al., 2008). This is especially important given that Mediterranean ecosystems are classified as biodiversity hotspots due to the high diversity of plant species and high endemism in these areas (Cowling et al., 1996; Myers et al., 2000). Therefore, conservation programs are necessary to preserve the biodiversity of these ecosystems.

The Mediterranean climate is suitable for viticulture; subsequently vineyards are becoming important components of Mediterranean landscapes due to the development of the wine industry in these regions (Hannah et al., 2013; Viers et al., 2013). Land occupied by vineyards increased by 70% between 1988 and 2010 in New World Mediterranean zones (Chile, California, Australia, and South Africa) (Viers et al., 2013). By replacing natural landscapes and by simplifying the structure and composition of ecological communities (Viers et al., 2013), the expansion of vineyards threatens Mediterranean ecosystems. In addition, agricultural management (e.g., tillage, pesticide, and fertilizer applications) directly affects soil biodiversity by altering the physical and chemical properties of soil (Pampulha & Oliveira, 2006; Jangid et al., 2008). For instance, organically managed vineyards have higher soil microbial biomass and nematode densities than conventional vineyards (Coll et al., 2011). On the other hand, Bevivino et al. (2014) report that undisturbed soils have more stable bacterial communities throughout the change of seasons than do vineyards. This suggests that natural habitats are more resilient to environmental or human perturbations. Furthermore, soil biodiversity is very significant to wine production, which relies on soil and climatic conditions as key components of *terroir* (van Leeuwen et al., 2004; Gilbert, van der Lelie & Zarraonaindia, 2014).

The soil horizon is one of the most diverse environments on Earth; currently it is estimated that thousands of different microbial species inhabit 1 g of soil (Delmont et al., 2011; Xu et al., 2014). There is abundant evidence confirming the important role played by soil microorganisms in several ecosystem services such as erosion control, soil formation, nutrient cycling, and plant health (Tiedje et al., 1999; Nannipieri et al., 2003; Garbeva, van Veen & van Elsas, 2004; Gardi et al., 2009). However, soil microbial communities are not static and can change across agricultural practices and environmental gradients (Bevivino et al., 2014; García-Orenes et al., 2013). For instance, the addition of organic matter to managed soils increases fungal abundance and causes the microbial community structure to resemble that of undisturbed forest soil (García-Orenes et al., 2013). In addition, Corneo et al. (2013) report that microbial communities change across altitudinal gradients, where soil physical (e.g., soil moisture, clay content) and chemical (e.g., aluminum, magnesium, molybdenum, and boron) properties explain most of the altitudinal variation in soil communities.

The recent development of high-throughput sequencing techniques has allowed a deeper understanding of the microbial diversity of vineyard soils in different wine-producing regions around the world (Corneo et al., 2013; Fujita et al., 2010; Zarraonaindia et al., 2015). Although Mediterranean Chile is one of the most important wine-producing regions and the area occupied by vineyards in Chile has rapidly expanded (Viers et al., 2013), there are very few studies exploring the microbial diversity of these vineyard soils (but see Aballay, Maternsson & Persson, 2011; Castañeda et al., 2015). Recently, Castañeda et al. (2015) have shown that the soil bacterial communities in native forests and vineyards are similar, whereas the fungal communities differ between the habitats. This study employed T-RFLPs, which are reliable technique but do not provide deep taxonomic resolution or information about the metabolic functioning of the microbial community. In the present study, our goal is to describe the taxonomic diversity as well as the metabolic functions of bacterial and fungal communities present in forest and vineyard soils in Chile. To accomplish this goal, we assessed the taxonomic and metabolic diversity of soil samples from three organic vineyards in central Chile; we employed a shotgun sequencing approach, paying particular attention to species associated with viticulture and wine making. The organic vineyards sampled are relatively young (<10 years old) and are surrounded by natural landscapes. The surrounding natural landscapes are dominated by native sclerophyllous forests and shrubs, thus these ecosystems likely represent the soil characteristics of the area before the establishment of the vineyard. The knowledge of the soil microbial communities of native habitats could provide valuable information for the conservation management of vulnerable ecosystems (Heilmann-Clausen et al., 2014) such as for the Chilean Mediterranean region (Mittermeier et al., 2011; Hannah et al., 2013; Viers et al., 2013). Knowledge of microbial community dynamics found in this biome is scarce and metagenomic studies could provide a starting point for the conservation of microbial diversity and for the preservation of ecosystem functions provided by natural habitats (Gardi et al., 2009).

MATERIALS AND METHODS

Sampling

Soil samples were collected from three different organic vineyards and from neighboring sclerophyllous forest patches in central Chile; Ocoa (32°52′S–71°7′W), Leyda (33°34′S–71°22′W), and Colchagua (34°36′S–71°7′W). Samples were collected in March (during the harvest season) of 2012. The owners of the vineyards and the surrounding native forest patches granted all necessary permits to access the sampling sites: Seña Vineyards in Ocoa (Chile), Cono Sur Vineyards in Leyda (Chile), and Emiliana Vineyards in Colchagua (Chile; Table 1). The vineyards contain woody-perennial monocultures of *Vitis vinifera*, whereas the forest patches mainly contain *Cryptocarya alba*, *Peumus boldus*, *Quillaja saponaria*, *Lithraea caustica*, and *Acacia caven*, among other tree and shrub species.

In each vineyard, a plot near the forest patch was randomly selected. In each vineyard plot, five vines each separated from the other by 3.5 m were randomly selected. One bulk soil sample was collected at a distance of 5 cm from each vine stem; the soil samples were taken from the first 15 cm of the soil horizon using soil cores. This depth was chosen because the majority of microbial activity is thought to occur within the upper 15 cm (O'Brien et al., 2005). The same procedure was performed in the adjacent forest patch, where five native trees and corresponding soil samples were randomly selected and collected as previously described.

All collected samples were stored in a sterile bag and placed in a cooler with ice packs. During the same day, the 30 soil samples were transported to the laboratory where they were individually homogenized, sieved, and stored at −80 °C until DNA extraction was performed.

Table 1: Descriptive Information of Each Sampling Site

	Ocoa, Chile	Leyda, Chile	Colchagua, Chile
Latitude	32°52′S	33°34′S	34°36′S
Longitude	71°7′W	71°22′W	71°7′W
Altitude (m)	307	216	268
Mean temperature (°C)	14.7	16.2	14.6
Precipitation (mm)	354	457	731
pH forest soil[1]	7.87	6.86	6.34
pH vineyards soil[2]	8.1 ± 0.1	7.8 ± 0.5	7.5 ± 0.4
Forest soil content (sand, silt, and clay) (%)	73–16–11	67–22–11	47–37–15
Vineyard soil content (sand, silt and clay) (%)	56–38–16	61–26–13	61–27–12
Soil taxonomy	Alfisol	Alfisol	Alfisol
Vine variety	Cabernet Sauvignon	Sauvignon Blanc	Syrah
Planting year (±SD)	2002 ± 3	2006 ± 1	2001 ± 4

Notes:

[1] pH in forests was determined from a single soil sample.

[2] pH in vineyards was determined in each plot and the mean (±SD) is shown.

Metagenomic Sequencing

For a total of 30 soil samples (three vineyards × two habitats × five soil samples), DNA was extracted using the Power Soil DNA isolation kit (MoBio Laboratories Inc., Carlsbad, CA, USA) following the manufacturer's instructions. The quality of the DNA extracted was determined by electrophoresis using a 0.8% agarose gel. Furthermore, the DNA was quantified using a nanospectrophotometer (NanoDrop Technologies Inc., Wilmington, DE, USA).

For sequencing, the DNA extractions from each habitat (five samples) were pooled into one sample. Thus, one pooled vineyard sample and one pooled forest sample were sequenced for each vineyard (six samples in total). The concentration of DNA was assessed by fluorescence using the Quant-iT PicoGreen dsDNA kit (Invitrogen, Carlsbad, CA, USA); fluorescence was measured on a DQ 300 fluorometer (Hoefer Scientific Instruments, San Francisco, CA, USA). Following this, each metagenomic library was prepared using the 454 GS Junior Titanium Rapid DNA library preparation kit according to the manufacturer's instructions. Emulsion PCR (emPCR) was performed according to the Amplification Method Manual using a Lib-L kit. All steps involved in massive DNA sequencing were performed in the AUSTRAL-omics Core-Facility (Facultad de Ciencias, Universidad Austral de Chile) in a 454 GS Junior Titanium Series (Roche, Branford, CT, USA) following the standard protocol from Roche.

Data Analysis

The raw sequences of each of the six metagenomes were uploaded to the MG-RAST server at http://metagenomics.anl.gov (Meyer et al., 2008). The number of uploaded sequences ranged from 141,694 to 195,138 sequences for the forest soil samples and from 189,372 to 208,095 for the vineyard soil samples. After quality control was performed using MG-RAST, the number of retained sequences for the forest soil samples ranged from 114,120 to 131,618 with an average length of 442.7 bp, whereas 108,385–138,101 sequences with an average length of 445.3 bp were retained for the vineyard soil samples. Taxonomic assignments were performed using the SEED database, and metabolic assignments were performed using the Subsystems database. For both types of assignments, we employed a maximum *e*-value of 1*e*–5, a minimum identity of 60%, and a maximum alignment length of 15 bp. The accession numbers for the metagenomes in the MG-RAST server (http://metagenomics.anl.gov/metagenomics.cgi?page=MetagenomeProject&project=8742) are: 4565458.3, 4565459.3, 4565460.3, 4565461.3, 4565462.3, and 4565463.3.

For taxonomic analysis, the OTU table was downloaded from the MG-RAST server and analyzed in QIIME v1.9.1 (Caporaso et al., 2010). OTUs that matched the following criteria were removed from the OTU table: (1) OTUs matched to mitochondria, chloroplast, plant, or animal sequences; (2) OTUs observed fewer than 10 times; and (3) OTUs observed in fewer than two samples. The resulting OTU table was analyzed employing the vegan (Oksanen et al., 2013) and phyloseq (McMurdie & Holmes, 2013) packages in R (R Development Core Team, 2016). To standardize the number of sequences between samples, they were rarefied to 289,800 sequences. Venn diagrams were made to visualize which OTUs were shared between forest and vineyard soils using Venny 2.1.0 (Oliveros, 2015). Species richness, Shannon diversity, and Pielou evenness indices were estimated for each of the samples, and these indices were compared between habitats using a Kruskal–Wallis test. Beta diversity was estimated using Bray–Curtis dissimilarity, employing the vegdist function of the vegan package in R. Then, a permutational multivariate analysis of variance (PERMANOVA) was used to compare the microbial community structure between forest and vineyard soils; this was performed with 999 permutations using the adonis function of the vegan package in R. Finally, a canonical correspondence analysis (CCA) conducted in vegan was used to visualize the community structure.

To analyze the metabolic profiles, the relative abundances of reads in forest and vineyard soils were compared via a White's non-parametric *t*-test (White, Nagarajan & Pop, 2009) using the software STAMP (Parks & Beiko, 2010). Comparisons of metabolic profiles between habitats were performed using a PERMANOVA analysis; this was done using the adonis function of vegan in R. Finally, CCA conducted in vegan was used to visualize the functional-based community structure and the relationship between soil samples and functional categories.

Finally, raw datasets and specific analyses are available at the Figshare server (https://dx.doi.org/10.6084/m9.figshare.2058060.v2).

RESULTS

Taxonomical Analysis

Metagenomic analyses using the SEED database showed that bacteria, followed by Eukaryota and Archaea, dominated the forest as well as the vineyard soil samples. The other sequences corresponded to viruses and unassigned sequences (Table 2). Among bacteria, Proteobacteria was the most abundant phylum both in forest soil as well as in vineyard soil; which was followed by Actinobacteria, Acidobacteria, Bacteroidetes, Firmicutes, and Planctomycetes (Table 2). However, we did not find significant differences in the relative abundances of these phyla (Table 2). By taking a closer look at the taxonomy, we found 4,104 bacterial OTUs (97% nucleotide ID) corresponding to 1,326 species, of which 87.1% were shared between habitats (Fig. 1). The 10 most abundant species were *Candidatus Solibacter usitatus* (pooled mean = 2.5%, $P = 0.83$), *Bradyrhizobium japonicum* (pooled mean = 2.5%, $P =$

0.51), *Rhodopseudomonas palustris* (pooled mean = 2.1%, *P* = 0.51), *Conexibacter woesei* (pooled mean = 1.9%, *P* = 0.83), *Candidatus Koribacter versatilis* (pooled mean = 1.7%, *P* = 0.83), *Gemmatimonas aurantiaca*(pooled mean = 1.5%, *P* = 0.28), *Sorangium cellulosum* (pooled mean = 1.4%, *P* = 0.83), *Mycobacterium tuberculosis* (pooled mean = 1.4%, *P* = 0.51), *Rhodopirellula baltica* (pooled mean = 0.9%, *P* = 0.83), and *Myxococcus xanthus* (pooled mean = 0.9%, *P* = 0.51). Nevertheless, there were no significant differences in the relative abundances of these dominant species in forest and vineyard soils. Conversely, the relative abundances of 36 OTUs were significantly different between habitats ($P < 0.05$); but all of these OTUs were found in very low relative abundances in both habitats. We also explored the presence of lactic acid (Lactobacillaceae and Leuconostocaceae) and acetic bacteria (Acetobacteraceae) in the forest and vineyard soils; lactic acid bacteria is known to positively affect wine production while acetic bacteria negatively affects production. Typically, these bacteria are found in low relative abundances in soil samples, but we expected they might be found in the sampled vineyards, being derived from the grape skins that are often used as compost. We found the presence of lactic acid bacteria including *Lactobacillus* (vineyard = 0.04%, forest = 0.03%, and *P* = 0.51) and acetic bacteria such as *Gluconobacter* (vineyard = 0.038%, forest = 0.041%, and *P* = 0.51) and *Acetobacter* (vineyard = 0.13%, forest = 0.12%, and *P* = 0.82).

Table 2: Abundances of Taxonomic Groups in Forest and Vineyard Soils

Taxa	Forest	Vineyard	*P*-value
Archaea	**0.454 ± 0.050**	**0.486 ± 0.077**	**0.83**
Crenarchaeota	0.050 ± 0.007	0.048 ± 0.003	0.51
Euryarchaeota	0.376 ± 0.040	0.395 ± 0.054	0.83
Korarchaeota	0.004 ± 0.002	0.004 ± 0.002	0.83
Thaumarchaeota	0.021 ± 0.015	0.036 ± 0.019	0.28
Unclassified	0.003 ± 0.001	0.004 ± 0.001	0.66
Bacteria	**90.34 ± 0.561**	**90.43 ± 0.073**	**0.72**
Acidobacteria	5.118 ± 0.748	5.080 ± 0.841	0.83
Actinobacteria	20.49 ± 2.532	20.17 ± 1.732	0.83
Aquificae	0.098 ± 0.005	0.105 ± 0.009	0.38
Bacteroidetes	3.049 ± 0.502	3.369 ± 0.576	0.51
Chlamydiae	0.051 ± 0.014	0.050 ± 0.013	0.83
Chlorobi	0.315 ± 0.030	0.321 ± 0.034	0.83

Chloroflexi	1.918 ± 0.304	2.017 ± 0.250	0.83
Chrysiogenetes	0.015 ± 0.005	0.016 ± 0.005	0.83
Cyanobacteria	1.774 ± 0.236	1.819 ± 0.190	0.51
Deferribacteres	0.048 ± 0.006	0.048 ± 0.005	0.83
Deinococcus–Thermus	0.574 ± 0.043	0.599 ± 0.049	0.51
Dictyoglomi	0.035 ± 0.010	0.046 ± 0.002	0.27
Elusimicrobia	0.012 ± 0.002	0.016 ± 0.004	0.27
Fibrobacteres	0.006 ± 0.001	0.010 ± 0.003	0.08
Firmicutes	2.945 ± 0.296	3.313 ± 0.295	0.13
Fusobacteria	0.028 ± 0.003	0.030 ± 0.011	0.51
Gemmatimonadetes	1.465 ± 0.208	1.542 ± 0.180	0.28
Lentisphaerae	0.030 ± 0.010	0.037 ± 0.003	0.28
Nitrospirae	0.200 ± 0.018	0.193 ± 0.037	0.51
Planctomycetes	3.001 ± 0.062	3.425 ± 0.617	0.13
Ptobacteria	0.019 ± 0.003	0.021 ± 0.007	0.83
Proteobacteria	46.12 ± 0.245	45.12 ± 1.481	0.28
Spirochaetes	0.235 ± 0.019	0.228 ± 0.015	0.83
Synergistetes	0.054 ± 0.003	0.062 ± 0.010	0.13
Tenericutes	0.001 ± 0.001	0.002 ± 0.001	0.27
Thermotogae	0.104 ± 0.010	0.134 ± 0.025	0.13
Verrucomicrobia	2.224 ± 0.656	2.210 ± 0.249	0.83
Unclassified	0.419 ± 0.040	0.439 ± 0.042	0.28
Eukaryota	**0.582 ± 0.162**	**0.434 ± 0.139**	**0.51**
Ascomycota	0.462 ± 0.102	0.339 ± 0.131	0.28
Basidiomycota	0.032 ± 0.013	0.026 ± 0.003	0.83
Unclassified	0.088 ± 0.056	0.069 ± 0.014	0.51
Viruses	**0.002 ± 0.008**	**0.001 ± 0.009**	**0.49**
Unassigned/unclassified	**8.624 ± 0.370**	**8.623 ± 0.115**	**0.51**

Note:
Values are shown as percentage of abundance in each habitat (mean ± standard deviation) *P*-values were derived from a Kruskal–Wallis test.

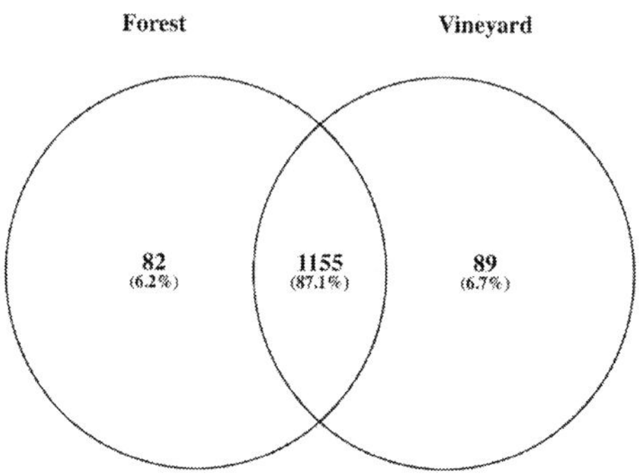

Figure 1: Numbers and percentage of bacterial species found in forest and vineyard soils.

For the Eukaryota domain, we focused on fungal OTUs, which were mainly related to the phyla Ascomycota and Basidiomycota (Table 2). We found 95 fungal OTUs (97% nucleotide ID) corresponding to 47 Ascomycota and 8 Basidiomycota species. Among the most abundant fungal-related OTUs, we found the Ascomycota *Gibberella zeae* (vineyard = 0.040% and forest = 0.0042%, $P = 0.83$), *Aspergillus fumigatus* (vineyard=0.03% and forest=0.05%, P=0.13), and *Neurospora crassa* (vineyard = 0.026% and forest = 0.029%, $P = 0.28$). Exploring the presence of fermenting yeasts in soil, we found some OTUs related to *Saccharomyces cerevisiae* (97% identity), which were significantly more abundant in forest (0.004%) than in vineyard (0.002%) soils ($P = 0.046$). Another important group found in both habitats was the domain Archaea represented by its five phyla: Crenarchaeota, Euryarchaeota, Korarchaeota, and Thaumarchaeota. Of these, the phylum Euryarchaeota was the most abundant, but no significant differences were found between the Archaea of forest and vineyard soils (Table 2).

Microbial community analyses showed that vineyards had higher species richness than forests, while the Shannon diversity, Shannon richness, and evenness indices were not significantly different between the habitats (Table 3A). We also found that the microbial community structure did not differ between habitats (PERMANOVA, $P = 0.45$); this is illustrated in the CCA plot (Fig. 2).

Table 3: Diversity indices for microbial communities from forest and vineyard soils

	Forest	Vineyard	*P*-value
(A) Index			
Richness	5101 ± 14.7	5251 ± 28.6	0.05
Shannon diversity	7.175 ± 0.054	7.209 ± 0.075	0.51
Shannon richness	517.0 ± 15.4	530.6 ± 32.5	0.83
Pielou evenness	0.840 ± 0.007	0.843 ± 0.009	0.51
(B) Shannon richness			
Nitrogen metabolism	287.8 ± 20.0	295.9 ± 11.2	0.51
Phosphorous metabolism	286.7 ± 11.9	258.5 ± 22.5	0.13
Potassium metabolism	183.6 ± 11.1	193.5 ± 12.7	0.28

Note:
Values are shown as percentage of abundance for each habitat (mean ± standard deviation) *P*-values were derived from a Kruskal–Wallis test.

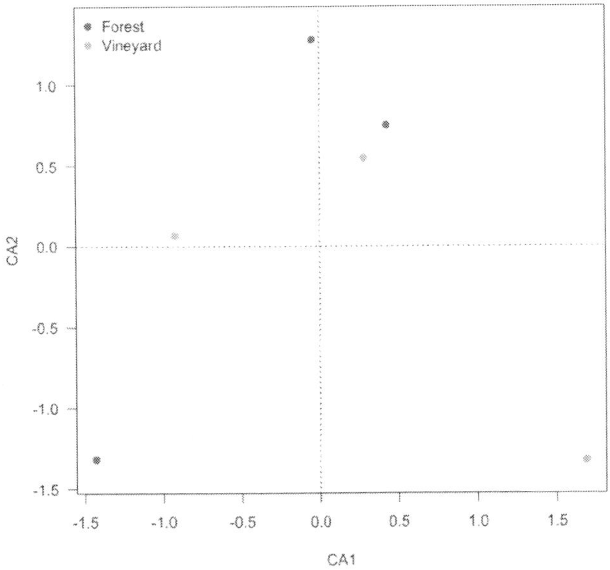

Figure 2: Ordination plot for microbial composition in soils.

Functional Analysis

Functional metabolic categories related to microorganisms found in forest and vineyard soils are represented in Fig. 3. The most abundant functional categories included sequences related to carbohydrate metabolism (forest mean = 14.4% and vineyard mean = 14.6%), functionally coupled genes but with unknown function (i.e., clustering-based on subsystems) (forest mean = 14.0% and vineyard mean = 14.2%), and metabolism of amino acids and their derivatives (forest mean = 10.8% and vineyard mean = 10.6%). We did not find differences between habitats in the microbial communities' functional profiles at the SEED level-1 gene annotation (PERMANOVA, $P = 0.80$, Fig. 3), and any functional categories showed a clear association to any habitat type (Fig. 3). Conversely, we found that genes related to metabolism of amino acids and their derivatives ($P = 0.007$), fatty acid and lipid metabolism ($P = 0.024$), nucleoside and nucleotide metabolism ($P = 0.045$), and secondary metabolism ($P = 0.011$) showed significantly higher relative abundances in forest than vineyard soils (Fig. 4). On the other hand, only genes related to miscellaneous functions ($P = 0.033$) showed a significantly higher relative abundance in vineyard than forest soils (Fig. 4). Therefore, we explored the functional profiles of the categories that showed significant abundance between habitats, but we did not find significant differences in the microbial communities' metabolic profiles: amino acid metabolism (PERMANOVA, $P = 0.22$), fatty acid metabolism (PERMANOVA, $P = 0.23$), nucleoside and nucleotide metabolism (PERMANOVA, $P = 0.25$), secondary metabolism (PERMANOVA, $P = 0.40$), and miscellaneous functions (PERMANOVA, $P = 0.23$).

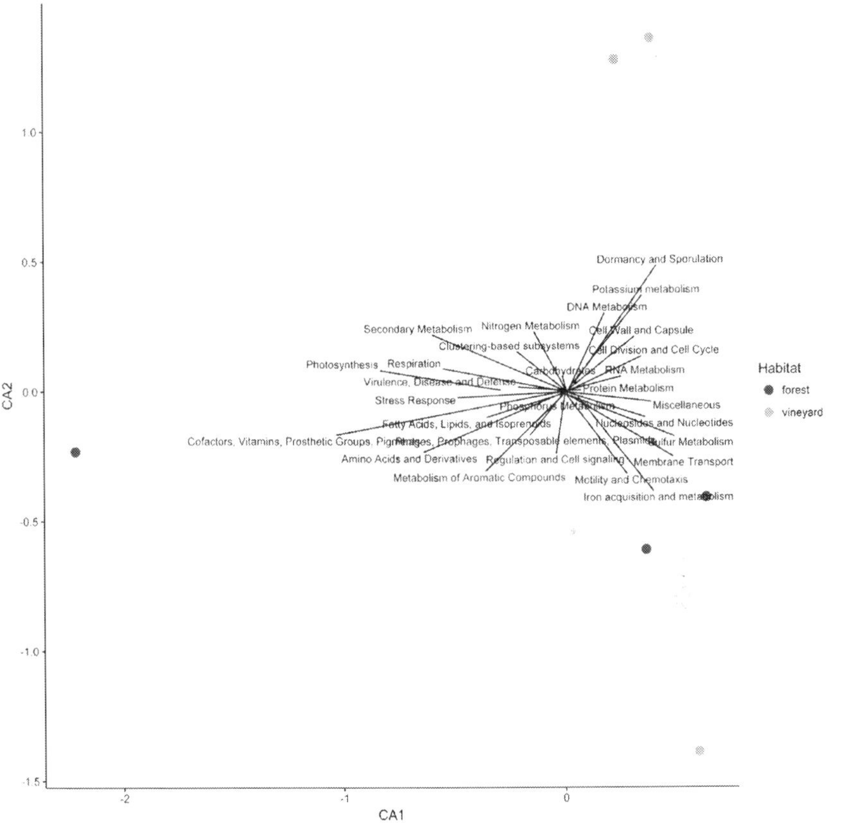

Figure 3: Ordination plot for metabolic categories in soils.

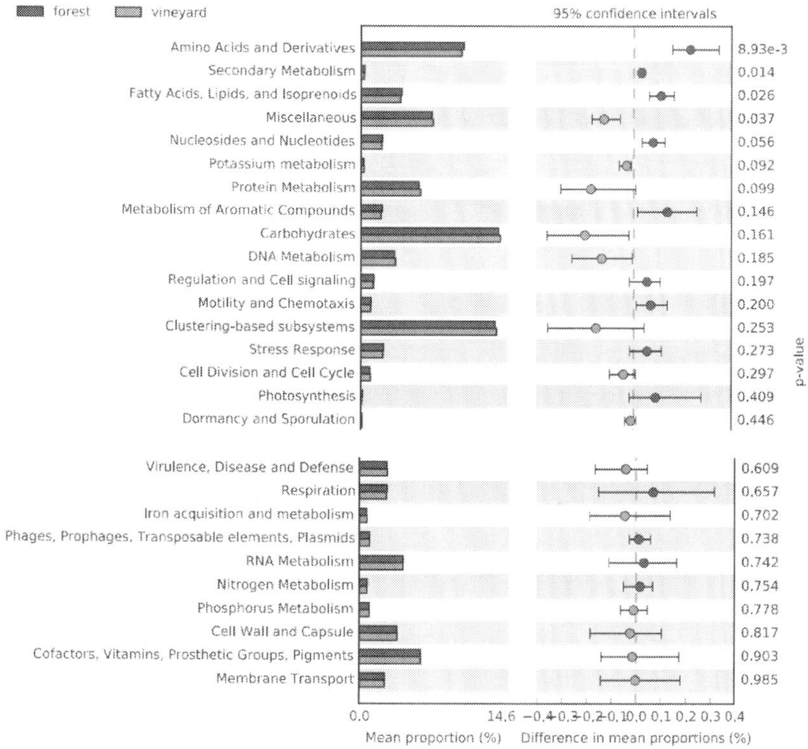

Figure 4: Functional categories found in soil microbial communities.

Additionally, we explored some functional categories that could be associated with nutrient cycling. From this, we found sequences related to sulfur metabolism (forest mean = 1.18% and vineyard mean = 1.15%), phosphorous metabolism (forest mean = 1.04% and vineyard mean = 1.05%), nitrogen metabolism (forest mean = 0.82% and vineyard mean = 0.80%), and potassium metabolism (forest mean = 0.30% and vineyard mean = 0.33%). The relative abundances of these functions were similar in forest and vineyard soils ($P > 0.1$). We also explored the SEED level-3 hierarchical gene annotation. In general, assimilation of inorganic sulfur (overall mean = 0.37%), phosphate metabolism (overall mean = 0.54%), phosphorous uptake (overall mean = 0.20%), ammonia assimilation (overall mean = 0.38%), nitrate and nitrite assimilation (overall mean = 0.15%), and potassium homeostasis (overall mean = 0.28%) were the most abundant level-3 functions related to nutrient cycling. However, the relative abundances of these functions were not significantly different between forest and vineyard soils. We also estimated the species richness of OTUs related with some specific functional categories (level 1 of SEED subsystems hierarchy) such as nitrogen metabolism, phosphorous metabolism, potassium metabolism, and defense-related genes. However, we did not find any significant differences in alpha-diversity between habitats (Table 3B).

DISCUSSION

Our analysis showed that bacterial-related OTUs had the highest relative abundance in both habitats. For soil environments, Uroz et al. (2013) have reported similar bacterial abundances in organic and mineral soils: bacterial sequences accounted for ca. 94% of the total sequences. Proteobacteria are very common in soil environments and are related to a wide variety of functions involved in carbon, nitrogen, and sulfur cycling (Spain, Krumholz & Elshahed, 2009). The relative abundances of Proteobacteria found in the present study are similar to those previously reported for other soil types such as crops, forests, and grasslands (ca. 40% according to Janssen (2006)). Participating in carbon cycling and producing secondary metabolites, Actinobacteria are also dominant in soils (Jenkins et al., 2010). In our study, the most abundant bacterial genera in the soil samples were *C. Solibacter*, *Bradyrhizobium*, *Conexibacter,* and *Rhodopseudomonas*, which have been previously reported as dominant genera in several types of soil (Delmont et al., 2011; Pearce et al., 2012). Comparing bacterial phyla and genera, we did not find differences in relative abundances between forest and vineyard soils. Additionally, we found that diversity indices and microbial community structure were similar between forest and vineyard soils; this agrees with our previous work performed using T-RFLPs (Castañeda et al., 2015). Conversely, these findings differ from previous evidence suggesting that bacterial communities differ between forest and managed soils (García-Orenes et al., 2013). However, the relationship between microbial diversity and habitat disturbance is very complex and can depend on the degree of disturbance; some disturbed habitats can even exhibit higher diversity than forest systems (Miura et al., 2016; Montecchia et al., 2015).

Similar to the bacterial community results, fungal communities did not differ between habitats. Most of the fungi-related sequences were assigned to Ascomycota, outweighing other groups such as Basidiomycota, which only represented a small fraction of the total fungal sequences. At the species level, the most abundant fungal species was *Gibberella zeae/Fusarium graminearum*, a well-known plant pathogen that attacks cereals (Bai & Shaner, 2004). From a comparative point-of-view, we found similar fungal abundance between forest and vineyard soils. Interestingly, our previous work employing T-RFLPs showed that fungal community structure differed between forest and vineyard soils (Castañeda et al., 2015); this agrees with another T-RFLP-based study that shows that fungal diversity differs between native *Eucalyptus* forests and *Pinus* plantations in Australia (Kasel, Bennett & Tibbits, 2008). It should be noted that the lack of differences in fungal abundances in the present study might be related to the low representation of fungal sequences in the soil samples. Additionally, differences in taxonomic abundance could be limited to differences in functional taxonomic groups because taxonomic assignment was based on the SEED non-redundant protein database (for additional information see Carrino-Kyker, Smemo & Burke, 2013). Therefore, complementary approaches such as metatranscriptomic or amplicon-sequencing approaches should be employed to study soil eukaryotic communities to gain a deeper understanding of the ecology of these communities.

Microbes play important roles in several stages of wine production (Mills et al., 2008). For instance, fermenting yeasts are involved in alcoholic fermentation (i.e., the conversion of sugar into ethanol and carbon dioxide), and lactic acid bacteria perform malolactic fermentation (i.e., the conversion of malate into lactate) (Fleet, 2003; Mills et al., 2008). Our data show the presence of lactic acid bacteria such as *Lactobacillus* and *Gluconobacter*, acetic bacteria such as *Acetobacter*, and the fermenting-yeast *S. cerevisiae* in the soil samples collected. Although these species are not usually common in soils, we decided to search for them because organic vineyards often use recycled grape skins (also known as pomace) as organic fertilizer. Knowing this, one would expect that some lactic acid bacteria and fermenting yeasts could colonize, or at least survive, in vineyard soils. Recently, Zarraonaindia et al. (2015) have reported that soil acts as a source of grape-associated bacteria, and thus with edaphic factors, soil can influence grapevine microbiota. However, the abundance of lactic acid, acetic, and fermenting microbes was relatively low compared to other dominant taxa. This suggests, contrary to what has been previously suggested (Bester, 2005; Chen, Yanagida & Shinohara, 2005; Zarraonaindia et al., 2015), that soil may not be a suitable ecological niche or reservoir for microorganisms important to wine production. It must be noted, however, that differences in methodological approaches may explain disparities between our findings and those previously reported: some studies have employed enrichment methods (Bester, 2005; Chen, Yanagida & Shinohara, 2005) or amplicon sequencing (Zarraonaindia et al., 2015), while shotgun sequencing (technique employed in the present study) could underestimate the abundance of fungal sequences. Future studies should evaluate the presence of enologically important microorganisms in surrounding native flora (i.e., leaves and fruits) to determine if these habitats are potential sources and/or reservoirs of microbial diversity relevant to wine production. This is particularly interesting due to the fact that high-quality wines are strongly associated with the concept of *terroir*, which encompasses regional characteristics such as climate and grape variety, and also gives special importance to soil and the interactions that occur with microorganisms (Anonymous, 2010). The fact that natural habitats can be potential reservoirs of microorganisms could safeguard the identity of *terroir* over time.

Most sequences obtained from forest and vineyard soils were related to the metabolism of carbohydrates and amino acids. This finding suggests that soil microbial communities are capable of degrading carbohydrates and playing an important role in the carbon cycle through organic matter and litter decomposition. Indeed, these results agree with the high relative abundance (ca. 12%) of genes related to carbohydrate metabolism previously reported for organic soils (Uroz et al., 2013; Paula et al., 2014). On the other hand, land-use changes alter the community structure of soil microorganisms and can have profound effects on ecosystem functions and processes (Griffiths & Philippot, 2013; Paula et al., 2014). In this sense, it has been reported that the land conversion of primary forests to long-term pastures has changed the microbial functional diversity of Amazon soils and especially so for genes related to carbon and nitrogen cycling (Paula et al., 2014). In the present study, we found differences in the abundance of genes related to the metabolism of amino acids and their derivatives, fatty acid and lipid metabolism, nucleoside and nucleotide metabolism, secondary metabolism, and miscellaneous functions. However, from a community perspective, we did not detect differences in the metabolic profiles nor did we find a difference in the number of microbial species related with each metabolic function. We also explored functional categories of genes related to nutrient metabolism and potentially involved in nutrient cycling (Fierer et al., 2012). For instance, nitrogen-related genes represented 0.8% of the total functional reads, and the abundances of these genes did not differ between forest and vineyard soils. These abundance values are in concordance with previous studies, including environments enriched with nitrogen-fixing bacteria such as in soybean crops (Mendes et al., 2014). A plausible explanation for the lack of differences between habitats is that organic agriculture supplies nitrogen in its organic form (e.g., compost and manure) similarly to what occurs in forests; thus nitrogen could be available in the same chemical form in both habitats only in higher quantities in vineyards (NH_4 vineyard = 9.2 mg/kg and NH_4 forest = 4.2 mg/kg; NO_3 vineyard = 11.1 mg/kg and NO_3 forest = 7.2 mg/kg).

CONCLUSION

We explored the taxonomic and functional diversity of microbial communities in Chilean vineyards using shotgun sequencing. We also analyzed the taxonomic and functional diversity of microbial communities in forest soils of the Chilean Mediterranean biome, one of the most threatened biodiversity hotspots in the world (Myers et al., 2000; Viers et al., 2013). Our metagenomic analyses revealed that the soil microbial communities of organic vineyards and native forests are similar, suggesting that taxonomic composition does not significantly differ between habitats. Conversely, some functional categories differed between forest and vineyard soils. These results could suggest that native forests surrounding vineyards act as microbial reservoirs buffering land conversion. However, additional research is needed to explore the role of landscape complexity and agriculture management on microbial communities in forest–vineyard agroecosystems. Finally, we propose that the implementation of environmentally friendly practices by the wine industry may help to maintain the microbial diversity and ecosystem functions related to natural habitats. This will not only preserve biodiversity but also help to maintain the typicity of wine, which is a valuable cultural and commercial characteristic.

ACKNOWLEDGMENTS

We thank Marlene Manzano and Fernando Alfaro for collecting soil samples and DNA extraction, Andrea Silva for advice during the metagenome sequencing, Juan Opazo for exploratory analysis on the sequencing data, and Juan Ugalde and Toshiko Miura for their advice on the metagenomic analysis. We also thank Elizabeth Cook and Emily Giles for their valuable suggestions and English editing of the manuscript draft. Finally, we thank Keith Crandall and the anonymous reviewers for their comments and suggestions that improved the quality of our manuscript.

FUNDING STATEMENT

This work was funded by CONICYT PFB 23/2008 and ICM P05-002 through Instituto de Ecología and Biodiversidad (IEB-Chile). Luis E. Castañeda was partially supported by FONDECYT 1140066. There was no additional external funding received for this study. The funders had no role in study design, data collection and analysis, decision to publish, or preparation of the manuscript.

AUTHOR CONTRIBUTIONS

Luis E. Castañeda analyzed the data, wrote the paper, prepared figures and/or tables, and reviewed drafts of the paper.

Olga Barbosa conceived and designed the experiments, performed the experiments, contributed reagents/materials/analysis tools, and reviewed drafts of the paper.

REFERENCES

1. Aballay, Maternsson & Persson (2011) Aballay E, Maternsson A, Persson P. Screening of rhizosphere bacteria from grapevine for their suppressive effect on *Xiphinema index* Thorne & Allen on in vitro grape plants. Plant and Soil. 2011;347(1–2):313–325. doi: 10.1007/s11104-011-0851-6.
2. Anonymous (2010) Anonymous . Definition of Viticultural 'Terroir.'

Resolution OIV/VITI 333/2010.Tbilisi: International Organization of Vine and Wine General Assembly (OIV); 2010. [6 June 2016].

3. Bai & Shaner (2004) Bai G, Shaner G. Management and resistance in wheat and barley to *Fusarium* head blight. Annual Review of Phytopathology. 2004;42(1):135–161. doi: 10.1146/annurev.phyto.42.040803.140340.
4. Bester (2005) Bester R. Stellenbosch: University of Stellenbosch; 2005. [6 June 2016]. Growth and survival of Saccharomyces cerevisiae in soil. MSc thesis.
5. Bevivino et al. (2014) Bevivino A, Paganini P, Bacci G, Florio A, Pellicer MS, Papaleo MC, Mengoni A, Ledda L, Fani R, Benedetti A, Dalmastri C. Soil bacterial community response to differences in agricultural management along with seasonal changes in a Mediterranean region. PLoS ONE. 2014;9(8):e105515 doi: 10.1371/journal.pone.0105515.
6. Caporaso et al. (2010) Caporaso JG, Kuczynski J, Stombaugh J, Bittinger K, Bushman FD, Costello EK, Fierer N, Gonzalez-Peña A, Goodrich JK, Gordon JI, Huttley GA, Kelley ST, Knights D, Koenig JE, Ley RE, Lozupone CA, McDonald D, Muegge BD, Pirrung M, Reeder J, Sevinsky JR, Turnbaugh PJ, Walters WA, Wildmann J, Yatsunenko T, Zaneveld J, Knight R. QIIME allows analysis of high-throughput community sequencing data. Nature Methods. 2010;7:335–336. doi: 10.1038/nmeth.f.303.
7. Carrino-Kyker, Smemo & Burke (2013) Carrino-Kyker SR, Smemo KA, Burke DJ. Shotgun metagenomic analysis of metabolic diversity and microbial community structure in experimental vernal pools subjected to nitrate pulse. BMC Microbiology. 2013;13(1):78. doi: 10.1186/1471-2180-13-78.
8. Castañeda et al. (2015) Castañeda LE, Manzano M, Godoy K, Marquet PA, Barbosa O. Comparative study between soil microbial structure communities from vineyards and sclerophyllous forest in central Chile. Ecology and Evolution. 2015;5(18):3857–3868. doi: 10.1002/ece3.1652.
9. Chen, Yanagida & Shinohara (2005) Chen YS, Yanagida F, Shinohara T. Isolation and identification of lactic acid bacteria from soil using enrichment procedure. Letters in Applied

Microbiology. 2005;40(3):195–200. doi: 10.1111/j.1472-765x.2005.01653.x.

10. Cincotta, Wisnewski & Engelman (2000) Cincotta RP, Wisnewski J, Engelman R. Human population in biodiversity hotspots. Nature. 2000;404(6781):990–992. doi: 10.1038/35010105.
11. Coll et al. (2011) Coll P, Le Cadre E, Blanchart E, Hinsinger P, Villenave C. Organic viticulture and soil quality: a long-term study in Southern France. Applied Soil Ecology. 2011;50:37–44. doi: 10.1016/j.apsoil.2011.07.013.
12. Corneo et al. (2013) Corneo PE, Pellegrini A, Cappellin L, Roncador M, Chierici M, Gessler C, Pertot I. Microbial community structure in vineyard soils across altitudinal gradients and in different seasons. FEMS Microbiology Ecology. 2013;84(3):588–602. doi: 10.1111/1574-6941.12087.
13. Cowling et al. (1996) Cowling RM, Rundel PW, Lamont BB, Arroyo MK, Arianoutsou M. Plant diversity in Mediterranean climate regions. Trends in Ecology and Evolution. 1996;11(9):362–366. doi: 10.1016/0169-5347(96)10044-6.
14. Delmont et al. (2011) Delmont TO, Prestat E, Keegan KP, Faubladier M, Robe P, Clark IM, Pelletier E, Hirsch PR, Meyer F, Gilbert JA, Le Paslier D, Simonet P, Vogel TM. Structure, fluctuation and magnitude of a natural grassland soil metagenome. ISME Journal. 2011;6(9):1677–1687. doi: 10.1038/ismej.2011.197.
15. Fierer et al. (2012) Fierer N, Leff JW, Adams BJ, Nielsen UN, Bathes ST, Lauber CL, Owens S, Gilbert GA, Wall DH, Caporaso JG. Cross-biome metagenomic analyses of soil microbial communities and their functional attributes. Proceedings of the National Academy of Sciences of the United States of America. 2012;109(52):21390–21395. doi: 10.1073/pnas.1215210110.
16. Fleet (2003) Fleet GH. Yeast interactions and wine flavour. International Journal of Food Microbiology. 2003;86(1–2):11–22.
17. Fujita et al. (2010) Fujita K, Furuya S, Kohno M, Suzuki S, Takayanagi T. Analysis of microbial community in Japanese vineyard soils by culture-independent molecular approach. International Journal of Wine Research. 2010;2:75–104. doi: 10.2147/ijwr.s13008.

18. Garbeva, van Veen & van Elsas (2004) Garbeva P, van Veen JA, van Elsas JD. Microbial diversity in soil: selection microbial populations by plant and soil type and implications for disease suppressiveness. Annual Review of Phytopathology. 2004;42(1):243–270. doi: 10.1146/annurev.phyto.42.012604.135455.

19. García-Orenes et al. (2013) García-Orenes F, Morugán-Coronado A, Zornoza R, Scow K. Changes in soil microbial community structure influences by agricultural management practices in a Mediterranean agro-ecosystem. PLoS ONE. 2013;8(11):e80522 doi: 10.1371/journal.pone.0080522.

20. Gardi et al. (2009) Gardi C, Montanarella L, Arrouays D, Bispo A, Lemanceau P, Jovilet C, Mulder C, Ranjard L, Römbke J, Rutgers M, Menta C. Soil biodiversity monitoring in Europe: ongoing activities and challenges. European Journal of Soil Science. 2009;60(5):807–819. doi: 10.1111/j.1365-2389.2009.01177.x.

21. Gilbert, van der Lelie & Zarraonaindia (2014) Gilbert JA, van der Lelie D, Zarraonaindia I. Microbial *terroir* for wine grapes. Proceedings of the National Academy of Sciences of the United States of America. 2014;111(1):5–6. doi: 10.1073/pnas.1320471110.

22. Griffiths & Philippot (2013) Griffiths BS, Philippot L. Insights into the resistance and resilience of the soil microbial community. FEMS Microbiology Reviews. 2013;37(2):112–139. doi: 10.1111/j.1574-6976.2012.00343.x.

23. Hannah et al. (2013) Hannah L, Roehrdanz PR, Ikegami M, Shepard AV, Shaw MR, Tabor G, Zhi L, Marquet PA, Hijmans RJ. Climate change, wine, and conservation. Proceedings of the National Academy of Sciences of the United States of America. 2013;110:6907–6912.

24. Heilmann-Clausen et al. (2014) Heilmann-Clausen J, Barron ES, Boddy L, Dahlber A, Griffith GW, Nordén J, Ovaskainen O, Perini C, Senn-Irlet B, Halme P. A fungal perspective on conservation biology. Conservation Biology. 2014;29(1):61–68. doi: 10.1111/cobi.12388.

25. Jangid et al. (2008) Jangid K, Williams MA, Franzluebbers AJ, Sanderlin JS, Reeves JH, Jenkins MB, Endale DM, Coleman DC, Whitman WB. Relative impacts of land-use, management intensity and fertilization upon soil microbial community structure in agricultural

systems. Soil Biology and Biochemistry. 2008;40(11):2843–2853. doi: 10.1016/j.soilbio.2008.07.030.

26. Janssen (2006) Janssen PH. Identifying the dominant soil bacterial taxa in libraries of 16S rRNA and 16S rRNA genes. Applied and Environmental Microbiology. 2006;72(3):1719–1728. doi: 10.1128/aem.72.3.1719-1728.2006.
27. Jenkins et al. (2010) Jenkins SN, Waite IS, Blackburn A, Husbund R, Rushton SP, Manning DC, O'Donnell AG. Actinobacterial community dynamics in long term managed grasslands. Antonie van Leeuwenhoek. 2010;95(4):319–334. doi: 10.1016/j.funbio.2011.02.002.
28. Kasel, Bennett & Tibbits (2008) Kasel S, Bennett LT, Tibbits J. Land use influences soil fungal community composition across central Victoria, south-eastern Australia. Soil Biology and Biochemistry. 2008;40(7):1724–1732. doi: 10.1016/j.soilbio.2008.02.011.
29. Lauber et al. (2008) Lauber CL, Strickland MS, Bradford MA, Fierer N. The influence of soil properties on the structure of bacterial and fungal communities across land-use types. Soil Biology and Biochemistry. 2008;40(9):2407–2415. doi: 10.1016/j.soilbio.2008.05.021.
30. McMurdie & Holmes (2013) McMurdie P, Holmes S. Phyloseq: an R package for reproducible interactive analysis and graphics of microbiome census data. PLoS ONE. 2013;8:e61217 doi: 10.1371/journal.pone.0061217.
31. Mendes et al. (2014) Mendes LW, Kuramae EE, Navarrete AA, van Veen JA, Tsai SM. Taxonomical and functional microbial community selection in soybean rhizosphere. ISME Journal. 2014;8(8):1577–1587. doi: 10.1038/ismej.2014.17.
32. Meyer et al. (2008) Meyer F, Paarman D, D'Souza M, Olson R, Galss EM, Kubal M, Paczian T, Rodriguez A, Steens R, Wilke A, Wilkening J, Edwards RA. The metagenomics RAST server: a public resource for the automatic phylogenetic and functional analysis of metagenomes. BCM Bioinformatics. 2008;9(1):386. doi: 10.1186/1471-2105-9-386.
33. Mills et al. (2008) Mills DA, Phister T, Neeley E, Johannsen E.

Wine fermantation. In: Cocolin L, Ercolini D, editors. Molecular Techniques in the Microbial Ecology of Fermented Foods. Berlin: Springer-Verlag; 2008. pp. 162–192.

34. Mittermeier et al. (2011) Mittermeier RA, Turner WR, Larsen FW, Brooks TM, Gascon C. Global biodiversity conservation: the critical role of hotspots. In: Zachos FE, Hable JC, editors. Biodiversity Hotspots. Berlin: Springer-Verlag; 2011. pp. 3–22.

35. Miura et al. (2016) Miura T, Niswati A, Swibawa IG, Haryani S, Gunito H, Arai M, Yamada K, Shimano S, Kaneko N, Fujie K. Shifts in the composition and potential functions of soil microbial communities responding to a no-tillage practice and bagasse mulching on a sugarcane plantation. Biology and Fertility of Soils. 2016;52(3):307–322. doi: 10.1007/s00374-015-1077-1.

36. Montecchia et al. (2015) Montecchia MS, Tosi M, Soria MA, Vogrig JA, Sydorenko O, Correa OS. Pyrosequencing reveals changes in soil bacteria communities after conversion of Yungas forests to agriculture. PLoS ONE. 2015;10(3):e0119426 doi: 10.1371/journal.pone.0119426.

37. Myers et al. (2000) Myers N, Mittermeier RA, Mittermeier CG, da Fonseca GA, Kent J. Biodiversity hotspots for conservation priorities. Nature. 2000;403:853–858. doi: 10.1038/35002501.

38. Nannipieri et al. (2003) Nannipieri P, Ascher J, Ceccherini MT, Landi L, Pietramellara G, Renella G. Microbial diversity and soil functions. European Journal of Soil Science. 2003;54(4):655–670. doi: 10.1046/j.1351-0754.2003.0556.x.

39. O'Brien et al. (2005) O'Brien HE, Parrent JL, Jackson JA, Moncalvo J-M, Vilgalys R. Fungal community analysis by large-scale sequencing of environmental samples. Applied and Environmental Microbiology. 2005;71(9):5544–5550. doi: 10.1128/aem.71.9.5544-5550.2005.

40. Oksanen et al. (2013) Oksanen J, Blanchet FG, Friendly M, Kindt R, Legendre P, McGlinn D, Minchin PR, O'Hara RB, Simpson GL, Solymos P, Stevens MHH, Szoecs E, Wagner H. vegan: Community Ecology PackageR package version 2.0-10https://cran.r-project.org/ 2013

41. Oliveros (2015) Oliveros JC. Venny. An interactive tool for

comparing lists with Venn's diagrams. 2015. http://bioinfogp.cnb.csic.es/tools/venny/index.html. [23 June 2016]. http://bioinfogp.cnb.csic.es/tools/venny/index.html

42. Pampulha & Oliveira (2006) Pampulha ME, Oliveira A. Impact of an herbicide combination of bromoxynil and prosulfuron on soil microorganisms. Current Microbiology. 2006;53(3):238–243. doi: 10.1007/s00284-006-0116-4.
43. Parks & Beiko (2010) Parks DH, Beiko RG. Identifying biologically relevant differences between metagenomic communities. Bioinformatics. 2010;26(6):715–721. doi: 10.1093/bioinformatics/btq041.
44. Paula et al. (2014) Paula FS, Rodrigues JLM, Zhou J, Wu L, Mueller RC, Mirza BS, Bohannan BJM, Nüsslein K, Deng Y, Tiedje JM, Pellizari VH. Land use change alters functional gene diversity, composition and abundance in Amazon forest soil microbial communities. Molecular Ecology. 2014;23(12):2988–2999. doi: 10.1111/mec.12786.
45. Pearce et al. (2012) Pearce DA, Newsham KK, Thorne MAS, Calvo-Bado L, Krsek M, Laskaris P, Hodson A, Wellington EM. Metagenomic analysis of a southern maritime Antarctic soil. Frontiers in Microbiology. 2012;3:403. doi: 10.3389/fmicb.2012.00403.
46. R Development Core Team (2016) R Development Core Team . R: A language and environment for statistical computing. Vienna: R Foundation for Statistical Computing; 2016.
47. Spain, Krumholz & Elshahed (2009) Spain AM, Krumholz LR, Elshahed MS. Abundance, composition, diversity and novelty of soil Proteobacteria. ISME Journal. 2009;3(8):992–1000. doi: 10.1038/ismej.2009.43.
48. Tiedje et al. (1999) Tiedje JM, Asuming-Brempong S, Nusslein K, Marsh TL, Flynn SJ. Opening the black box of soil microbial diversity. Applied Soil Ecology. 1999;13(2):109–122. doi: 10.1016/s0929-1393(99)00026-8.
49. Underwood et al. (2008) Underwood EC, Viers JH, Klausmeyer KR, Cox RL, Shaw MR. Threats and biodiversity in the Mediterranean biome. Diversity and Distributions. 2008;15(2):188–197. doi: 10.1111/j.1472-4642.2008.00518.x.

50. Uroz et al. (2013) Uroz S, Ioannidis P, Lengelle J, Cébron A, Morin E, Buée M, Martin F. Functional assays and metagenomic analysis reveals differences between the microbial communities inhabiting the soil horizons of a Norway spruce plantation. PLoS ONE. 2013;8(2):e55929 doi: 10.1371/journal.pone.0055929.
51. van Leeuwen et al. (2004) van Leeuwen C, Friant P, Chone C, Tregoat O, Koundouras S, Dubourdieu D. Influence of climate, soil, and cultivar on terroir. American Journal of Enology and Viticulture. 2004;55:207–217.
52. Viers et al. (2013) Viers JH, Willimas JN, Nicholas KA, Barbosa O, Kotzé I, Spence L, Webb LB, Merenlender A, Reynols M. Vinecology: pairing wine with nature. Conservation Letters. 2013;6:287–299. doi: 10.1111/conl.12011.
53. Vitousek et al. (1997) Vitousek PM, Mooney HA, Lubchenco J, Melillo JM. Human domination of Earth's ecosystems. Science. 1997;227(5325):194–199. doi: 10.1126/science.277.5325.494.
54. White, Nagarajan & Pop (2009) White JR, Nagarajan N, Pop M. Statistical methods for detecting differentially abundant features in clinic metagenomic samples. PLoS Computational Biology. 2009;5(4):e1000352 doi: 10.1371/journal.pcbi.1000352.
55. Xu et al. (2014) Xu Z, Hansen MA, Hansen LH, Jacquiod S, Sorensen S. Bioinformatic approaches reveal metagenomic characterization of soil microbial community. PLoS ONE. 2014;9(4):e93445 doi: 10.1371/journal.pone.0093445.
56. Zarraonaindia et al. (2015) Zarraonaindia I, Owens SM, Welsenhorn P, West K, Hampton-Marcell J, Lax S, Bokulich NA, Mills DA, Martin G, Taghavi S, van der Lelie D, Gilbert JA. The soil microbiome influences grapevine-associated microbiota. mBio. 2015;6(2):e02527 doi: 10.1128/mbio.02527-14.

CHAPTER 12

16S RDNA PYROSEQUENCING ANALYSIS OF BACTERIAL COMMUNITY IN HEAVY METALS POLLUTED SOILS

Marcin Gołębiewski[1, 2, 3], Edyta Deja-Sikora[1], Marcin Cichosz[5], Andrzej Tretyn[1, 2], Borys Wróbel[3, 4]

[1]Chair of Plant Physiology and Biotechnology, Nicolaus Copernicus University, Lwowska 1, 87-100 Toruń, Poland

[2]Centre of Modern Interdisciplinary Technologies, Nicolaus Copernicus University, Wilenska 4, 87-100 Toruń, Poland

[3]Institute of Oceanology PAS, Powstańców Warszawy 55, 81-712 Sopot, Poland

[4]Evolutionary Systems Laboratory, Adam Mickiewicz University, Umultowska 89, Room S1, 61-614 Poznań, Poland

[5]Department of Chemical Pro-ecological Processes, Nicolaus Copernicus University, Gagarina 7, 87-100 Toruń, Poland

ABSTRACT

Soil contamination with heavy metals is a widespread problem, especially prominent on grounds lying in the vicinity of mines, smelters, and other industrial facilities. Many such areas are located in Southern Poland; they are polluted mainly with Pb, Zn, Cd, or Cu, and locally also with Cr. As for now, little is known about most bacterial species thriving in such soils and even less about a core bacterial community—a set of taxa common to polluted soils. Therefore, we wanted to answer the question if such a set could be found in samples differing physicochemically and phytosociologically. To answer the question, we analyzed bacterial communities in three soil samples contaminated with Pb and Zn and two contaminated with Cr and lower levels of Pb and Zn. The communities were assessed with 16S rRNA gene fragments pyrosequencing. It was found that the samples differed significantly and Zn decreased both diversity and species richness at species and family levels, while plant species richness did not correlate with bacterial diversity. In spite of the differences between the samples, they shared many operational taxonomic units (OTUs) and it was possible to delineate the core microbiome of our sample set. The core set of OTUs comprised members of such taxa as *Sphingomonas*, *Candidatus Solibacter*, or *Flexibacter* showing that particular genera might be shared among sites ~40 km distant.

INTRODUCTION

Analysis of biodiversity in areas under anthropogenic pressure (e.g., pollution or climate change) allows to quantify the possibly detrimental effects of human activity on particular taxonomic groups. Due to the rapid response, microbial biodiversity may also be used to monitor the outcome of bioremediation/bioreclamation efforts [1].

A number of existing studies showed varying degree of heavy metals impact on soil bacteria. Frostegård and colleagues [2] found that Cd, Cu, Ni, Pb, and Zn influenced patterns of phospholipid fatty acids (PLFA) and decreased biomass and activity (measured as respiration) of microbial communities in forest and arable soils. However, determination which bacterial species or higher taxonomic units were affected was beyond the capabilities of this methodology. rRNA-based methods allowed for more detailed analyses; certain groups reported differential influence of metals on culturable and unculturable fractions of soil bacterial community [3, 4]. Total community richness was found to be only moderately affected by heavy metals [5, 6], but Gans et al. found that the diversity in heavily polluted soils may constitute only 1 % of that in pristine soils [7].

It is estimated that only 0.1–1 % of soil bacteria are readily culturable [8, 9]. This leads to the conclusion that culture-independent methods should be preferred, although some researchers disagree with this view claiming that the culturable subcommunity is the most active, and therefore the most important [4, 10, 11].

Among the methods that can assess both culturable and unculturable bacteria, the majority is based on marker genes, such as 16S rRNA or *recA* genes [12, 13]. Next generation sequencing (NGS) quickly becomes the method of choice, due to the high data yield and relatively low costs of data generation. To date, several reports on soil bacterial communities analyzed via 16S rRNA gene fragments pyrosequencing were published [14, 15, 16, 17], but only a few of them were devoted to metals-contaminated soils. The dependence of bacterial diversity and species richness on the contamination was found in Cd, Pb, and Zn [18, 19] as well as Cr and As [20] and Cu [19] contaminated soils. Metals caused decrease of diversity, increase of Proteobacteria, and decrease of Acidobacteria and Actinobacteria levels. However, when considered with other factors, they were not the most important variables shaping the community structure [19].

We wanted to see if bacterial communities in two soils, differently contaminated with heavy metals and varying in other physicochemical parameters as well as phytosociologically but coming from the same geographical region were similar. Basing on the shared operational taxonomic units (OTUs) set, we delineated a core microbiome of our samples, defined as the set of OTUs evenly distributed among all the samples and present at the percentage greater than 0.1 %.We looked also at the diversity, the species richness, and the community structure. To study the parameters of interest, barcoded libraries of V3–V4 16S rRNA gene fragments were pyrosequenced and between 15,000 and 20,000 raw reads from each sample were generated. We developed a bioinformatic pipeline based on MOTHUR [21] and used it to analyze the parameters mentioned above.

METHODS

Soil Samples

Samples were collected at two sites in Southern Poland: near Olkusz (50°17′ N, 19°30′ E), in the vicinity of a lead and zinc ore enrichment facility (specifically its flotation waste collector), and in Alwernia (50°02′ N, 19°31′ E), near a chromium green-producing facility. The sampling points were located on public grounds, not protected by nature reserve or national park, and therefore no specific permission was required. The fieldwork did not involve any endangered or protected species.

Phytosociological analysis of the sampling sites was performed according to the Braun-Blanquet method [22]. Briefly, one phytosociological picture was taken at each site, then the plant species were identified and their coverage (in standard six-point Braun-Blanquet scale: +—rare species, 1—abundant species, coverage lower than 20 %, 2—very abundant species with coverage <20 % or abundant one with coverage between 20 and 25 %, 3—25–50 % coverage, 4—50–75 % coverage, 5—75–100 % coverage) was estimated.

Top soil from the 5–30-cm layer was collected in July 2009. Five subsamples were taken at each site at the four vertices and in the center of a 2.5-m square. After visible plant fragments and large stones were removed, the subsamples were combined, and the sample was thoroughly mixed with a spade. After sampling, the soils were transferred to plastic bags which were sealed and stored in a cold room (at 9 °C) until analysis. DNA isolation (details below) was performed within 5 days, other analyzes until the end of 2009. The soil texture was assessed with the laser diffraction method on Analysette 22 (Fritsch, Germany). The organic matter and carbonates content were estimated by loss on ignition at 550 and 950 °C, respectively. The pH was measured in a 1:10 soil-water slurry with a digital pH meter (Elmetron, Poland). The metals content was analyzed by X-ray fluorescence (XRF), and their bioavailable concentrations were determined with atomic absorption spectrometry (AAS) after DETAPAC extraction [23]. The samples characteristics are summarized in Table 1, and the phytosociological analysis can be found in Table 2.

Table 1: Soil samples characteristics

Name	GPS coordinates	Soil type	OM (%)	C_{inorg}[a] (%)	P_2O_5 (%)	H_2O (%)	pH	Fe_2O_3 (%)	Cd_t[b] (mg/kg)	Cd_{ba}[c] (mg/kg)	Pb_t (mg/kg)	Pb_{ba} (mg/kg)	Zn_t (mg/kg)	Zn_{ba} (mg/kg)	Cr_t (mg/kg)	Cr_{ba} (mg/kg)
A1	50° 02' 54.8" N, 19° 31' 34.2" E	Sand	2.82	0.12	0.02	2.18	7.0	2.23	4.2±0.2	2.5±0.1	204±5	104±5	749±5	521±5	740±10	492±10
A2	50° 02' 40.4" N, 19° 31' 38.4" E	Silt	4.62	0.04	0.09	2.77	6.5	3.23	3.5±0.2	2.1±0.1	203±5	99±5	767±5	538±5	760±10	501±10
O1	50° 17' 21,6" N, 19° 30' 48,1" E	Sand	19.90	0.30	0.13	5.60	6.7	2.02	3.4±0.2	2.1±0.1	1,378±5	730±5	1,247±5	863±5	100±10	63±10
O2	50° 17' 17,2" N, 19° 30' 28,8" E	Sand	16.23	0.62	0.05	2.95	6.5	1.06	3.9±0.2	2.3±0.1	461±5	241±5	957±5	692±5	180±10	121±10
O3	50° 17' 15,2" N, 19° 30' 11,7" E	Sand	3.82	30.37	0.01	1.07	7.6	7.43	3.7±0.2	2.2±0.1	1,097±5	563±5	2,002±5	1,478±5	300±10	196±10

[a] Inorganic carbon content
[b] Total concentration
[c] Bioavailable concentration

Table 2: Phytosociological analysis of sampling points

Sample	Phytosociological classification	Plant sp. richness	Trees coverage	Herbaceous plants coverage	Main species
A1	No-rank *Equisetum arvense* Class: *Agropyretea intermedio-repentis*	12	0 %	15 %+ 10 % mosses	*Equisetum arvense*
A2	Assoc.: *Phleum pratense* Alliance: *Cynosurion* Order: *Arrhenatheretalia* Class: *Molinio-Arrhenatheretea*	36	0 %	100 %	*Phleum pratense*
O1	Assoc.: *Querco roboris-Pinetum* Alliance: *Dicrano-Pinion* Order: *Piceetalia abietis* Class: *Vaccinio-Piceetaea*	35	65 %	80 %	*Pinus sylvestris, Festuca ovina, Leontodon hispidus, Rumex thyrsiflorus*
O2	Assoc.: *Spergulo vernalis-Corynephoretum* Alliance: *Corynephorion canescentis* Order: *Corynephoretalia canescentis* Class: *Koelerio-Corynephoretea*	14	5 % (shrubs)	20 %	*Pinus sylvestris, Corynephorus canescens*

O3	No-rank *Gypsophila fastigiata* Class: *Koelerio-Corynephoretea*	14	0 %	15 %	*Gypsophila fastigiata*

Analysis of 357f and 786r Primers Coverage and Specificity

Primers' coverage (percentage of bacterial sequences in a database from which the primer set generates amplicons) and specificity (percentage of nonbacterial sequences amplified by the primers) were assessed by TestProbe from the SILVA website (http://beta.arb-silva.de/search/testprobe; [24]) run on the SSU r114 database and by Ribosomal Database Project's tool ProbeMatch (http://rdp.cme.msu.edu/probematch/search.jsp; [25]) run against the RDP 10 update 31 database. Both tools were run with two mismatches allowed and otherwise default settings, with the exception of using weighted mismatches in TestProbe. The parameters for the pair of primers were also assessed with ProbeMatch of RDP and TestPrime from SILVA [26].

Isolation of Metagenomic DNA

Metagenomic DNA was isolated from 30 g of homogenized soils with the method developed by Zhou and colleagues [27] modified to increase the yield of DNA and the spectrum of lysed bacteria. The modifications involved adding an enzymatic digestion (lysozyme and achromopeptidase) step before the SDS lysis and lowering the lysis temperature (55 °C instead of 65 °C) [28]. Subsequently, the DNA was purified in two agarose gel electrophoresis steps, first using 0.7 %, and then in 1 % agarose. Pure DNA was recovered from the gel with an agarose gel extraction kit (Roche). The quality of the preparations was assessed spectrophotometrically on NanoDrop ND-1000 (NanoDrop, Wilmington, DE).

PCR Amplification of 16S rRNA Gene Fragments and Pyrosequencing

A fragment of the 16S rRNA gene encompassing the V3 and V4 variable regions [29] was amplified with primers 357f (5′ CCT ACG GGA GGC

AGC AG 3′, [30]) and 786r (5′ AC CAG GGT ATC TAA WCC 3′, this work) with overhangs required for pyrosequencing. The forward primer was used in three versions differing in the multiplex identifier (MID) sequence between the A adaptor and the actual primer. The reaction mix contained the following: 1 ng of template DNA, 10 pmol primers, 0.2 mM dNTP, 1.5 mM $MgCl_2$, 0.2 U of Taq polymerase, and the appropriate 1× buffer (Fermentas, Lithuania). The cycling conditions were as follows: 95 °C for 5 min, 30 cycles of: 95 °C for 30 s, 52 °C for 30 s, 72 °C for 30 s; and finally 72 °C for 2 min. The reactions were carried out in MasterCycler thermocycler (Eppendorf, Germany). Eight independent reactions were performed for each sample.

PCR products were purified with the DNA CleanUp kit (A&A Biotechnology, Poland), pooled in equimolar amounts, vacuum dried, and either sent for sequencing at Max Planck Molecular Genetics Institute sequencing service (Berlin, Germany—Olkusz samples) or sequenced at the Chair of Genetics, University of Silesia (Katowice, Poland—Alwernia ones). Pyrosequencing was performed with the use of Titanium chemistry on GS FLX and GS Junior instruments (Roche, USA) as described in the manufacturer's protocols.

Bioinformatic Analysis

We developed a bash script that allowed a streamlined analysis of sequence data using MOTHUR v.1.30.2 [21] and R [31] and by calling other scripts in bash. The script took as the input an .sff file(s) (either a single one with all samples or many single-sample files) and a file containing information on the primers used for amplicons generation with optional barcodes (in the case of single .sff) or, in the case of multiple .sff files, additional file with sample names. The computations were parallelized where possible to speed up the analysis. The limits on the number of processors and the amount of RAM to be used in the clustering step could be set to allow running the script on multi-processors machines without consuming all the resources available.

The script can be run on any POSIX-compatible operating system, where installation of bash shell is possible. We ran the analyzes on a 48-core 256 GB RAM machine under control of the 64-bit version of Fedora 14 Linux, using eight cores and with limit of RAM amount set to 10 GB. All the procedures described below were pipelined in this script, available from the authors on request. The actual parameters for all steps of the pipeline can be found in Supplementary Materials.

Pyrosequencing adaptor and MID sequences were removed from the Olkusz reads by the sequencing center, and quality trimming was performed for separate sample files. For the Alwernia samples, a single .sff file was obtained, and the trimming was done accordingly. Quality trimming was performed with the AmpliconNoise algorithm [32] implemented in MOTHUR's shhh.flows. The reads were then dereplicated, aligned against the SILVA template alignment, and screened for those covering the desired alignment region. The gap-only and terminal gap-containing columns were filtered out of the alignment and putative chimeras were identified with the uchime algorithm [33] and removed. Finally, the sequencing noise was further reduced via single-linkage pre-clustering [34]. Then, the high-quality reads were down-sampled to the smallest sample size and (1) classified with naïve Bayesian classifier implemented in MOTHUR (classify.seqs) using SILVA taxonomy provided by the authors of the program on the webpage www.mothur.org/w/images/9/98/Silva.bacteria.zip, as well as (2) a distance matrix was calculated, and OTUs were constructed by average neighbor clustering. Sequences were regarded successfully classified at a given level if their bootstrap support was greater than 80 %.

Species Richness and Diversity Analyzes

Species richness was assessed with Chao1 [35] and ACE [36] indices, as well as with catchall v3.0 [37]. Phylogenetic diversity was measured according to the methodology of Faith and coworkers [38] (MOTHUR's command "phylo.diversity"). The estimators were calculated for five subsamples containing 1,500 sequences drawn randomly from each group and averaged to yield the final values.

An averaged shared OTU table was prepared from the subsamples (for details, see Supplementary Materials).

Morisita-Horn [39] and Bray-Curtis [40] dissimilarities were calculated for all the pairs of samples via dist.shared command of MOTHUR. They were visualized as heatmaps with the use of the heatmap.sim command. Venn diagrams were prepared from the shared OTU table using a custom Perl script.

To check if the communities differed significantly, the parsimony [41] and weighted UniFrac [42] tests were employed. The MOTHUR commands parsimony and unifrac.weighted were used, respectively. The tests were run on a phylogenetic tree generated from the same distance matrix that served for the OTU construction, with the use of the relaxed neighbor joining algorithm of clearcut [43] implemented in MOTHUR.

Statistical Analysis

Statistical analysis was performed in R 2.15.2 [31], using vegan [44] (redundancy analysis (RDA) with significance testing by permutation—ANOVA) and Hmisc [45] packages (Spearman's correlation in rcorr).

Core Microbiome Delineation

The "core microbiome," as defined by Turbaugh and coworkers [46], would consist of the OTUs that were present in all the samples. We added two conditions here: (1) the OTUs should be present in roughly equal amounts and (2) the frequency of reads has to be >0.1 % in each sample (>2 reads). OTUs meeting the conditions were identified by calculation of Shannon's evenness, treating the samples as taxa and reads as individuals and eliminating those with two or less reads in any sample. The OTUs with the value greater than 0.9 were regarded "evenly distributed" and assigned to the core microbiome.

RESULTS

Sites and Soils Characteristics

In this study, we analyzed the bacterial community structure and species richness in three soil samples from Olkusz and two from Alwernia, both towns located in the Malopolska province of Poland. The soils came from the area lying in the vicinity of a flotation waste pond of a zinc and lead ore enrichment facility (Olkusz) and from the grounds adjacent to a factory producing chromium-based pigments (Alwernia).

The samples differed from each other, both in terms of plant species richness (A1 being the least rich, and A2 being the most) and phytosociological classification (Table 2). Also, the ground coverage by plants varied greatly, from almost 0 % for O3 to 100 % for A2. Two samples were classified as no-rank associations (A1—no rank *Equisetum arvense*, O3—no rank *Gypsophila fastigiata*), while the remaining three belonged to syntaxons well placed in the hierarchy (Table 2).

The three Olkusz samples appeared to be sandy (Table 1), the organic matter content varied between 19.90 % (O1) and 3.82 % (O3), and the inorganic carbon (carbonates) content was below 1 % in O2 and O3, while it was over 30 % in O3. High C_{org} content in the O1 and O2 samples was due to the addition of peat to the upper layer of soil in an effort to immobilize heavy metal ions to allow land reclamation, which was done in 1990s [47]. High level of carbonates in O3 was brought about by dolomite, which was the main component of flotation waste. The pH was close to neutral in all samples. Slightly alkaline pH in the O3 sample might have resulted from the high dolomite content and the remnants of the flotation reagents (xanthates) leaking from the pond. The organic matter content was 2.82 and 4.62 % in A1 and A2, respectively, the carbonates were low, and the pH was close to neutral.

The samples were only moderately contaminated with Cd (3.4–4.2 mg kg^{-1} total conc., 2.1–2.5 mg kg^{-1} bioavailable conc.—Table 1), and the sites did not differ much in the concentration of this metal. The levels of Pb and, particularly, Zn were high in the Olkusz samples, reaching 1,097 mg kg^{-1} (Pb) and 2,002 mg kg^{-1} (Zn) total concentration at O3, and were much lower in Alwernia ones. The opposite pattern was found for Cr, whose concentration was highest in A2 (760 mg kg^{-1}). Bioavailable concentrations of the metals were at the level of 70 % of the total concentrations. It seems that in the Olkusz samples, Pb and Zn were the main contaminants which might have influenced bacterial communities. In the Alwernia ones, Cr and Zn were the dominant pollutants, while Cd could be neglected in all the samples due to its low concentration.

Primers' Coverage and Specificity

The 357f primer was highly specific, as only 2.5 % of matches came from Archaea according to RDP and 20.6 % were not from bacteria according to SILVA. It had also high coverage, as it amplified over 99.4 % (RDP) and 96.7 % (SILVA) of bacterial sequences. The 786r primer also performed well, having 99.3 and 97.2 % of coverage among bacteria (according to RDP and SILVA, respectively) and slightly lower specificity (5.4 and 26.1 % of matches were nonbacterial in RDP and

SILVA). Both tools reported a fair coverage of Archaea by the primers (up to 86.6 % for Euryarchaeota), but when no mismatches were allowed, these matches disappeared.

As a pair, the primers were highly specific for bacteria (2.9 and 25.7 % nonbacterial matches in RDP and SILVA, respectively) and the coverage was also high (98.8 and 88.7 %). The values reported were calculated excluding sequences lacking the 357–786 region.

Sequencing

The DNA preparations were of high quality, as their A260/280 ratios were ~1.8 for all samples and A260/230 at ~2 indicating high purity. All of them performed well as PCR templates. In total, 66,385 reads were obtained (Table 3), out of which 25,808 (38.9 %) met the quality criteria. The 2, 376 potential chimeras were removed, and 23,432 (35.3 % of the total number) high-quality nonchimeric reads were included in the downstream analysis. The median of read length was 250 bp and the mean was 255 bp. All the reads were successfully classified at the domain (all of them came from bacteria) and phylum level (bootstrap support >80 %), for lower taxonomic levels, the percentage of unclassified sequences increased, to reach 50.5 % for genus. The reads generated in this project were deposited in the NCBI's SRA database under SRA IDs SRA024301.1 (Olkusz) and SRA059341.1 (Alwernia).

Table 3: Sequencing statistics

Sample	Raw reads no.	High-quality reads		Chimeras		High-quality non-chimeric reads	
		Number	%[a]	Number	%[b]	Number	%[a]
A1	4,235	3,054	72.1 %	422	13.8 %	2,632	62.1 %
A2	5,947	4,229	71.1 %	783	18.5 %	3,446	57.9 %
O1	15,967	7,474	46.8 %	412	5.5 %	7,062	44.2 %
O2	17,522	6,215	35.5 %	363	5.8 %	5,852	33.4 %
O3	22,714	4,836	21.2 %	396	8.2 %	4,440	19.5 %
Total	**66,385**	**25,808**	**38.9 %**	**2 376**	**9.2 %**	**23,432**	**35.3 %**

[a]Percent of raw reads

[b]Percent of high-quality reads

Community Species Richness and Diversity Indices

The communities were moderately sampled, indicating that much more reads would be required to capture all the diversity (Fig. 1). The Alwernia samples were more diverse than the Olkusz ones, which was particularly visible at the 0.10 level (Table 4). Bacterial diversity in the Olkusz soils decreased with increasing concentration of Pb and Zn (O2 > O1 > O3), while A1 was more diverse than A2 in spite of the higher contamination level. The rarefaction analysis (Fig. 1) gave results similar to the comparison of Shannon's H′ (Table 5). Evenness in all samples was close to 0.9 for the 0.03 clustering level and in the range 0.7–0.8 for the 0.10 level, the highest values were observed in A1.

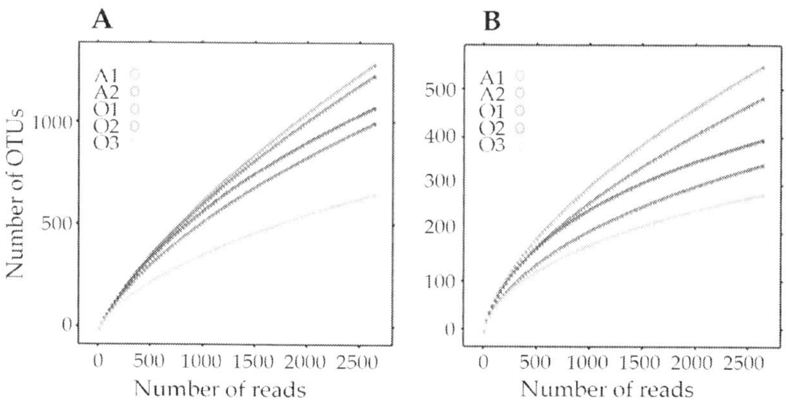

Figure 1: Rarefaction curves. a 0.03 dissimilarity level, b 0.10 dissimilarity level.

Table 4: Species richness estimators

Sample	OTUs observed		Chao1		ACE		Catchall[b]	
	0.03[a]	0.10[a]	0.03	0.10	0.03	0.10	0.03	0.10
A1	**858**	**386**	**3,025**(2,536–3,657)	**981**(790–1,261)	**5,188**(4,747–5,679)	**1,554**(1,381–1,758)	**8,637**	1,747
A2	826	342	2,530 (2,150–3,019)	881 (697–1,160)	4,515 (4,139–4,934)	1,347 (1,195–1,527)	6,662	**1,900**
O1	699	263	1,589 (1,383–1,858)	436 (372–536)	2,596 (2,361–2,864)	576 (504–670)	2,558	610
O2	749	312	1,728 (1,507–2,013)	523 (448–640)	2,723 (2,485–2,993)	637 (569–723)	3,239	900
O3	476	224	1,069 (903–1,301)	402 (330–521)	1,768 (1,589–1,977)	571 (495–667)	1,934	546

Each value is a mean of five obtained for random subsamples of equal size, 95 % confidence intervals are in parentheses. Highest values are in boldface

[a]Dissimilarity level for OTU construction

[b]Due to a bug in catchall there is no confidence intervals for the estimations

Table 5: Diversity and evenness estimators

Sample	Shannon's H′		Shannon's evenness		Phylodiversity
	0.03[a]	0.10[a]	0.03	0.10	
A1	**6.367 (6.309–6.425)**	**4.984 (4.908–5.060)**	**0.943**	0.837	**53,34**
A2	6.309 (6.249–6.368)	4.800 (4.723–4.878)	0.939	0.823	49.14
O1	5.978 (5.909–6.047)	4.280 (4.194–4.366)	0.913	0.768	40,32
O2	6.196 (6.137–6.254)	4.810 (4.738–4.883)	0.936	**0.838**	45,63
O3	5.321 (5.247–5.395)	4.169 (4.087–4.251)	0.863	0.771	30,66

Each value is an average of five coming from random subsamples of size 1,500. 95 % confidence intervals are given in parentheses

[a]Dissimilarity level for OTU construction

Both the observed and estimated total richness was higher in the Alwernia communities than in the Olkusz ones (Table 4). At the 0.03 dissimilarity level, all the methods predicted the highest richness in A1, while at the 0.10 level, only catchall predicted a greater number of OTUs in A2 than in A1. The parametric methods, such as catchall, should be independent of the community coverage, therefore they could be trusted more than the nonparametric estimators. The collectors curves for the Chao1 and ACE indices showed no saturation, so one would expect that the values of these estimators would increase with the sampling effort. In other words, they should not be treated as the true richness estimations.

The estimated number of OTUs in the least rich sample (O3) was about five times lower than in the richest one (A1 or A2). As all the methods used predicted that the richness was smallest in the most contaminated sample (O3), it might mean that species richness was affected by heavy metals contamination. Indeed, regression analysis showed that the Zn concentration negatively correlated with all estimators of species richness and diversity, and the correlation seemed to be nonlinear (Spearman's rank correlation, $p<0.05$, Table 6). However, due to the low number of samples, such results should be treated cautiously, in spite of their high statistical significance. No other variables, including the phosphorus concentration and the plant species richness, correlated with bacterial diversity (RDA model's $p>0.05$) or species richness measures (Spearman's rank correlation, $p>0.05$).

Table 6: Correlation of zinc concentration with bacterial species richness and diversity estimators

	Spearman's correlation coefficient and *p* value
Sp. observed (OTUs 0.03)	$\rho=-1, p=0$
Chao1 (OTUs 0.03)	$\rho=-1, p=0$
ACE (OTUs 0.03)	$\rho=-1, p=0$
Shannon's H′ (OTUs 0.03)	$\rho=-1, p=0$

Sp. observed (OTUs 0.10)	$\rho=-1, p=0$
Chao1 (OTUs 0.10)	$\rho=-1, p=0$
ACE (OTUs 0.10)	$\rho=-1, p=0$
Shannon's H′ (OTUs 0.10)	$\rho=-0.90, p=0.03$
Phylodiversity	$\rho=-1, p=0$

Bacterial Community Structure

All the reads were assigned to bacteria. Proteobacteria were the most abundant phylum in all five samples (27.5–49.5 %—Fig. 2), in the Alwernia ones and O3, they were followed by Acidobacteria, while in O1 and O2 by Actinobacteria. Other major phyla were Bacteroidetes, Chloroflexi, and Gemmatimonadetes. The most abundant classes were Alphaproteobacteria, Actinobacteria, Acidobacteria, Gammaproteobacteria, Sphingobacteria, and Betaproteobacteria (Fig. 3b). The most abundant genera (>100 reads across all samples) collectively comprised 43.1 % of all reads, and their percentage varied from 35.8 % (O2) to 55.4 % (O3) (Fig. 3). The four most abundant ones were uncultured members of Acidobacteriaceae, Gemmatimonadaceae, Nitrosomonadaceae, and Xanthobacteraceae. The Alwernia samples containedmorerepresentativesofanunculturedgenusofAcidobacteriaceae family and *Geobacter* (Deltaproteobacteria, Geobacteraceae) than the Olkusz ones, while genus *Solirubrobacter* (Actinobacteria, Solirubrobacteraceae) was more abundant in the Olkusz soils. The O3 sample contained a high proportion of sulfur oxidizing bacteria of genera *Beggiatoa*(Gammaproteobacteria, Thiotrichaceae) and *Thiobacillus* (Betaproteobacteria, Hydrogenophilaceae) that probably utilized sulfides present in mine tailings, mainly pyrite, and possibly organosulfur compounds used in the flotation process (xanthates).

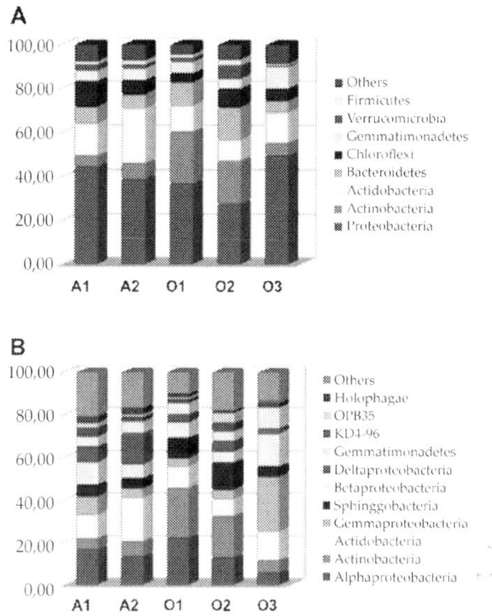

Figure 2: Bacterial community structure in samples. a Phylum level, b class level.

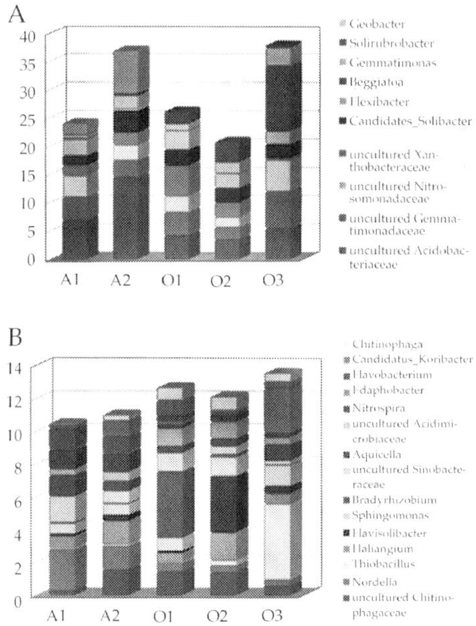

Figure 3: Most abundant genera in samples. a 10 genera of highest total abundance; **b** next 15 genera.

Influence of environmental variables on the community structure was assessed with RDA. Zn, Pb, Cr, and pH were found to be significantly correlated with the community structure both on the 0.10 and 0.03 clustering levels ($p < 0.05$). Notably, the P concentration and plant species richness did not influence the communities under study.

Comparison of Communities

To measure the "overlap" between the communities in the samples, an abundance-based index (Morisita-Horn—M-H), as well as an incidence-based one (Bray-Curtis—B-C) were used to calculate distances and community similarity heatmaps were generated (Fig. 4). At both clustering levels, the B-C similarities were smaller than the M-H ones, indicating that the structures of the abundant OTUs subcommunities were more similar than the rare OTUs communities structures. In other words, the rare OTUs were unique for each sample, while the abundant ones were more likely shared. Regardless of the clustering level and the similarity measure, O1 and O2 appeared to be the most similar samples, A1 and A2 formed another pair, while O3 showed little similarity to other ones.

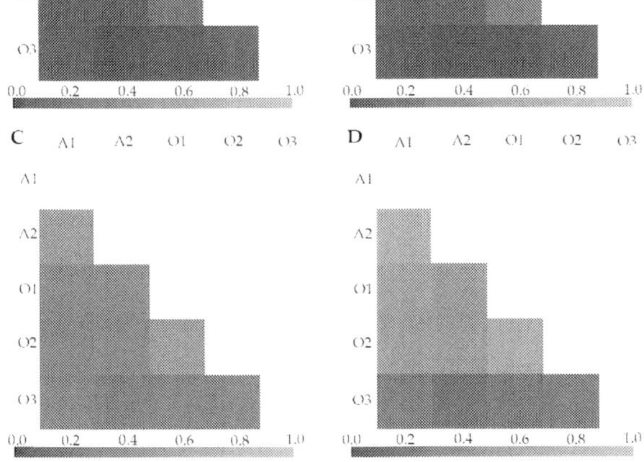

Figure 4: Community distance heatmaps. Darker shade means greater dis-

tance (lower similarity). *Light grey* small distances (greater similarity), *dark grey*—greater distances (lower similarity). Heatmaps based on Bray-Curtis dissimilarity a at 0.03 dissimilarity level and b at 0.10 dissimilarity level; Heatmaps based on Morisita-Horn dissimilarity c at 0.03 dissimilarity level and d at 0.10 dissimilarity level.

The community overlaps were depicted as Venn diagrams (Fig. 5). The results corroborated analyzes of the M-H and B-C similarities presented above: O1 and O2 shared almost identical number of OTUs as A1 and A2. Surprisingly, in spite of differences in the physicochemical features and the plant vegetation between sampling sites, many of the OTUs were shared by all five samples, particularly at the 0.10 clustering level (37 and 68 at 0.03 and 0.10 level, respectively). The percentage of sequences belonging to these OTUs reached 12 % at the 0.03 level and almost 60 % at the 0.10 level.

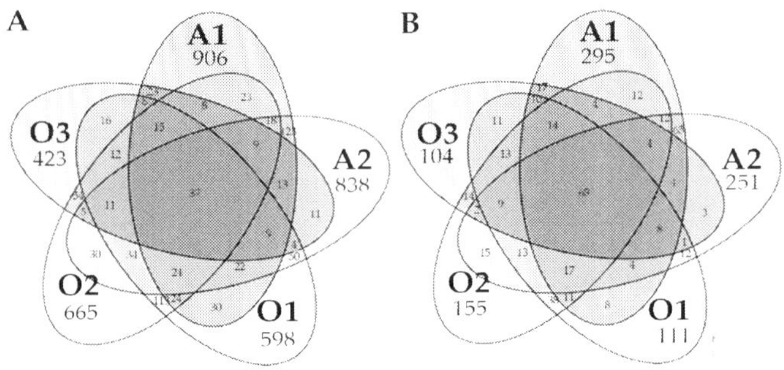

Figure 5: Venn diagrams of shared OTUs. a 0.03 dissimilarity level, b 0.10 dissimilarity level.

The statistical significance of differences in community structure were assessed via the parsimony and weighted UniFrac tests. Both methods showed that all the pairs of samples differed significantly ($p < 0.001$).

Identification of OTUs Shared by all Samples—"Core Microbiome"

To learn which taxons were shared, consensus taxonomy was constructed for each OTU and the ones present in all the samples were identified with an in-house developed Perl script.

At the 0.03 level, the shared OTUs belonged mainly to Alphaproteobacteria and Acidobacteria, with KD4-96 (Chloroflexi), Actinobacteria, and Betaproteobacteria being less frequent. The largest shared OTUs were Otu5473 (an uncultured member of Xanthobacteraceae), Otu5214 (*Sphingomonas*), and Otu4909 (*Candidatus Solibacter*).

At the 0.10 level, the picture was similar; however, members of the Bacteroidetes phylum (order Sphingomonadales) were more numerous than Chloroflexi. When a definition of core microbiome given in "Methods" was used, at the 0.03 level, there were four "core" OTUs (Table 7), two belonging to Alphaproteobacteria (genera *Sphingomonas* and *Holospora*), one to Acidobacteria (unclassified Acidobacteriaceae) and one to Chloroflexi (members of KD4-96). At the higher level, there were 21 core OTUs (Table 8), coming mostly from Sphingobacteria (mainly *Flexibacter* and members of Chitinophagaceae), Alphaproteobacteria (members of Sphingomonadaceae and Rhodospirillales), KD4-96 (phylum Chloroflexi), and Betaproteobacteria (members of Nitrosomonadaceae, Comamonadaceae, TRA3-20 and SC-I-84).

Table 7: Consensus taxonomies for 0.03 level core microbiome OTUs

	A1	A2	O1	O2	O3	E[a]	Total	Taxonomy
Otu4707	4	3	4	6	6	0,979	23	Proteobacteria; Alphaproteobacteria; Rickettsiales; Holosporaceae; Holospora;
Otu5214	13	26	24	36	41	0,960	140	Proteobacteria; Alphaproteobacteria; Sphingomonadales; Sphingomonadaceae; Sphingomonas;
Otu3722	11	3	12	4	9	0,925	39	Chloroflexi;KD4-96;
Otu4868	6	17	18	15	3	0,903	59	Acidobacteria; Acidobacteria; Acidobacteriales; Acidobacteriaceae;

The numbers of reads in each sample and the total number are reported

[a]Shannon evenness showing how evenly the reads were distributed among samples

Table 8: Consensus taxonomies for 0.10 level core microbiome OTUs

	A1	A2	O1	O2	O3	E[a]	Total	Taxonomy
Otu1237	4	8	9	9	8	0.979	38	WCHB1-60;
Otu1558	15	6	8	7	6	0.956	42	Proteobacteria; Gammaproteobacteria; Legionellales; Coxiellaceae; Aquicella;
Otu1464	9	21	3	16	12	0.912	61	Proteobacteria; Deltaproteobacteria; Myxococcales; Nannocystineae; Haliangiaceae; Haliangium;
Otu1410	11	10	9	7	4	0.969	41	Proteobacteria; Deltaproteobacteria;GR-WP33-30;uncultured;
Otu1537	21	3	13	20	14	0.921	71	Proteobacteria; Betaproteobacteria; TRA3-20;
Otu1534	11	22	7	10	15	0.953	65	Proteobacteria; Betaproteobacteria;SC-I-84;
Otu1572	41	21	23	25	53	0.957	163	Proteobacteria; Betaproteobacteria; Nitrosomonadales; Nitrosomonadaceae; uncultured;
Otu1489	11	19	18	4	6	0.909	58	Proteobacteria; Betaproteobacteria; Burkholderiales; Comamonadaceae;
Otu1428	37	46	36	74	48	0.977	241	Proteobacteria; Alphaproteobacteria; Sphingomonadales; Sphingomonadaceae;
Otu1248	10	5	3	5	13	0.919	36	Proteobacteria; Alphaproteobacteria; Rickettsiales; Holosporaceae; Holospora;
Otu1434	35	38	45	36	8	0.938	162	Proteobacteria; Alphaproteobacteria; Rhodospirillales;
Otu1574	7	18	13	26	13	0.948	77	Proteobacteria; Alphaproteobacteria; Rhodospirillales;
Otu1524	12	23	19	19	3	0.916	76	Proteobacteria; Alphaproteobacteria; Rhizobiales; Methylocystaceae; uncultured;
Otu1557	10	13	18	17	3	0.926	61	Proteobacteria; Alphaproteobacteria; Caulobacterales; Caulobacteraceae;
Otu1531	7	4	4	4	3	0.973	22	Chloroflexi;TK10;
Otu1522	28	19	11	14	6	0.928	78	Chloroflexi;S085;
Otu1594	101	56	45	59	55	0.973	316	Chloroflexi;KD4-96;

Otu1504	20	20	10	20	25	0.977	95	Candidate_division_TM7;
Otu1589	74	11	96	61	58	0.915	300	Bacteroidetes; Sphingobacteria; Sphingobacteriales; Cytophagaceae; Flexibacter;
Otu1573	23	79	79	140	40	0.902	361	Bacteroidetes; Sphingobacteria; Sphingobacteriales; Chitinophagaceae;
Otu1470	42	109	90	72	54	0.966	367	Acidobacteria; Acidobacteria; Acidobacteriales; Acidobacteriaceae; Candidatus_Solibacter;

The numbers of reads in each sample and total number are reported

[a]Shannon's evenness showing how evenly the reads were distributed among samples

DISCUSSION

In spite of the high level of soils contamination with Pb and Zn, the species richness, diversity, and evenness of bacterial communities under study were similar or slightly lower than those found in studies of unpolluted soils (e.g., [16, 17]). Hur and coworkers [18] found much lower numbers of OTUs in Cd, Pb, and Zn polluted soils (less than 300 observed OTUs at the 0.03 clustering level); however, the soils they analyzed were much more contaminated with Cd (>300 mg kg^{-1}) and less with Zn. In spite of the small number of samples, we found a significant negative correlation between bioavailable Zn concentration and all measures of species richness, Shannon's diversity as well as evenness, which was also similar to Hur's results [18]. Interestingly, Berg and colleagues [48] found that a long term Cu exposure changes the bacterial community structure, but not its diversity and species richness, demonstrating that various metals influence the communities differently. Similar observations were made by Chodak et al. [20].

Other metals did not correlate with the diversity variables, which might have been caused by the low concentration (cadmium) or the relatively low toxicity (chromium and lead). We did not measure Cr(VI) level, which is the most toxic species of this metal, but the lack of correlation with diversity and richness might indicate that the concentration of chromates was low in our samples. Sheik and coworkers [20] found that Cr(VI) significantly impacted both the diversity and the community structure, lowering the former as well as the species richness.

Interestingly, the plant species richness did not seem to correlate with the bacterial diversity and richness. However, it might be hypothesized that specific plant species influenced these parameters. Which species did that and the kind of the influence remains to be elucidated.

The species richness and the diversity was higher in the Alwernia soils than in the Olkusz ones. The reason for this might be the soil texture, as there are evidences that higher clay content promotes bacterial diversity [49, 50], or OM and Cr, which seem to be positively correlated in our samples.

The samples were taken in a way to avoid sampling rhizosphere communities (depth 5–30 cm, roots removed with soil sticking to them); however, in the case of the Olkusz ones, it was more difficult (sandy soils, rhizosphere not sticking to roots), which could have caused certain amount of the rhizospheric soil to contaminate the samples. Certain reports [51] showed that the rhizosphere harbored less species than the bulk soil. It is also possible that plants that were common at the Olkusz sites (e.g., *Pinus sylvestris*) but absent from the Alwernia ones favored specific communities with lower diversity. There are papers showing that plant species together with the soil type determine the shape of the microbial community in the rhizosphere [52, 53].

The bacterial communities at the level of phyla were similar in all our samples, in spite of the distance separating the sampling sites and the differences in other parameters such as the plant cover and the contamination. They were generally similar to those found in other soils analyzed by the pyrosequencing-based methods, even though different primer systems were used. Proteobacteria were found to be the dominant phylum both in grassland (e.g., [16, 17, 18]) and forest soils [16], just as in the Olkusz and Alwernia samples. Other abundant phyla were Actinobacteria and Acidobacteria followed by Bacteroidetes and Firmicutes. The three first phyla were also present in high numbers in our samples; however, the last one was relatively rare. Thus, the level of phylum does not seem to be adequate for assessing differences in soil bacterial communities. The same, to a lesser extent, applies also to the class level. Lower levels should therefore be used if the effect of heavy metals is to be demonstrated. In our case, both the levels chosen (0.03—roughly species, 0.10—family) allowed the demonstration of Zn influence on bacterial communities. It seems that the lower the level, the more specific effects can be seen, but on the other hand going below the 0.03 dissimilarity is difficult due to the sequencing noise, therefore constructing the OTUs at the 0.03 level is now the best choice. For large studies, it might be that higher levels would be more appropriate, as the variability at the strain or species level may be too high.

As we were interested in defining the core microbiome of our samples, we decided to use the level of species and families (0.03 and 0.10 OTUs, respectively). We chose to assign evenly distributed OTUs with over two reads in each sample to the core microbiome to lower the probability that rare sequences that were present in one or more samples purely by chance (e.g., due to the erroneous reading of a MID sequence) would be taken as shared. Core microbiomes were to date determined for various parts of human body in the Human Microbiome Project (e.g., [46]) and outside of HMP [54]; however, to our knowledge, this is the first attempt to define the core microbiome of soil samples. Our samples' core microbiome appeared to be fairly large and even at the 0.03 level, a single OTU (Otu5214, *Sphingomonas*) constituted over 0.5 % of total reads number in each sample. At the level of 0.10 dissimilarity, there were six such OTUs, some of them present at the level over 1.71 %. In the case of human body core microbiomes, they were impossible to define at low phylogenetic levels and no abundant OTUs (over 0.5 % of reads from each sample) were universally shared [54]. As our samples were diverse in terms of physicochemical properties except for the Zn contamination, it seems that the core OTUs represent bacteria that are adapted to heavy metals, albeit various ones (Zn and Pb in Olkusz samples, while Zn and Cr in Alwernia). It would be interesting to see if more geographically distant samples also share so many phylotypes at low phylogenetic level and if the core genes set could be defined. The latter question could be addressed by deep sequencing of metagenomic DNA, assembly, and analysis of genes encoded on the assembled contigs.

In conclusion, our work showed that OTUs constructed at the 0.03 and 0.10 levels allow assessing the influence of heavy metals on both bacterial diversity and community structure. Zn turned out to be the factor that had the greatest impact on these parameters, other were less important; in particular, no correlation between plant diversity and bacterial diversity was observed. We defined our samples core microbiome comprising *Holospora* and *Sphingomonas*(Alphaproteobacteria) as well as KD4-96 (Chloroflexi) and an unclassified member of Acidobacteriaceae at the species (0.03) level and Cytophagaceae, Chitinophagaceae, Acidobacteriaceae, KD4-96, Sphingomonadaceae, and Nitrosomonadaceae at the family level (0.10).

ACKNOWLEDGMENTS

This work was supported by the Nicolaus Copernicus University (grant no. 2009/312-B to MG) and Polish Ministry of Science and Higher Education (grant no. N N302 094834 to AT, ED-S, and MG, additional support was provided by grants AODP/09/2010/0 and N304 03832/1819 to BW). We are grateful to Dr. Marcin Chodak for helpful discussions and to the anonymous reviewers for their help in improving the manuscript.

REFERENCES

1. Baek KH, Yoon BD, Kim BH, Cho DH, Lee IS et al (2007) Monitoring of microbial diversity and activity during bioremediation of crude oil-contaminated soil with different treatments. J Microbiol Biotechnol 17:67–73
2. Frostegård Å, Tunlid A, Bååth E (1993) Phospholipid fatty acid composition, biomass, and activity of microbial communities from two soil types experimentally exposed to different heavy metals. Appl Environ Microbiol 59:3605–3617
3. Sandåå R-A, Torsvik V, Enger Ø (2001) Influence of long term heavy-metal contamination on microbial communities in soil. Soil Biol Biochem 33:287–295
4. Ellis RJ, Morgan P, Weightman AJ, Fry JC (2003) Cultivation-dependent and -independent approaches for determining bacterial diversity in heavy-metal-contaminated soil. Appl Environ Microbiol 69:3223–3230
5. Lorenz N, Hintemann T, Kramarewa T, Katayama A, Yasuta T et al (2006) Response of microbial activity and microbial community composition in soils to long-term arsenic and cadmium exposure. Soil Biol Biochem 38:1430–1437
6. Müller AK, Westergaard K, Christensen S, Sørensen SJ (2001)

The effect of long-term mercury pollution on the soil microbial community. FEMS Microbiol Ecol 36:11–19

7. Gans J, Wolinsky M, Dunbar J (2005) Computational improvements reveal great bacterial diversity and high metal toxicity in soil. Science 309:1387–1390
8. Hugenholtz P, Goebel BM, Pace NR (1998) Impact of culture-independent studies on the emerging phylogenetic view of bacterial diversity. J Bacteriol 180:4765–4774
9. Handelsman J (2004) Metagenomics: application of genomics to uncultured microorganisms. Microbiol Mol Biol Rev 68:669–685
10. Bakken LR, Olsen RA (1987) The relationship between cell-size and viability of soil bacteria. Microb Ecol 13:103–114
11. Bakken LR (1997) Culturable and nonculturable bacteria in soil. In: van Elsas JD, Trevors JT, Wellington EMH (eds) Modern soil microbiology. Marcel Dekker, New York, pp 47–61
12. Weisburg WG, Barns SM, Pelletier DA, Lane DJ (1991) 16S ribosomal DNA amplification for phylogenetic study. J Bacteriol 173:697–703
13. Scholtz HC, Al Dahouk S, Tomaso H, Neubauer H, Witte A et al (2008) Genetic diversity and phylogenetic relationships of bacteria belonging to the *Ochrobactrum–Brucella* group by *recA* and 16S rRNA gene-based comparative sequence analysis. Syst Appl Microbiol 32:1–16
14. Roesch LFW, Fulthorpe RR, Riva A, Casella G, Hadwin AKM et al (2007) Pyrosequencing enumerates and contrasts soil microbial diversity. ISME J 1:283–290
15. Fulthorpe RR, Roesch LFW, Riva A, Triplett EW (2008) Distantly sampled soils carry few species in common. ISME J 2:901–910
16. Nacke H, Thürmer A, Wollher A, Will C, Hodac L et al (2011) Pyrosequencing-based assessment of bacterial community structure along different management types in German forest and grassland soils. PLoS ONE 6:e17000
17. Will C, Thürmer A, Wollherr A, Nacke H, Herold N et al (2010) Horizon-specific bacterial community composition of German grassland soils, as revealed by pyrosequencing-based analysis of

16S rRNA genes. Appl Environ Microbiol 76:6751–6759
18. Hur M, Kim Y, Song H-R, Kim JM, Choi YI et al (2011) Effect of genetically modified poplars on soil microbial communities during the phytoremediation of waste mine tailings. Appl Environ Microbiol 77:7611–7619
19. Chodak M, Gołębiewski M, Morawska-Płoskonka J, Kuduk K, Niklińska M (2013) Diversity of microorganisms from forest soils differently polluted with heavy metals. Appl Soil Ecol 64:7–14
20. Sheik CS, Mitchell TW, Rizvi FZ, Rehman Y, Faisal M et al (2012) Exposure of soil microbial communities to chromium and arsenic alters their diversity and structure. PLoS ONE 7(6):e40059. doi:10.1371/journal.pone.0040059
21. Schloss PD, Westcott SL, Ryabin T, Hall JR, Hartmann M et al (2009) Introducing mothur: open-source, platform-independent, community-supported software for describing and comparing microbial communities. Appl Environ Microbiol 75:7537–7541
22. Dzwonko Z (2007) Guide to phytosociological analysis. Vademecum Geobotanicum. Sorus, Institute of Botany Jagiellonian University, Krakow-Poznan
23. Lindsay WL, Norvell WA (1969) Development of a DTPA micronutrient soil test. Agron Abstr 84
24. Quast C, Pruesse E, Yilmaz P, Gerken J, Schweer T et al (2013) The SILVA ribosomal RNA gene database project: improved data processing and web-based tools. Nucl Acids Res 41:D590–D596
25. Cole JR, Wang Q, Cardenas E, Fish J, Chai B et al (2009) The Ribosomal Database Project: improved alignments and new tools for rRNA analysis. Nucl Acids Res 37:D141–D145
26. Klindworth A, Pruesse E, Schweer T, Peplies J, Quast C et al (2013) Evaluation of general 16S ribosomal RNA gene PCR primers for classical and next-generation sequencing-based diversity studies. Nucl Acids Res 41:e1
27. Zhou J, Bruns MA, Tiedje JM (1996) DNA recovery from soils of diverse composition. Appl Environ Microbiol 62:316–322
28. Deja-Sikora E (2012) Search for bacterial cadmium, zinc, lead, copper and chromium resistance genes in metagenome of soils

polluted with heavy metals. PhD thesis, Nicolaus Copernicus University, Torun, Poland

29. Neefs J-M, Van De Peer Y, Hendriks L, De Wachter R (1990) Compilation of small ribosomal subunit RNA sequences. Nucleic Acids Res 18:2237–2317
30. Lane DJ (1991) 16S/23S rRNA sequencing. In: Stackebrandt E, Goodfellow MD (eds) Nucleic acid techniques in bacterial systematics. John Wiley and Sons, New York, pp 115–175
31. R Development Core Team (2011) R: A language and environment for statistical computing. R Foundation for Statistical Computing, Vienna, Austria. ISBN 3-900051-07-0, URL http://www.R-project.org/
32. Quince C, Lanzen A, Davenport RJ, Turnbaugh PJ (2011) Removing noise from pyrosequenced amplicons. BMC Bionformatics 12:38
33. Edgar RC, Haas BJ, Clemente JC, Quince C, Knight R (2011) UCHIME improves sensitivity and speed of chimera detection. Bioinformatics 27:2194–2200
34. Huse S, Huber J, Morrison H, Sogin M, Welch MD (2007) Accuracy and quality of massively parallel DNA pyrosequencing. Genome Biol 8:R143
35. Chao A (1984) Nonparametric estimation of the number of classes in a community. Scand J Stat 11:265–270
36. Chao A, Lee S-M (1992) Estimating the number of classes via sample coverage. J Am Stat Assoc 87:210–217
37. Bunge J (2011) Estimating the number of species with catchall. Pac Symp Biocomput 2011:121–130
38. Faith DP, Lozupone CA, Nipperess D, Knight R (2009) The cladistic basis for the phylogenetic diversity (PD) measure links evolutionary features to environmental gradient and supports broad applications of microbial ecology's "Phylogenetic Beta Diversity" framework. Int J Mol Sci 10:4723–4741
39. Horn H (1966) Measuring of "overlap" in comparative ecological studies. Am Nat 100:419–424
40. Bray JR, Curtis JT (1957) An ordination of upland forest communities of southern Wisconsin. Ecol Monogr 27:325–349

41. Schloss PD, Handelsman J (2006) Introducing TreeClimber, a test to compare microbial community structure. Appl Environ Microbiol 72:2379–2384
42. Lozupone CA, Hamady M, Kelley ST, Knight R (2007) Quantitative and qualitative beta diversity measures lead to different insights into factors that structure microbial communities. Appl Environ Microbiol 73:1576–1585
43. Sheneman L, Evans J, Foster JA (2006) Clearcut: a fast implementation of relaxed neighbor joining. Bioinformatics 22:2823–2824
44. Oksanen J, Blanchet FG, Kindt R, Legendre P, Minchin PR et al (2013) Vegan: Community ecology package. R package version 2.0-7. http://CRAN.R-project.org/package=vegan
45. Harrell FE et al (2013) Hmisc: Harrell miscellaneous. R package version 3.10-1.1. http://CRAN.R-project.org/package=Hmisc
46. Turnbaugh PJ, Ley RE, Hamady M, Fraser-Liggett CM, Knight R et al (2007) The human microbiome project. Nature 449:804–810
47. Cabała J (2009) Heavy metals in ground soil environment of Olkusz area of Zn-Pb ore exploitation. University of Silesia Press, Katowice
48. Berg J, Brandt KK, Al-Soud WA, Holm PE, Hansen LH et al (2012) Selection for Cu-tolerant bacterial communities with altered composition, but unaltered richness, via long-term Cu exposure. Appl Environ Microbiol 78:7438–7446
49. Chau JF, Bagtzoglou AC, Wilig MR (2011) The effect of soil texture on richness and diversity of bacterial communities. Environ Forensics 12:333–341
50. Carson JK, Gonzalez-Quinones V, Murphy DV, Hinz C, Shaw JA et al (2010) Low pore connectivity increases bacterial diversity in soil. Appl Environ Microbiol 76:3936–3942
51. Uroz S, Buee M, Murat C, Frey-Klett P, Martin F (2010) Pyrosequencing reveals a contrasted bacterial diversity between oak rhizosphere and surrounding soil. Environ Microbiol Rep 2:281–288
52. Marschner P, Yang C-H, Lieberei R, Crowley DE (2001) Soil and plant specific effects on bacterial community composition in the rhizosphere. Soil Biol Biochem 33:1437–1445

53. Ladygina N, Hedlund K (2010) Plant species influence microbial diversity and carbon allocation in the rhizosphere. Soil Biol Biochem 42:162–168

54. Ravel J, Gajer P, Abdo Z, Schneider GM, Koenig SS et al (2011) Vaginal microbiome of reproductive-age women. PNAS 108(S1):4680–4687

CHAPTER 13

MICROBIAL COMMUNITIES ASSOCIATED WITH THE ANTHROPOGENIC, HIGHLY ALKALINE ENVIRONMENT OF A SALINE SODA LIME, POLAND

Agnieszka Kalwasińska [1], Tamás Felföldi[3], Attila Szabó[3], Edyta Deja-Sikora[1], Przemysław Kosobucki[2], and Maciej Walczak[1]

[1]Department of Environmental Microbiology and Biotechnology, Faculty of Biology and Environmental Protection, Nicolaus Copernicus University, Torun´, Poland

[2]Chair of Environmental Chemistry and Bioanalytics, Faculty of Chemistry, Nicolaus Copernicus University, Torun´, Poland

[3]Department of Microbiology, Eo¨tvo¨s Lora´nd University, Budapest, Hungary

ABSTRACT

Soda lime is a by-product of the Solvay soda process for the production of sodium carbonate from limestone and sodium chloride. Due to a high salt concentration and alkaline pH, the lime is considered as a potential habitat of haloalkaliphilic and haloalkalitolerant microbial communities. This artificial and unique environment is nutrient-poor and devoid of vegetation, due in part to semi-arid, saline and alkaline conditions. Samples taken from the surface layer of the lime and from the depth of 2 m (both having pH ~11 and EC_e up to 423 dS m^{-1}) were investigated using culture-based (culturing on alkaline medium) and culture-independent microbiological approaches (microscopic analyses and pyrosequencing). A surprisingly diverse bacterial community was discovered in this highly saline, alkaline and nutrient-poor environment, with the bacterial phyla Proteobacteria (representing 52.8% of the total bacterial community) and Firmicutes (16.6%) showing dominance. Compared to the surface layer, higher bacterial abundance and diversity values were detected in the deep zone, where more stable environmental conditions may occur. The surface layer was dominated by members of the genera *Phenylobacterium*, *Chelativorans* and *Skermanella*, while in the interior layer the genus *Fictibacillus* was dominant. The culturable aerobic, haloalkaliphilic bacteria strains isolated in this study belonged mostly to the genus *Bacillus* and were closely related to the species *Bacillus pseudofirmus*, *B. cereus*, *B. plakortidis*, *B. thuringensis* and *B. pumilus*.

Keywords: Alkaliphiles, Halophiles, Saline soda (Solvay) lime, 16S rRNA gene pyrosequencing

INTRODUCTION

Saline soda lime is a by-product of the Solvay soda process for the production of sodium carbonate from limestone and sodium chloride. The multistage Solvay process comprises: (I) sodium bicarbonate precipitation from brine saturated with ammonia and carbon dioxide, (II) conversion of sodium bicarbonate into the final product–sodium carbonate, and (III) regeneration of ammonia from ammonium chloride solution. Alkaline soda lime is generated as a by-product of the third stage of process. It has the form of a thick sludge and consists mainly of calcium carbonate (constitutes over 90% of the lime), amorphous calcium hydroxide (represents around 6%), and sodium and calcium chlorides (Ziółkowska et al. 2013). The remaining constituents include silicon dioxide and magnesium and aluminium salts. Since soda production started in Janikowo (Poland) in 1957, the alkaline lime has been deposited in the repository ponds called the "white seas" (Fig. 1). This unique, artificial environment is a potential habitat of haloalkaliphilic and haloalkalitolerant microorganisms due to high pH (~11) and high salt concentration (up to 423 dS m^{-1}). It can be expected that these environmental conditions favour the selection of microorganisms capable of growing both in the presence of high salt concentrations and in conditions with highly alkaline pH (Ma et al. 2010).

Figure 1: The surface of the saline soda lime deposited in the repository in Janikowo.

Microbiomes of the natural saline-alkaline environments have been thoroughly investigated in recent years (Qazi 2013). However, studies have been focused mainly on microorganisms of soda lakes (Duckworth et al. 2000; Pikuta et al. 2003; Sorokin et al. 2011; Borsodi et al. 2008), while the microbial communities inhabiting natural and man-made soda (solonchak) soils and deserts have not been thoroughly studied (Sorokin et al. 2008; Pan et al. 2010). Microorganisms of unique and underexplored terrestrial saline and alkaline habitats require more consideration as knowledge of them is really limited. They live under adverse and unstable environmental conditions. Apart from extremely high concentrations of salt and alkaline pH, the microbes are exposed to seasonal or constant drought, excessively high or low temperatures, and oligotrophy (Pan et al. 2010). It is to be expected that haloalkaliphilic taxa predominating in dry or semi-arid land habitats would be similar to those previously found in soda lakes (Sorokin et al. 2008); however, there is still little evidence for complex microbial communities functioning under such harsh conditions.

Saline soda lime (an artificial, man-made environment) was chosen by us as a model of a nutrient-poor, semi-arid (annual precipitation less than 500 mm, Kędziora 2011), saline-alkaline habitat, resembling soda soils in such parameters like alkalinity, solidity, dryness and low nutrient content. The surface of the lime is permanently devoid of vegetation. It loses water seasonally due to strong evaporation and becomes hard and cracked (Fig. 1).

The aim of our study was to determine the abundance, biodiversity and community composition of bacteria inhabiting saline soda lime and to compare the communities of this unique and extreme environment with the microbiota inhabiting natural saline and alkaline environments. Two layers of the lime, from the surface and from a depth of two meters, corresponding to environments 1 and 10 years old, respectively, and differing in physicochemical properties were sampled and analysed.

MATERIALS AND METHODS

Study Area

Saline soda lime has been deposited in Janikowo (Kuyavia, Central Poland; 52°46′45.0″N 18°07′30.0″E) since 1957. The repository of the lime covers an area of 200 ha, has the capacity of over 13,000,000 Mg and is composed of several ponds, which are separated by causeways. This area rises up to 16 m above the ground level (Regional Inspectorate for Environmental Protection 2013). Most of the ponds contain desalted lime (with chlorides concentration around 30 g kg^{-1}), which is used in agriculture for de-acidification of soil. The particular pond from which samples were taken contains increased concentration of salts due to the deposition of lime sludge after cleaning a technological line used in wet and dry vacuum salt production.

Saline Soda Lime Samples

Samples of the saline soda lime were collected in September 2013 from the surface layer of the deposit (depth of 0–20 cm) and from a depth of 1.8–2.0 m using a telescopic Edelman mud auger (GeomorTechnik, Szczecin, Poland). The alkaline lime, a product of chemical processes, is more homogenous than soil. Therefore only ten samples were taken: five from the surface layer and five samples from the deep layer. The surface layer was the most recent, compared to the bottom layer, and the age of the two layers was estimated at 1 and 10 years, respectively. For each sampled layer, five separate subsamples were taken in the corners and in the centre of a 2.5 m square using a hand coring device (4 cm in diameter; Pepper and Gerba 2014). Samples (50 cm^3) were placed in sterile glass jars and they were transported to the laboratory in a portable ice bag at 4 °C. In the laboratory, subsamples (10 g) from a particular layer were mixed thoroughly for subsequent microbial and physicochemical tests. All analyses started within 4 h after sample collection.

Chemical Analyses of Soda Lime

As a pre-treatment, solid and dried samples were hammered to reduce agglomeration and then homogenized in a roller mill. Gravimetric moisture of the saline soda lime was determined by the method of mass loss in drying. To avoid the known problem of imprecise pH measurement due to high salinity, the pH of the samples was determined by the alkacymetric titration of the aqueous lime suspensions (using 0.1:100 lime: solvent ratio, w/v). The suspensions were titrated until the indicator dye was stable (Ziółkowska et al. 2013). Electrical conductivity (EC) in extracts of the water-saturated lime paste was measured by the conductometric method. The concentration of chloride ions was determined by the argentometric titration with silver nitrate. Calcium and sodium ion concentrations were analysed with BWB XP (Halstead, England) flame photometer. Total organic carbon (TOC) was measured in a TOC 5000 analyzer combined with a SSM-5000A module Shimadzu (Kyoto, Japan). TOC values were calculated by subtracting inorganic carbon concentration (IC) from total carbon concentration (TC). Organic nitrogen concentration was calculated by subtracting ammonia nitrogen from total Kjeldahl nitrogen (TKN). TKN was determined in a Turbotherm system (Gerhardt, Konigswinter, Germany). Ammonia nitrogen concentration was measured using the Nessler method (Spectroquant Merck SQ 118, Darmstadt, Germany) after distillation in a Vapodest 20 system (Gerhardt, Konigswinter, Germany). Nitrate concentration was determined using the spectrophotometric method (Spectroquant Merck SQ 118, Darmstadt, Germany) after extracting nitrate from the sample with 2.0 M KCl. Total phosphorus content was measured with the spectrophotometric molybdenum blue method (Spectroquant Merck SQ 118, Darmstadt, Germany) after sample digestion with aqua regia. All analyses were performed in triplicate; the mean values with standard deviations (SD) are given in Table 1.

Table 1: Chemical properties of the saline soda lime

	Surface	Interior
pH	10.81 ± 0.10	10.84 ± 0.06
EC_e (dS m^{-1})	378 ± 20.20	423 ± 13.00
Chlorides (g kg^{-1})	108.64 ± 3.10	122.02 ± 2.36
Sodium (g kg^{-1})	56.22 ± 3.31	63.21 ± 2.84
Calcium (g kg^{-1})	23.97 ± 0.25	24.28 ± 0.18
Water content (%)	27.80 ± 1.39	68.85 ± 4.68
Total organic carbon (g kg^{-1})	2.94 ± 0.96	3.23 ± 0.82
Organic nitrogen (mg kg^{-1})	<LOD	32.97 ± 5.24
Nitrates (mg kg^{-1})	<LOD	<LOD
Ammonia nitrogen (mg kg^{-1})	<LOD	3.9 ± 0.20
Total phosphorus (mg kg^{-1})	29.63 ± 6.32	27.63 ± 5.98

The values are given as means ± SD; for organic nitrogen and ammonia nitrogen LOD (limit of detection) <0.01 mg kg^{-1}, for nitrates LOD <0.001 mg kg^{-1}

Isolation and Enumeration of Bacteria

To isolate culturable bacteria and to estimate their number, 10 g of lime sample was added to 90 ml of NaCl solution (a concentration corresponding to the chloride content in a given sample). Mixtures were blended thoroughly, tenfold serial dilutions were prepared, and 0.1 ml of each suspension was spread onto three replicate plates. An isolation medium consisted of the lime extract (100 g dissolved in 1 l of distilled water, mixed for 20 min, filtered and adjusted to pH 10 using the buffer solution (Na_2CO_3–K_2HPO_4) according to Ntougias et al. (2006), enriched with glucose (0.5 g l^{-1}) and yeast extract (0.2 g l^{-1}), and solidified with agar (1.5%). The cultivation was performed in three parallel repetitions.

All plates were incubated up to 21 days at 25° C. The number of bacteria was expressed as colony forming units (CFU) per g dry weight of the saline soda lime.

Isolates differing in appearance, colour, size and shape of bacterial colonies on the medium were chosen for subsequent taxonomic identification. Colonies were subcultured in a medium containing glucose (5 g l^{-1}), yeast extract (1 g l^{-1}), $MgSO_4$ anhydrous (0.01 g l^{-1}), and NaCl (60 g l^{-1}), supplied with the buffer as given above. Then glycerol stocks of isolates were prepared and stored at −80 °C.

The total number of bacteria in alkaline lime samples was determined by the acridine orange direct count technique. 1 g of the lime (fixed with formaldehyde in 3.8% v/v final concentration, sterilized with MF-Millipore Membrane Filters, 0.22 μm) was placed into glass tubes containing 9 ml of filter-sterilized citric acid solution (final concentration 0.6 M). Citric acid was added in order to dissolve calcium carbonate present in the samples. For each suspension, three subsamples were stained with acridine orange (final concentration 0.01%) for 3 min and then the slides were immediately viewed under a microscope (Nicon Eclipse E200, Japan).

Identification of Bacterial Strains based on the 16S rRNA Gene

Total genomic DNA was extracted from the strains using the Gene MATRIX Bacterial & Yeast Genomic DNA Purification Kit (EURx, Gdańsk, Poland) following the instructions given by the manufacturer. The 16S rRNA gene was amplified by the PCR in a reaction mixture containing 1 U *Taq*DNA polymerase (EURx), 0.2 mM dNTP mixture (EURx), 1× Polbuffer B with 1.5 mM $MgCl_2$ (EURx), 0.25 μM of universal primers 27F (5′-AGA GTT TGA TCM TGG CTC AG-3′, (Lane 1991)) and 1492R (5′-TAC GGY TAC CTT GTT ACG ACT T-3′ (Polz and Cavanaugh 1998)), and 1 μl of genomic DNA in a total volume of 20 μl. The thermal profile consisted of an initial denaturation at 95 °C for 3 min, 35 amplification cycles (denaturation at 95 °C for 30 s, annealing at 53 °C for 20 s and extension at 72 °C for 1 min 40 s) and a final extension at 72 °C for 5 min. PCR amplicons were checked on a 1 w/v % agarose gel stained with Midori Green DNA Stain (Nippon Genetics Europe GmbH, Dueren, Germany). Sequencing reactions of PCR products were performed with primer 27F, using a Big Dye Terminator v 3.1 Cycle Sequencing Kit (Applied Biosystems, Thermo Fisher Scientific, Waltham, MA, USA) following the instructions given by the manufacturer. Capillary electrophoresis was performed by the Oligo sequencing laboratory (IBB PAS, Warsaw, Poland). Manual correction of automatic base calling on chromatograms was carried out using MEGA 6.0 software (Tamura et al. 2013). Nucleotide sequences of the bacterial strains obtained in this study were deposited to GenBank under the accession numbers KT387570-KT387597.

The taxonomic affiliation of bacterial strains was determined using the EzTaxon online tool (Kim et al. 2012). Additional closest relatives of isolates were identified with BLASTN and these sequences were also used in the phylogenetic analysis. Sequence alignment was performed with the SINA Aligner (Pruesse et al. 2012). Phylogenetic analysis including the search for the best-fit model was conducted with the MEGA 6.0 software (Tamura et al. 2013).

Pyrosequencing Analysis of the Total Bacterial Community

Environmental genomic DNA was isolated from samples homogenized as follows: 10 g of sample was added to 90 ml citric acid (final concentration 0.6 M) in order to dissolve calcium carbonate, the suspension was centrifuged (Rotina, Hettich, Germany 4000×g, 20 C, 10 min), and washed twice with pH 8 phosphate buffer.

DNA extraction, PCR amplification, pyrosequencing and subsequent sequence analysis was performed as generally described by Felföldi et al. (2015). Bacteria-specific primers S-D-Bact-0341-b-S-17 and S-D-Bact-0785-a-A-21 (Klindworth et al. 2013) were applied for the triplicate amplification of the 16S rRNA gene using 0.4 U Phusion High Fidelity DNA polymerase (Thermo Fisher Scientific), 1× Phusion HF Buffer (Thermo Fisher Scientific), 0.2 mM dNTP (Fermentas, Vilnius, Lithuania), 0.4 µg/µl BSA (Fermentas), 0.5 µM of both primers and 1.0 µl of genomic DNA in a final volume of 20 µl, applying the thermal profile of initial denaturation at 98 °C for 5 min, 25 cycles (with denaturation at 95 °C for 40 s, annealing at 55 °C for 2 min and extension at 72 °C for 1 min) and final extension at 72 °C for 10 min. Sequencing was performed with the GS Junior platform (Roche, Basel, Switzerland). Data were processed with mothur v 1.33 (Schloss et al. 2009); briefly, sequences were quality filtered, denoised, singletons and putative chimeras were removed; and subsequent taxon identification was performed with the SINA aligner tool (Pruesse et al. 2012) using the ARB-SILVA SSU NR reference database–SILVA Release 119 (Quast et al. 2013). After the removal of archaeal, chloroplast and mitochondrial sequences from the dataset, operational taxonomic units (OTUs) were assigned at 97% similarity level, representing bacterial species. The number of obtained sequences was normalized to the sample having the lowest sequence count prior to the calculation of richness estimators and diversity indices in mothur. Association networks of major bacterial OTUs were constructed with Cytoscape 3.2.1 (Shannon et al. 2003) based on the 50 OTUs having the highest abundance in the total sequencing dataset.

Since plastid 16S rRNA gene references are underrepresented in the ARB-SILVA database, chloroplast sequences were analysed separately: the closest related sequences were retrieved from GenBank using BLASTN, and following an alignment with SINA, the subsequent phylogenetic analysis was performed with MEGA 6.0 (Tamura et al. 2013).

The NCBI Sequence Read Archive accession codes for the unique sequences obtained with pyrosequencing are SRS1279117 and SRS1279119.

RESULTS

Lime Characteristics

The physicochemical characteristics of the saline soda lime samples are presented in Table 1. The surface layer of alkaline lime and the layer from the depth of 2.0 m had similar alkaline pH, measured in a water suspension. The EC in saturated lime extracts was very high, up to 423 dS m^{-1}. The water content was significantly higher in the deep layer of the lime compared to the surface layer. The samples contained a low concentration of nitrogen and organic carbon. Ammonia nitrogen was only detected at a depth of 2.0 m, and nitrates were not detected. In contrast to this, total phosphorus concentrations were rather high. The interior samples of the lime (Fig. 2b) were more compact compared to the surface layer (Fig. 2a).

Figure 2: Physical structure of the saline soda lime a less compact surface layer, b more compact internal layer, c, d *arrows* point putative bacterial cells, covered with salts).

Total Number of Bacteria and Number of Culturable Bacteria

The total number of bacteria determined by direct cell counting was much lower in the surface layer than in the lime at a depth of 2.0 m (Table 2). A similar trend was observed in the number of culturable bacteria. The number of colony forming units was one order of magnitude lower in the surface layer as compared with the deep layer. Both the direct count of bacteria and the number of culturable bacteria in the surface and the interior layer of saline-alkaline lime, differed significantly ($p < 0.01$).

Table 2: Abundance of microorganisms in the saline soda lime

	Surface	Interior
Culturable bacteria (CFUg^{-1} dry mass, 10^3)	1.44 ± 0.12	34.82 ± 3.30
Total number of bacteria (cells g^{-1} dry mass, 10^6)	7.52 ± 2.03	245.20 ± 95.55

The values are given as means (n = 3) ± SD

Community structure

In total, 10,702 high-quality 16S rRNA gene sequence reads obtained by pyrosequencing were clustered at the 97% similarity level, resulting in 422 operational taxonomic units (corresponding to the bacterial species; Table 3), out of which 82 OTUs (19.4%) were shared among the samples. Compared to the deep layer, the rarefaction curve of the surface layer reached a higher level of saturation. The two datasets shared only a 0.177 Bray-Curtis similarity based on the abundance of the bacterial OTUs, which showed that the bacterial communities in the two layers were highly dissimilar.

Table 3: Pyrosequencing data statistics of the saline soda lime samples

Sample	Total	Surface	Interior
Number of sequences after quality filtration	10702	4263	6439
Bacterial sequences	6894	2150	4744
Archaeal sequences	1217	70	1147
Chloroplast sequences	2471	1959	512
Mitochondrial sequences	120	84	36
Number of bacterial OTUs (97% cut off)[a]	422	190	321
Richness estimators[b]			
Chao 1	–	190 (190; 190)	385 (360; 427)
ACE	–	190 (190; 190)	414 (384; 459)
Diversity indices[b]			
Shannon-Wiener	–	3.97 (3.90; 4.04)	4.37 (4.29; 4.46)
Inverse Simpson's (1/D)	–	18.7 (17.0; 20.7)	15.1 (13.4; 17.4)

[a] The number of OTU found in the subset of sequences normalized to the sample with the lowest sequence count

[b] Numbers in parentheses are 95% confidence intervals

The applied 16S rRNA gene amplification protocol was designed for the domain Bacteria. Most of the obtained reads (62%) were assigned to Bacteria (including chloroplast and mitochondrial sequences, which are incorporated into the phylogenetic tree of bacteria due to the endosymbiotic origin of these organelles), however 9.7% of all reads belonged to Archaea. Most archaeal reads were identified in the interior sample (17.8% of total reads) compared to the surface (1.6%), indicating indirectly that Archaea were relatively numerous in the deep layer of the lime. The Archaeal community was dominated by members of genera "*Candidatus Halobonum*" and *Halorubrum*, however it should be noted that the revealed structure could be highly distorted since the applied primers were specific for the domain Bacteria (Klindworth et al. 2013). The estimated number of bacterial species was also much higher in the interior sample, based on both Chao 1 and the Abundance-based Coverage Estimator (ACE) species richness estimators. This value was less than half in the surface sample compared to the deep layer sample. Bacterial diversity was also higher in the deep layer.

An analysis of the distribution of the most abundant bacterial phyla revealed that the saline soda lime samples were dominated by Proteobacteria, Firmicutes, Bacteroidetes and Actinobacteria (all having >5% contribution to total bacterial reads, excluding chloroplast and mitochondrial 16S rRNA gene sequences; Fig. 3). Proteobacteria was the most abundant phylum in the surface layer of the lime (65.2%) followed by Acidobacteria (7.8%) and Bacteroidetes (7.7%). In the interior sample, 40.3% of the total bacterial sequences belonged to phylum Proteobacteria, followed by Firmicutes (31.9%), Bacteroidetes (10.1%) and Actinobacteria (7.2%). Within the Proteobacteria phylum, the most abundant class was Alphaproteobacteria in the case of the surface sample (representing 85.4% of proteobacterial sequences), while in the deep layer sample, Alphaproteobacteria (46.8%) and Gammaproteobacteria (45.2%) were the most abundant classes.

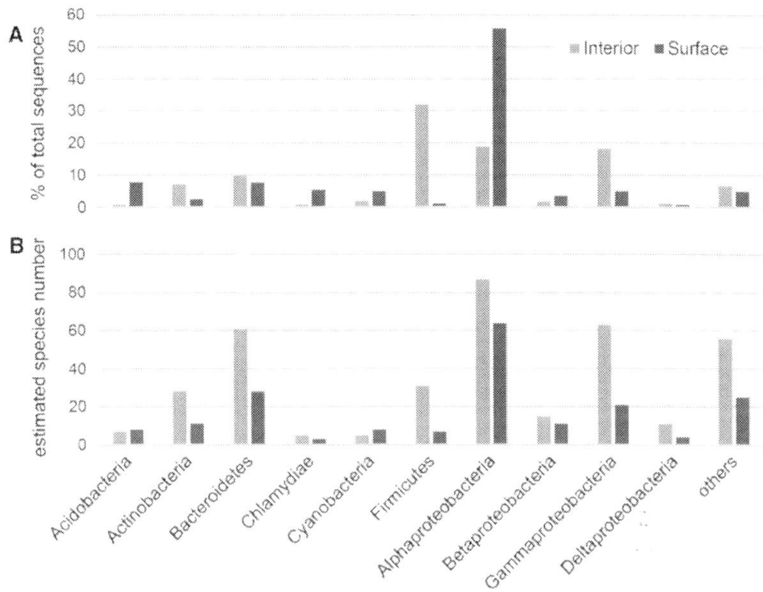

Figure 3: Distribution of bacterial sequences retrieved by pyrosequencing from the interior and the surface saline soda lime samples. Contribution of taxa to the total bacterial community is given based on their relative abundance (a) and the distribution of species (b). Phylum-level distribution (except for Proteobacteria, where class-level distribution is shown) was expressed as a percentage of total sequences and as estimated species number based on the number of observed OTUs (assigned at 97% similarity level) per group. Species numbers were calculated assuming 97% sequence similarity threshold based on the 16S rRNA gene. Only taxa having contribution higher than 5% at least in one sample are shown.

Differences in overall community composition are presented as an association network of major bacterial OTUs (Fig. 4). OTUs belonging to *Phenylobacterium*, *Skermanella*, *Bryobacter*, *Simkania*, *Salinibacter*, *Psychromonas*, *Halomonas*, *Synechococcus*, *Haloferula* and *Euhalothece* were detected in both layers of the alkaline lime, contributing in different measure to the total bacterial community. On the other hand, the major OTU related to the genus *Chelativorans* was detected exclusively in the surface sample, while *Fictibacillus* and other bacilli (*Bacillus* OTU1-OTU3) were present only in the deep layers of the lime.

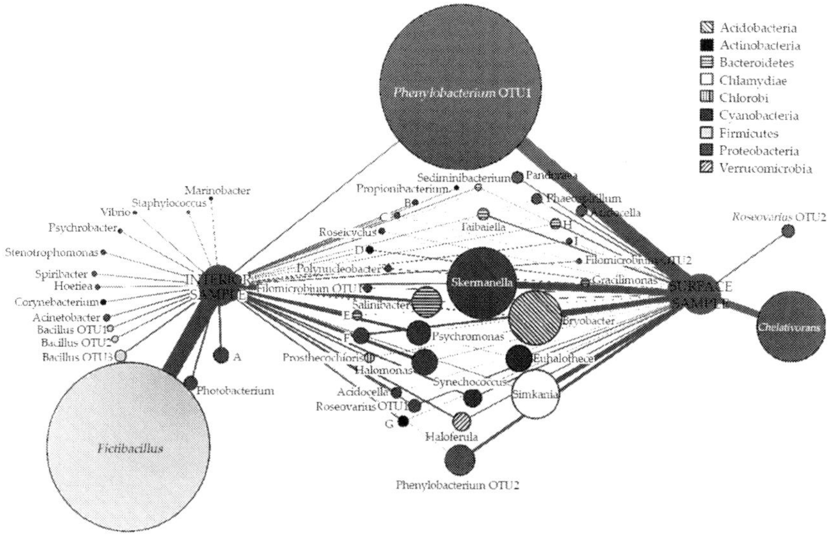

Figure 4: Association network of major bacterial OTUs in the saline soda lime samples based on pyrosequencing data. OTUs were calculated assuming 97% species-level sequence similarity threshold based on the 16S rRNA gene. Top 50 most abundant bacterial OTUs are shown as circles with size corresponding to relative abundance in the total dataset, while the width of the edges corresponds with the relative abundance of an OTU in a particular sample. Circles are color-coded according to phylogenetic affiliation and genus-level identity of each OTU is also shown. OTUs distantly related to cultured bacteria were marked with letters as follows: a uncultured Rhodobacteraceae, b unclassified DB1-14 Alphaproteobacteria, c uncultured Rhodobacteraceae, d unclassified Elev-16S-976 Actinobacteria, e unclassified E6aC02 Sphingobacteriales, f unclassified Rhizobiales alphal cluster, g unclassified Microbacteriaceae, h unclassified ML602 M-17 Bacteriodetes, i unclassified NKB5 Gammaproteobacteria.

A significant portion of the sequences constituted chloroplasts of algae in both samples (27%, on average). They were more abundant in the surface layer of the saline-alkaline lime (46% of total reads) than in the interior sample (8%). Chloroplast sequences formed three major clades in the phylogenetic tree presented in Fig. 5. The most abundant were reads related to the genus *Ochromonas* (Chrysophyceae, Stramenopiles;

representing 68.9% of the total number of chloroplast reads), while the others were related to the genera *Mesotaenium* (Zygnemophyceae, Streptophyta; 14.8%) and *Picochlorum* (Trebouxiophyceae, Chlorophyta; 13.4%). Other detected genera included *Dunaliella* (Chlorophyceae, Chlorophyta), *Chaetoceros*(Bacillariophyta) and *Euglena* (Euglenozoa).

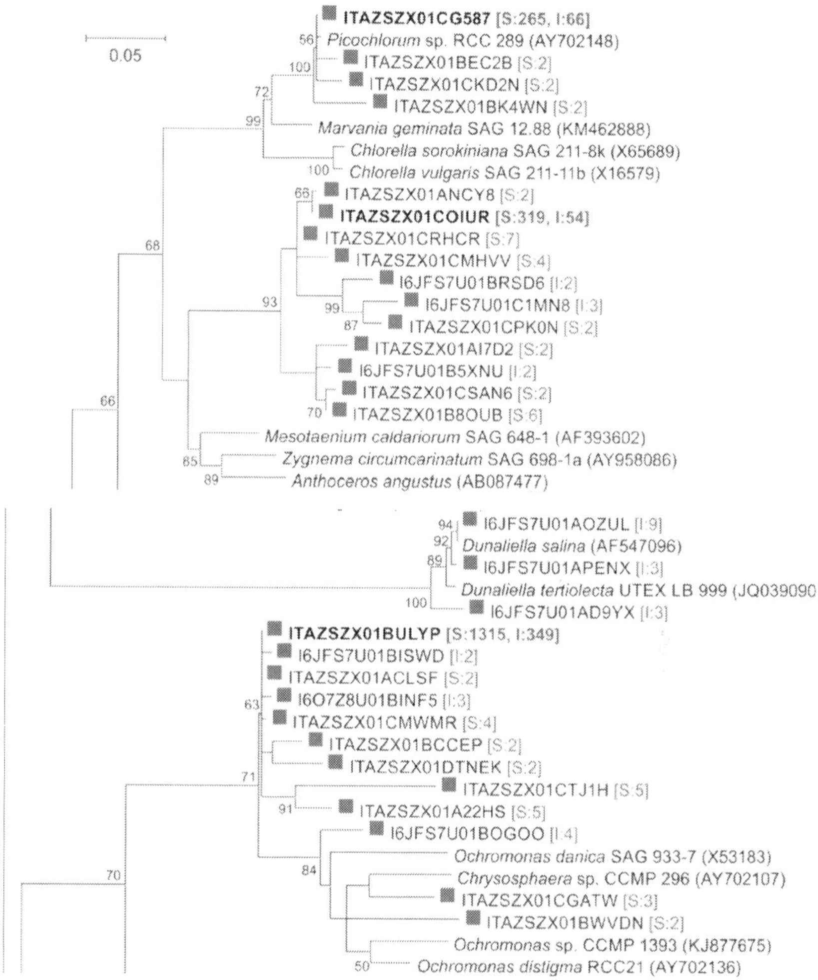

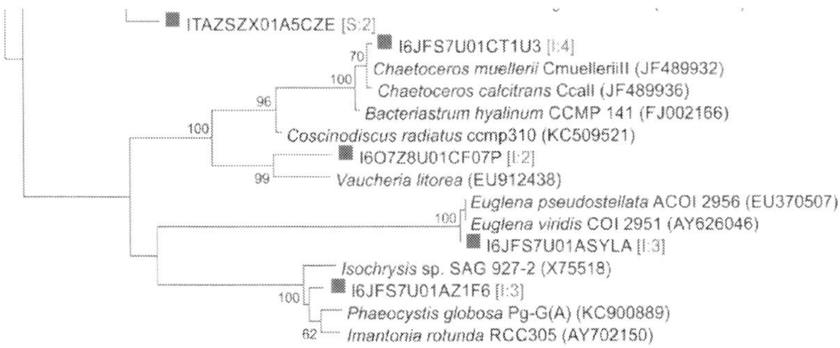

Figure 5: Phylogenetic tree of chloroplast genotypes obtained from the saline soda lime samples based on the pyrosequencing of the 16S rRNA gene. Tree was constructed using the Maximum Likelihood method with the General Time Reversible nucleotide substitution model and is based on 334 nucleotide positions. GenBank accession numbers are given in parentheses. Representative sequences obtained by pyrosequencing are marked with grey squares. In this case, sequences differing only in one position are represented with a single sequence, number of sequences is given in square brackets (I interior and S surface sample) and major phylotypes representing at least 0.5% of total chloroplast sequences are highlighted with bold letters.

Bacterial Isolates

Eleven isolates were retrieved from the surface (marked MS, strains MS1–MS11) and seventeen from the interior sample (marked MI, strains MI1–MI17). All of the bacterial isolates stained Gram-positive. Molecular taxonomic identification of the isolates (Fig. 6) revealed a very low diversity level of culturable bacteria. The strains belonged to three genera: *Bacillus, Micrococcus* (one strain), and *Nocardiopsis* (one strain). The closest relative of 18 strains (MS1–7, MI1–9, MI13, MI14) was *Bacillus* sp. A73, first isolated from a hypersaline lake in Iran (Amoozegar et al. 2014). Strain MI12 was identified as alkalitolerant *Bacillus plakortidis*, which was first isolated from a sponge in the Norwegian Sea (Borchert et al. 2007). The closest relative of isolate MI17 was *Bacillus* sp. ANL-isoa2 and its isolation source was a solonchak soda soil in the Kulunda Steppe, Mongolia (Sorokin et al. 2007). The closest relative of isolate MI10 was a moderately halotolerant and obligately alkaliphilic *Bacillus pseudofirmus* strain H13, isolated from the lime and described previously (Kalwasińska et al. 2015). *Bacillus cereus* was the closest relative of strain MI11, and *Bacillus thuringiensis* was the closest relative of strain MI15; however, it is not possible to distinguish between these two species on the basis of gene sequences that are not related to the production of entomopathogenic toxins (Sorokin et al. 2006). Isolate MI16 was identified as *Bacillus pumilus*, another alkalitolerant and common soil species. The closest relative of strain MS8 was *Nocardiopsis xinjangensis*, a halophilic actinomycete isolated from a saline soil in China (Li et al. 2003) and the closest relative of strain MS9 was *Micrococcus luteus*, which is ubiquitous in nature.

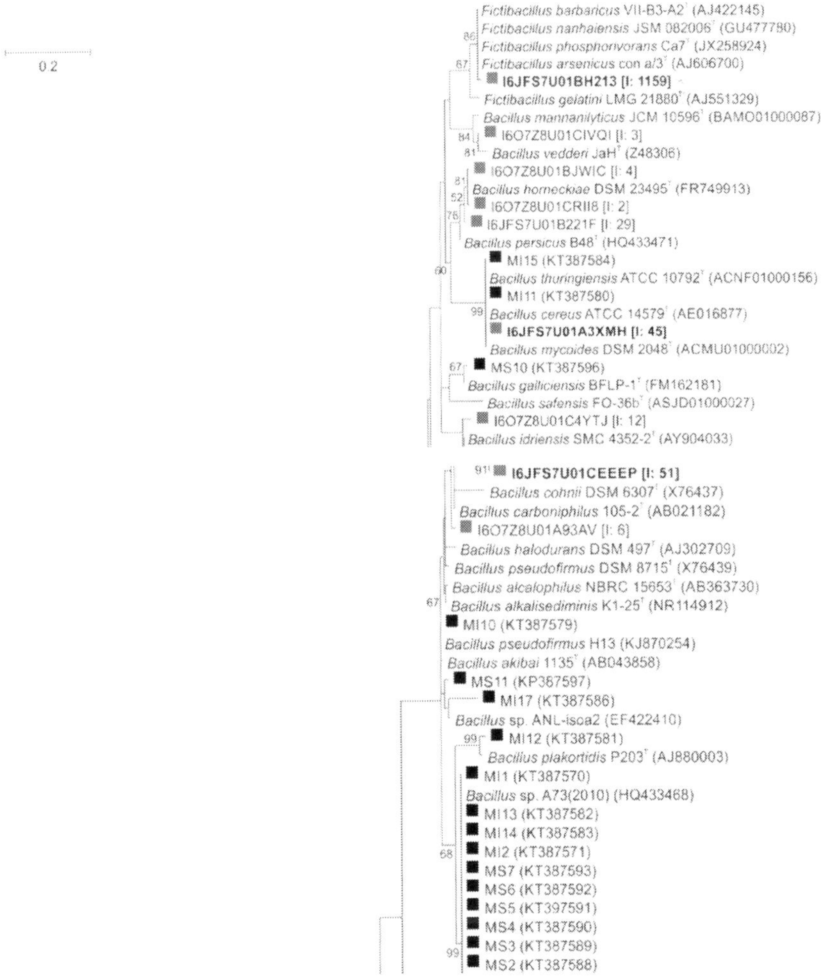

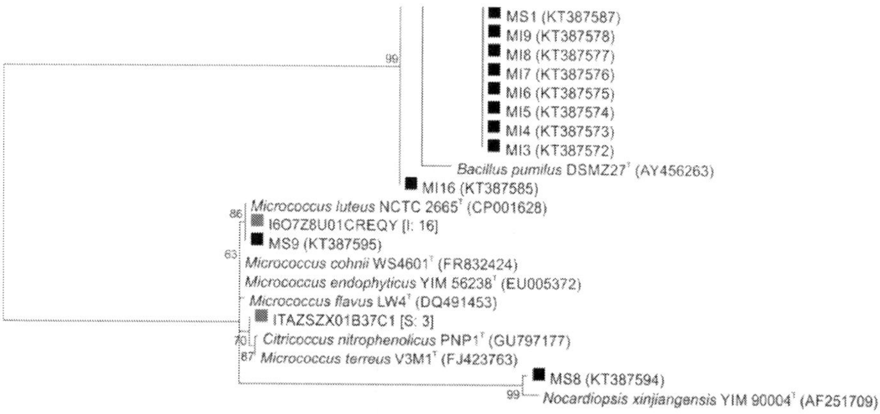

Figure 6: Phylogenetic tree of the saline soda lime bacterial isolates and closely related genotypes from pyrosequencing data based on the 16S rRNA gene. Tree was constructed using the Maximum Likelihood method with the Kimura 2-Parameter nucleotide substitution model and is based on 356 nucleotide positions. Type strains are marked with superscript T, GenBank accession numbers are given in parentheses. Bacterial isolates are marked with black squares, while representative sequences obtained by pyrosequencing are marked with grey squares. In the latter case, sequences differing only in one position are represented with a single sequence, number of sequences is given in square brackets (I, interior and S, surface sample) and major phylotypes representing at least 0.5% of total bacterial sequences are highlighted with bold letters.

DISCUSSION

The saline soda lime is an unusual environment created by humans as a result of the chemical processes of soda production. This extreme environment, consisting almost entirely of calcium carbonate, is highly alkaline, saline and nutrient-poor, especially in terms of nitrogen and organic carbon content. Most of the organic matter present in this artificial environment originates from allochtonous airborne material and atmospheric deposition on the lime surface. A comparison of the bacterial diversity observed in this study with that of other studies is difficult because of the lack of papers describing similar environments. The high alkalinity caused by the presence of carbonates, bicarbonates and hydroxides, as well as the high concentration of salts, are thought to be the key factors that shape the structure of microbial communities in this environment. The influence of pH and salinity on bacterial and archaeal community composition in various ecosystems was extensively investigated by other researchers (Liu et al. 2015; Wu et al. 2017; Maturrano et al. 2006; Zhang et al. 2013; Canfora et al. 2014). These studies have definitively shown that pH and salinity are the most important factors controlling microbial abundance, metabolic activity, and diversity.

The total number of bacteria in the lime (10^6–10^8 cells g^{-1}) was low compared to dolomites and soil, where it ranged from 10^8 to 10^9 cells g^{-1} (Stres 2007; Sheibani et al. 2013). This may be connected with the very harsh physicochemical conditions mentioned above. In general, a more extreme environment is expected to maintain a lower abundance of microorganisms by selecting a small group of highly specialized organisms able to survive and grow under such demanding conditions (Rothschild and Mancinelli 2001). On the other hand, the observed bacterial counts were quite similar to those obtained for highly alkaline and saline Mono Lake (Humayoun et al. 2003) or cave environments made of limestone (Barton and Jurado 2007; Wu et al. 2015).

The higher quantity of culturable bacteria and the total cell counts, as well as the higher community species richness and the diversity detected in the interior, compared to the surface layer of the saline soda lime, may be related to the more stable environmental conditions existing in the deeper layer of the lime. The flat, vegetation-free, semisolid surface layer of the lime is subjected to periodic fluctuations in moisture and temperature, and as a consequence, in salinity and pH. Due to the high evaporation rate and low precipitation rate (below 500 mm $year^{-1}$) in the Kuyavia region, Central Poland (Kędziora 2011), the studied area is similar to cold, semi-arid habitats. Microorganisms in this environment must adapt to changing environmental conditions such as long dry periods during late spring and summer. Desiccation cracking of the surface layer of the lime can substantially reduce microbial activity and growth (Pesaro et al. 2004) and change the microbial community structure (Thion and Prosser 2014). Furthermore, microbial biomass and activity in the lime may decrease with the increasing number of drying and rewetting cycles, as is observed in soils (Mikha et al. 2005; Wu and Brookes 2005).

Due to the high alkalinity connected with the presence of $Ca(OH)_2$, the saline soda lime increases the pH of aqueous solutions up to 11–12 within a few minutes (Jeliński et al. 2011). The repeated shifts between moderate and highly alkaline conditions, as well as the presence of drying and rewetting cycles, make the surface of the lime more changeable and selective for microorganisms than the deeper layers of the lime. A similar increase in diversity in the interior layer, compared to the surface, was observed in other extreme, alkaline environments like submarine ikaite columns (made of calcium carbonate hexahydrate, low salinity) located in Greenland (Glaring et al. 2015) and in alkaline-saline Mono Lake, California, and Soap Lake, Washington, when going down from the light-exposed, oxic surface waters to the anoxic bottom (Humayoun et al. 2003; Dimitriu et al. 2008).

The distribution of the most abundant bacterial phyla in the alkaline, saline soda lime was similar to that observed for submarine ikaite columns (Glaring et al. 2015) and saline desert Kutch (Pandit et al. 2014) and slightly different than that found in soda lakes, however that may be due to the different methodology applied. Most of the previous studies of solonchaks or soda lakes have been based on the construction of 16S rDNA clone libraries and subsequent sequencing of individual 16S rDNA clones. Since only limited amounts of clones were sequenced, the complexity of the intrinsic community could not be revealed. Pyrosequencing technology provides a large number of sequence reads in a single run, resulting in deep datasets and allowing for the detection not only the most abundant microbial community members but also the low abundance taxa (Pershina et al. 2012; Dudhagara et al. 2015). Nevertheless, it should be mentioned that pyrosequencing of 16S rRNA amplicons does not imply that the taxa found were actively growing in the saline soda lime, since the detected DNA could be derived from dormant or dead microorganisms.

The most abundant heterotrophic bacterial OTU present in the saline soda lime belonged to the genus *Phenylobacterium* (Alphaproteobacteria) and it was dominant in the surface layer. Members of this genus are strict aerobes or facultative anaerobes and have been isolated from various environments, including highly alkaline, oligotrophic groundwater (Tiago et al. 2005). However, thus far there has been no evidence of their presence in other caustic, saline, or lime habitats. Some members of the genus *Phenylobacterium*are xenobiotic compound-degrading bacteria, and some of them can reduce nitrate to nitrite (Oh and

Roh 2012). Others have a unique adaptation that allows them to thrive in oligotrophic environments (Abraham et al. 2008). This last feature of the representatives of the genus *Phenylobacterium* is certainly important in terms of their abundance in the nutrient-poor lime under study.

Autochthonous microorganisms can survive in extreme environment owing to their ability to form spores. The second most abundant bacterial OTU in the lime belonged to the genus *Fictibacillus* (phylum Firmicutes), which contains species recently reclassified from the genus *Bacillus* (Glaeser et al. 2013). Representatives of the phylum Firmicutes were also relatively abundant in metagenomes from saline and alkaline desert Kutch, India (Pandit et al. 2014), a saline and alkaline Mojave Desert playa lake (Navarro et al. 2009), and sediments of hyper saline and hyper alkaline Lonar Lake (Paul et al. 2016) and much less abundant in a limestone cave environment (Wu et al. 2015). One of the *Fictibacillus* representatives, halotolerant and alkaliphilic *Fictibacillus solisalsi*, was isolated from saline soil (Liu et al. 2009) resembling soda lime in terms of salinity and alkalinity. Surprisingly, in the soda lime, *Fictibacillus* was detected only in the interior samples. Nevertheless, alkaliphilic spore-forming Bacilli show a worldwide distribution and are likely to be spread widely by the wind and migratory birds (Jones and Grant 2000).

Representatives of *Chelativorans* (order Rhizobiales) and *Skermanella* (order Rhodospirillales) within the class Alphaproteobacteria were also among the abundant OTUs found in the surface layer of the lime. These Gram-negative, strictly aerobic bacteria were previously isolated from various environments, including nutrient-poor environments (air, desert soil, and coal mine soil (Weon et al. 2007; An et al. 2009; Luo et al. 2012). However there is no evidence of their presence in alkaline environments like soda lakes or limestone.

Members of phylum Acidobacteria were also present in the saline soda lime, especially in its surface layer. This is not surprising since Acidobacteria is the second most abundant taxon in some soils (Janssen 2006) and its representatives may be easily spread by wind. This group of bacteria is known to be more abundant in environments with low carbon availability (Fierer et al. 2007) and many members of Acidobacteria have been found to be abundant in alkaline soils (Dunbar et al. 2002; Rousk et al. 2010). The detected genus *Bryobacter* includes a type strain that was first isolated from Sphagnum-dominated acidic wetlands (Kulichevskaya et al. 2010), suggesting the presence of micro-habitats in the lime matrix that allow organisms sensitive to high pH to survive in this alkaline environment.

Members of the genus *Halomonas* (Gammaproteobacteria), encompassing moderately halophilic and halotolerant bacteria (Wang et al. 2014), were detected in the saline, alkaline soda lime, especially in the interior layer. Their presence in the lime is fully justified, as they show widespread distribution in saline and also alkaline environments like solonchaks and soda lakes (Shapovalova et al. 2009; Duckworth et al. 2000).

The most surprising is the presence of the genus *Simkania* (phylum Chlamydiae) in the surface layer of the highly alkaline and saline lime. Its representatives are obligate intracellular pathogens (Michel et al. 2005), however it is difficult to identify their host in this extremely harsh environment.

Hypersaline environments are generally dominated by halophilic Archaea, which belong to the order Halobacteriales (Ventosa and Nieto 1995). Since the applied PCR primers were designed to amplify members of the domain Bacteria, archaeal sequences obtained by pyrosequencing may present a highly distorted community structure. However, archaeal reads belonging to "*Candidatus Halobonum*" and *Halorubrum* (Halobacteriaceae) were still conspicuously present in the saline, alkaline lime deposit, especially in its deep layer.

Apart from Archaea, the bacterial genus *Salinibacter*, an extremely halophilic taxon within the Bacteroidetes with archaeal-like properties was detected in the saline, alkaline soda lime, especially in the interior layer. *Salinibacter* shares many features with the family Halobacteriaceae: both groups use KCl for osmotic adjustment of their cytoplasm, both mainly possess salt-requiring enzymes with a large excess of acidic amino acids, and both contain different retinal pigments (Oren 2013).

Primary producers present in the alkaline, saline lime were both prokaryotic cyanobacteria and eukaryotic algae. Cyanobacteria are among the most alkaliphilic microbes, and they frequently dominate alkaline environments, such as soda lakes and microbial mats (Somogy et al. 2010; Dadheech et al. 2013). They also constitute an important group of phototrophs in saline desert soils (Li et al. 2013; Pandtit et al. 2014). Therefore, their presence in the alkaline and saline surface layer of the soda lime can be expected. In particular, *Euhalothece* and *Synechococcus* phylotypes, detected in the lime, are among the most prominent inhabitants of hypersaline and alkaline soda environments (Garcia-Pichel et al. 1998; Feloföldi et al. 2011).

Most of the chloroplast sequences in the surface layer of the lime were distantly related (<93% pairwise sequence similarity) to the genera *Ochromonas* (Chrysophyceae) and *Mesotaenium* (*Zygnemophyceae*). Therefore, eukaryotic algal phylotypes detected in the studied samples may represent novel phylotypes, and most probably belong to genera which have yet to be described. Picoeukaryotic and phototrophic members of the genus *Picochlorum*, which carry sequences that were found in the saline soda lime, were also reported in various saline environments, both marine (Henley et al. 2004) and continental (Máthé et al. 2014). This is probably due to the broad halotolerance of some *Picochlorum* representatives (e.g. *P. oklahomense*, Henley et al. 2002). Plastid sequences related to *Dunaliella* were also found in the lime and this is consistent with common distribution of this genus in saline environments, such as the Dead Sea, Israel, solar salterns (Oren 2010) and the Great Salt Plains, USA (Buchheim et al. 2010).

The number of culturable, aerobic bacteria grown on an alkaline medium (10^3–10^4 CFU g^{-1} dry mas) was within the range observed for weathered limestone (Mulec et al. 2001), soils (Horikoshi 1991), soda lakes (Joshi et al. 2008) and salterns (Maturrano et al. 2006). However, it was lower than in alkaline alpeorujo (10^5 CFU g^{-1}), an organic sludge-like by-product of olive oil extraction (Ntougias et al. 2006) probably due to the lower amount of nutrients present in the lime and its high salinity. The low content of organic carbon and nitrogen are factors significantly limiting the growth of any heterotrophic community of microbes.

Most of the bacteria isolated from the alkaline and saline lime were alkaliphiles and moderate halophiles. Molecular phylogenetic identification of the strains revealed that culturable diversity was very low compared to other alkaline and saline environments (Antony et al. 2013; Grant and Jones 2016) since members of only three different genera, namely *Bacillus*, *Nocardiopsis,* and *Micrococcus*, were retrieved. A previous study of this environment, focusing on alkaline lime ponds of lower salinity, revealed the presence of other culturable genera: *Halomonas*, *Planococcus*, and *Microcella* (Kalwasińska et al. 2015). Nevertheless, the observed diversity of isolates from alkaline lime was still low and may reflect an inability to create the optimal growth conditions for microorganisms derived from an extreme habitat in the laboratory. On the other hand, the observations performed during the present investigation remain consistent with those of other researchers. Reported species, especially *Bacillus*, *Halomonas*, and *Micrococcus* are described to be alkali- and halotolerant and they exhibit a broad environmental distribution, therefore their common occurrence in alkaline and saline artificial lime is to be expected (Kamekura 2013; Grant and Jones 2016).

CONCLUSIONS

The data presented shows that surprisingly diverse, distinct bacterial communities are present in the highly alkaline, saline and nutrient-poor environment of soda lime (Poland). The examined artificial environment maintains extremophilic bacterial taxa adapted to thrive in this unusual ecosystem. Most of the detected bacterial and plastid sequences were related to microorganisms found in various alkaline and saline, nutrient-poor environments such as solonchaks, desert soils, limestone, soda lakes and ponds, oceans, saline springs and alkaline ground water or even air. The phylogenetic groups identified in the surface and in the interior layers of the lime were distinct. Higher bacterial abundance and diversity was detected in the deep zone of the lime than in the surface layer, possibly due to both geochemical and temporal stability differences between these two environments. Molecular phylogenetic identification of the isolates revealed a very low diversity of culturable bacteria, which belonged mainly to the genus *Bacillus*, which is widely distributed in different saline and alkaline habitats with high pH and a variable salt concentrations.

ACKNOWLEDGEMENT

This research was partially supported by the Grants KMOP-4.2.1/B-10-2011-0002 and TÁMOP-4.2.2/B-10/1-2010-0030 and by funds provided by Nicolaus Copernicus University (Toruń, Poland) to maintain research potential. T. F. was supported by the János Bolyai Research Scholarship of the Hungarian Academy of Sciences. The authors are grateful to Balázs Vajna for his useful comments.

REFERENCES

1. Abraham WR, Macedo AJ, Lunsdorf H, Fischer R, Pawelczyk S, Smit J, Vancanneyt M. Phylogeny by a polyphasic approach of the order Caulobacterales, proposal of *Caulobacter mirabilis* sp. nov., *Phenylobacterium haematophilum* sp. nov. and *Phenylobacterium conjunctum* sp. nov., and emendation of the genus *Phenylobacterium*. Int J Syst Evol Microbiol. 2008;58:1939–1949. doi: 10.1099/ijs.0.65567-0.
2. Amoozegar MA, Bagheri M, Makhdoumi-Kakhki A, Didari M, Schumann P, Nikou MM, Sanchez-Porro C, Ventosa A. *Aliicoccus persicus* gen. nov., sp. nov., a halophilic member of the Firmicutes isolated from a hypersaline lake. Int J Syst Evol Microbiol. 2014;64:1964–1969. doi: 10.1099/ijs.0.058545-0.
3. An H, Zhang L, Tang Y, Luo X, Sun T, Li Y, Wang Y, Dai J, Fang C. *Skermanella xinjiangensis* sp. nov., isolated from the desert of Xinjiang. China Int J Syst Evol Microbiol. 2009;59:1531–1534. doi: 10.1099/ijs.0.003350-0.
4. Antony CP, Kumaresan D, Hunger S, Drake HL, Murrell JC, Shouche YS. Microbiology of Lonar Lake and other soda lakes. ISME J. 2013;7:468–476. doi: 10.1038/ismej.2012.137.
5. Barton HA, Jurado V. What's up down there? microbial diversity in caves. Microbe. 2007;2:132–138.
6. Borchert MS, Nielsen P, Graeber I, Kaesler I, Szewzyk U, Pape T, Antranikian G, Schafer T. *Bacillus plakortidis* sp. nov. and *Bacillus murimartini* sp. nov., novel alkalitolerant members of rRNA group 6. Int J Syst Evol Microbiol. 2007;57:2888–2893. doi: 10.1099/ijs.0.65177-0.
7. Borsodi AK, Marialigeti K, Szabo G, Palatinszky M, Pollak B, Keki Z, Kovacs AL, Schumann P, Toth EM. *Bacillus aurantiacus* sp. nov., an alkaliphilic and moderately halophilic bacterium isolated from Hungarian soda lakes. Int J Syst Evol Microbiol. 2008;58:845–851. doi: 10.1099/ijs.0.65325-0.

8. Buchheim MA, Kirkwood AE, Buchheim JA, Verghese B, Henley WJ. Hypersaline soil supports a diverse community of *Dunaliella* (Chlorophyceae) J Phycol. 2010;46:1038–1047. doi: 10.1111/j.1529-8817.2010.00886.x.
9. Canfora L, Bacci G, Pinzari F, Lo Papa G, Dazzi C, Benedetti A. Salinity and bacterial diversity: to what extent does the concentration of salt affect the bacterial community in a saline soil? PLoS ONE. 2014;9:e106662. doi: 10.1371/journal.pone.0106662.
10. Dadheech PK, Glockner G, Casper P, Kotut K, Mazzoni CJ, Mbedi S, Krienitz L. Cyanobacterial diversity in the hot spring, pelagic and benthic habitats of a tropical soda lake. FEMS Microbiol Ecol. 2013;85:389–401. doi: 10.1111/1574-6941.12128.
11. Dimitriu PA, Pinkart HC, Peyton BM, Mormile MR. Spatial and temporal patterns in the microbial diversity of a meromictic soda lake in Washington State. Appl Environ Microbiol. 2008;74:4877–4888. doi: 10.1128/AEM.00455-08.
12. Duckworth AW, Grant WD, Jones BE, Meijer D, Marquez MC, Ventosa A. *Halomonas magadii* sp. nov., a new member of the genus *Halomonas*, isolated from a soda lake of the East African Rift Valley. Extremophiles. 2000;4:53–60.
13. Dudhagaraa P, Ghelanib A, Patelc R, Chaudharic R, Bhattc S. Bacterial tag encoded FLX titanium amplicon pyrosequencing (bTEFAP) based assessment of prokaryotic diversity in metagenome of Lonar sodalake, India. Genomics Data. 2015;4:8–11. doi: 10.1016/j.gdata.2015.01.010.
14. Dunbar J, Barns SM, Ticknor LO, Kuske CR. Empirical and theoretical bacterial diversity in four Arizona soils. Appl Environ Microbiol. 2002;68:3035–3045. doi: 10.1128/AEM.68.6.3035-3045.2002.
15. Felföldi T, Somogyi B, Márialigeti K, Vörös L. Notes on the biogeography of non-marine planktonic picocyanobacteria: re-evaluating novelty. J Plankton Res. 2011;33:1622–1626. doi: 10.1093/plankt/fbr051.
16. Felföldi T, Jurecska L, Vajna B, Barkács K, Makk J, Cebe G, Szabó A, Záray G, Márialigeti K. Texture and type of polymer fiber carrier determine bacterial colonization and biofilm properties

in wastewater treatment. Chem Eng J. 2015;264:824–834. doi: 10.1016/j.cej.2014.12.008.

17. Fierer N, Bradford MA, Jackson RB. Towards an ecological classification of soil bacteria. Ecology. 2007;88:1354–1364. doi: 10.1890/05-1839.
18. Garcia-Pichel F, Nubel U, Muyzer G. The phylogeny of unicellular, extremely halotolerant cyanobacteria. Arch Microbiol. 1998;169:469–482. doi: 10.1007/s002030050599.
19. Glaeser SP, Dott W, Busse HJ, Kampfer P. *Fictibacillus phosphorivorans* gen. nov., sp. nov. and proposal to reclassify *Bacillus arsenicus*, *Bacillus barbaricus*, *Bacillus macauensis*, *Bacillus nanhaiensis*, *Bacillus rigui*, *Bacillus solisalsi* and *Bacillus gelatini* in the genus *Fictibacillus*. Int J Syst Evol Microbiol. 2013;63:2934–2944. doi: 10.1099/ijs.0.049171-0.
20. Glaring MA, Vester JK, Lylloff JE, Al-Soud WA, Sorensen SJ, Stougaard P. Microbial diversity in a permanently cold and alkaline environment in Greenland. PLoS ONE. 2015;10:e0124863. doi: 10.1371/journal.pone.0124863.
21. Grant WD, Jones BE. Bacteria, archaea and viruses of Soda Lakes. In: Schagerl M, editor. Soda lakes of east Africa. Switzerland: Springer; 2016. pp. 97–149.
22. Henley WJ, Major KM, Hironaka JL. Response to salinity and heat stress in two halotolerant chlorophyte algae. J Phycol. 2002;38:757–766. doi: 10.1046/j.1529-8817.2002.01172.x.
23. Henley WJ, Hironaka JL, Guillou L, Buchheim MA, Buchheim JA, Fawley MW, Fawley KP. Phylogenetic analysis of the 'Nannochloris-like' algae and diagnoses of *Picochlorum oklahomensis*gen. et sp. nov. (Trebouxiophyceae, Chlorophyta) Phycologia. 2004;43:641–652. doi: 10.2216/i0031-8884-43-6-641.1.
24. Horikoshi K. Microorganisms in alkaline environments. Tokyo: Kodansha—VCH; 1991.
25. Humayoun SB, Bano N, Hollibaugh JT. Depth distribution of microbial diversity in Mono Lake, a meromictic soda lake in California. Appl Environ Microbiol. 2003;69:1030–1042. doi: 10.1128/AEM.69.2.1030-1042.2003.
26. Janssen PH. Identifying the dominant soil bacterial taxa in

libraries of 16S rRNA and 16S rRNA genes. Appl Environ Microbiol. 2006;72:1719–1728. doi: 10.1128/AEM.72.3.1719-1728.2006.

27. Jeliński T, Shyichuk O, Ziółkowska D. Alkalization properties of selected wastes from soda production by Solvay method. Arch Waste Manag. 2011;13:21–28.
28. Jones BE, Grant WD. Microbial diversity and ecology of alkaline environments. In: Seckbach J, editor. Journey to diverse microbial worlds: adaptation to exotic environments. Dordrecht: Springer; 2000. pp. 177–190.
29. Joshi AA, Kanekar PP, Kelkar AS, Shouche YS, Vani AA, Borgave SB, Sarnaik SS. Cultivable bacterial diversity of alkaline Lonar lake, India. Microb Ecol. 2008;55:163–172. doi: 10.1007/s00248-007-9264-8.
30. Kalwasińska A, Felföldi T, Walczak M, Kosobucki P. Physiology and molecular phylogeny of bacteria isolated from alkaline distillery lime. Pol J Microbiol. 2015;64:369–377. doi: 10.5604/17331331.1185236.
31. Kamekura M. Alkaliphilic microorganisms. In: Seckbach J, editor. Enigmatic microorganisms and life in extreme environments. Dordrecht: Springer Science+Buisness Media; 2013. pp. 479–487.
32. Kędziora A. Climatic conditions and water balance of the Kujawy Lakeland (in Polish) Rocz Glebozn—Soil Sci Ann. 2011;67:189–203.
33. Kim OS, Cho YJ, Lee K, Yoon SH, Kim M, Na H, Park SC, Jeon YS, Lee JH, Yi H, Won S, Chun J. Introducing EzTaxon-e: a prokaryotic 16S rRNA gene sequence database with phylotypes that represent uncultured species. Int J Syst Evol Microbiol. 2012;62:716–721. doi: 10.1099/ijs.0.038075-0.
34. Klindworth A, Pruesse E, Schweer T, Peplies J, Quast C, Horn M, Glockner FO. Evaluation of general 16S ribosomal RNA gene PCR primers for classical and next-generation sequencing-based diversity studies. Nucleic Acids Res. 2013;41:e1. doi: 10.1093/nar/gks808.
35. Kulichevskaya IS, Suzina NE, Liesack W, Dedysh SN. *Bryobacter*

aggregatus gen. nov., sp. nov., a peat-inhabiting, aerobic chemo-organotroph from subdivision 3 of the Acidobacteria. Int J Syst Evol Microbiol. 2010;60:301–306. doi: 10.1099/ijs.0.013250-0.

36. Lane D. 16S/23S rRNA sequencing. In: Stackebrandt E, Goodfellow M, editors. Nucleic acid techniques in bacterial systematics. New York: Wiley; 1991. pp. 115–175.
37. Li MG, Li WJ, Xu P, Cui XL, Xu LH, Jiang CL. *Nocardiopsis xinjiangensis* sp. nov., a halophilic actinomycete isolated from a saline soil sample in China. Int J Syst Evol Microbiol. 2003;53:317–321. doi: 10.1099/ijs.0.02055-0.
38. Li K, Liu R, Zhang H, Yun J. The diversity and abundance of bacteria and oxygenic phototrophs in saline biological desert crusts in Xinjiang, Northwest China. Microb Ecol. 2013;66:40–48. doi: 10.1007/s00248-012-0164-1.
39. Liu H, Zhou Y, Liu R, LaiR Zhang K-Y. *Bacillus solisalsi* sp. nov., a halotolerant, alkaliphilic bacterium isolated from soil around a salt lake. Int J Syst Evol Microbiol. 2009;59:1460–1464. doi: 10.1099/ijs.0.000653-0.
40. Liu S, Ren H, Shen L, Lou L, Tian G, Zheng P, Hu B. pH levels drive bacterial community structure in sediments of the Qiantang River as determined by 454 pyrosequencing. Front Microbiol. 2015;6:285.
41. Luo G, Shi Z, Wang H, Wang G. *Skermanella stibiiresistens* sp. nov., a highly antimony-resistant bacterium isolated from coal-mining soil, and emended description of the genus *Skermanella*. Int J Syst Evol Microbiol. 2012;62:1271–1276. doi: 10.1099/ijs.0.033746-0.
42. Ma Y, Galinski EA, Grant WD, Oren A, Ventosa A. Halophiles 2010: life in saline environments. Appl Environ Microbiol. 2010;76:6971–6981. doi: 10.1128/AEM.01868-10.
43. Máthé I, Borsodi AK, Tóth EM, Felföldi T, Jurecska L, Krett G, Kelemen Z, Elekes E, Barkács K, Márialigeti K. Vertical physico-chemical gradients with distinct microbial communities in the hypersaline and heliothermal Lake Ursu (Sovata, Romania) Extremophiles. 2014;18:501–514. doi: 10.1007/s00792-014-0633-1.
44. Maturrano L, Santos F, Rossello-Mora R, Anton J. Microbial diversity in Maras salterns, a hypersaline environment in the

Peruvian Andes. Appl Environ Microbiol. 2006;72:3887–3895. doi: 10.1128/AEM.02214-05.

45. Michel R, Muller K, Zoller L, Walochnik J, Hartmann M, Schmid E. Free-living amoebae serve as a host for the *Chlamydia*-like bacterium *Simkania negevensis*. Acta Protozool. 2005;44:113–121.
46. Mikha MM, Rice CW, Milliken GA. Carbon and nitrogen mineralization as affected by drying and wetting cycles. Soil Biol Biochem. 2005;37:339–347. doi: 10.1016/j.soilbio.2004.08.003.
47. Mulec J, Rupnik M, Zupan- HajnaN (2001) Comparison of the culturable bacterial flora from three microenvironments from Pečina v Borštu Cave(South–West Slovenia)13th International Congress of Speleology 4th Speleological Congress of Latin América and Caribbean 26th Brazilian Congress of Speleology, Brasília DF, pp 377-379
48. Navarro JB, Moser DP, Ross FA, Rosen MR, Dong H, Zhang G, HedlundBP. Bacterial succession within an ephemeral hypereutrophic Mojave Desert Playa Lake. Microb Ecol. 2009;57:307–320. doi: 10.1007/s00248-008-9426-3.
49. Ntougias S, Zervakis GI, Ehaliotis C, Kavroulakis N, Papadopoulou KK. Ecophysiology and molecular phylogeny of bacteria isolated from alkaline two-phase olive mill wastes. Res Microbiol. 2006;157:376–385. doi: 10.1016/j.resmic.2005.09.010.
50. Oh YS, Roh DH. *Phenylobacterium muchangponense* sp. nov., isolated from beach soil, and emended description of the genus *Phenylobacterium*. Int J Syst Evol Microbiol. 2012;62:977–983. doi: 10.1099/ijs.0.028902-0.
51. Oren A. Thoughts on the "missing link" between saltworks biology and solar salt quality. Global NEST J. 2010;12:417–425.
52. Oren A. *Salinibacter*: an extremely halophilic bacterium with archaeal properties. FEMS Microbiol Lett. 2013;342:1–9. doi: 10.1111/1574-6968.12094.
53. Pan H, Cheng Z, Zhang Y, Mu S, Qi X. Research progress and developing trends on microorganisms of Xinjiang specific environments. J Arid Land. 2010;2:51–56.
54. Pandit AS, Joshi MN, Bhargava P, Ayachit GN, Shaikh IM, Saiyed ZM, Saxena AK, Bagatharia SB. Metagenomes from the saline

desert of Kutch. Genome Announc. 2014;2:e00439-14. doi: 10.1128/genomeA.00439-14.

55. Paul D, Kumbhare SV, Mhatre SS, Chowdhury SP, Shetty SA, Marathe NP, Bhute S, Shouche YS. Exploration of microbial diversity and community structure of Lonar Lake: the only hypersaline meteorite crater lake within basalt rock. Front Microbiol. 2016;6:1553. doi: 10.3389/fmicb.2015.01553.

56. Pepper IL, Gerba CP. Environmental sample collection ans processing. In: Pepper IL, Gerba CP, Gentry TJ, editors. Environmental Microbiology. Oxford: Elsevier, Academic Press; 2014. pp. 157–175.

57. Pershina EV, Tamazyan GS, Dolnik AS, Pinaev AG, Sergaliev NH, Andronov EE. Studying the structure of soil microbial community in saline soils by high-throughput pyrosequencing. Ecol Genet. 2012;2:32–39.

58. Pesaro M, Nicollier G, Zeyer J, Widmer F. Impact of soil drying-rewetting stress on microbial communities and activities and on degradation of two crop protection products. Appl Environ Microbiol. 2004;70:2577–2587. doi: 10.1128/AEM.70.5.2577-2587.2004.

59. Pikuta EV, Hoover RB, Bej AK, Marsic D, Detkova EN, Whitman WB, Krader P. *Tindallia californiensis* sp. nov., a new anaerobic, haloalkaliphilic, spore-forming acetogen isolated from Mono Lake in California. Extremophiles. 2003;7:327–334. doi: 10.1007/s00792-003-0326-7.

60. Polz MF, Cavanaugh CM. Bias in template-to-product ratios in multitemplate PCR. Appl Environ Microbiol. 1998;64:3724–3730.

61. Pruesse E, Peplies J, Glockner FO. SINA: accurate high-throughput multiple sequence alignment of ribosomal RNA genes. Bioinformatics. 2012;28:1823–1829. doi: 10.1093/bioinformatics/bts252.

62. Qazi JI. Biotechnological potential and conservatory of extremophiles from climatically wide ranged developing countries: lesson from Pakistan. Crit Rev Microbiol. 2013;39:1–8. doi: 10.3109/1040841X.2012.679613.

63. Quast C, Pruesse E, Yilmaz P, Gerken J, Schweer T, Yarza P, Peplies

J, Glockner FO. The SILVA ribosomal RNA gene database project: improved data processing and web-based tools. Nucleic Acids Res. 2013;41:D590–D596. doi: 10.1093/nar/gks1219.

64. Regional Inspectorate for Environmental Protection (2013), 'State of the environment of the Kuyavian-Pomeranian voivodship, (Bydgoszcz: Library of the Environmental Monitoring), 186 (in Polish)

65. Rothschild LJ, Mancinelli RL. Life in extreme environments.Nature.2001;409:1092–1101.doi:10.1038/35059215.

66. Rousk J, Baath E, Brookes PC, Lauber CL, Lozupone C, Caporaso JG, Knight R, Fierer N. Soil bacterial and fungal communities across a pH gradient in an arable soil. ISME J. 2010;4:1340–1351. doi: 10.1038/ismej.2010.58.

67. Schloss PD, et al. Introducing Mothur: open-source, platform-independent, community-supported software for describing and comparing microbial communities. Appl Environ Microbiol. 2009;75:7537–7541. doi: 10.1128/AEM.01541-09.

68. Shannon P, Markiel A, Ozier O, Baliga NS, Wang JT, Ramage D, Amin N, Schwikowski B, Ideker T. Cytoscape: a software environment for integrated models of biomolecular interaction networks. Genome Res. 2003;13:2498–2504. doi: 10.1101/gr.1239303.

69. Shapovalova AA, Khijniak TV, Tourova TP, Sorokin DY. *Halomonas chromatireducens* sp. nov., a new denitrifying facultatively haloalkaliphilic bacterium from solonchak soil capable of aerobic chromate reduction microbiology. Microbiology. 2009;78:102–111. doi: 10.1134/S0026261709010135.

70. Sheibani S, Yanni SF, Wilhelm R, Whalen JK, Whyte LG, Greer CW, Madramootoo CA. Soil bacteria and archaea found in long-term corn (*Zea mays* L.) agroecosystems in Quebec, Canada. Can J Soil Sci. 2013;93:45–57. doi: 10.4141/cjss2012-040.

71. Somogy B, Felföldi T, Dinka M, Vörös L. Periodic picophytoplankton predominance in a large, shallow alkaline lake (Lake Fertő/ Neusiedlersee) Ann Limnol—Int J Limnol. 2010;46:9–19. doi: 10.1051/limn/2010001.

72. Sorokin A, Candelon B, Guilloux K, Galleron N, Wackerow-Kouzova N, Ehrlich SD, Bourguet D, Sanchis V. Multiple-locus sequence

typing analysis of *Bacillus cereus* and *Bacillus thuringiensis*reveals separate clustering and a distinct population structure of psychrotrophic strains. Appl Environ Microbiol. 2006;72:1569–1578. doi: 10.1128/AEM.72.2.1569-1578.2006.

73. Sorokin DY, van Pelt S, Tourova TP, Muyzer G. Microbial isobutyronitrile utilization under haloalkaline conditions. Appl Environ Microbiol. 2007;73:5574–5579. doi: 10.1128/AEM.00342-07.
74. Sorokin ID, Kravchenko IK, Doroshenko EV, Boulygina ES, Zadorina EV, Tourova TP, Sorokin DY. Haloalkaliphilic diazotrophs in soda solonchak soils. FEMS Microbiol Ecol. 2008;65:425–433. doi: 10.1111/j.1574-6941.2008.00542.x.
75. Sorokin DY, Kuenen JG, Muyzer G. The microbial sulfur cycle at extremely haloalkaline conditions of soda lakes. Front Microbiol. 2011;2:44. doi: 10.3389/fmicb.2011.00044.
76. Stres B. The relationship between total and culturable bacteria in cold soils. Acta Agric Slov. 2007;90:25–31.
77. Tamura K, Stecher G, Peterson D, Filipski A, Kumar S. MEGA6: molecular evolutionary genetics analysis version 6.0. Mol Biol Evol. 2013;30:2725–2729. doi: 10.1093/molbev/mst197.
78. Thion C, Prosser JI. Differential response of nonadapted ammonia-oxidising archaea and bacteria to drying-rewetting stress. FEMS Microbiol Ecol. 2014;90:380–389.
79. Tiago I, Mendes V, Pires C, Morais PV, Verssimo A. *Phenylobacterium falsum* sp. nov., an Alphaproteobacterium isolated from a nonsaline alkaline groundwater, and emended description of the genus *Phenylobacterium*. Syst Appl Microbiol. 2005;28:295–302. doi: 10.1016/j.syapm.2005.02.005.
80. Ventosa A, Nieto JJ. Biotechnological applications and potentialities of halophilic microorganisms. World J Microbiol Biotechnol. 1995;11:85–94. doi: 10.1007/BF00339138.
81. Wang YX, Xiao W, Dong MH, Zhao Q, Li ZY, Lai YH, Cui XL. *Halomonas qiaohouensis* sp. nov., isolated from salt mine soil in southwest China. Antonie Leeuwenhoek. 2014;106:253–260. doi: 10.1007/s10482-014-0189-1.
82. Weon HY, Kim BY, Hong SB, Joa JH, Nam SS, Lee KH, Kwon

SW. *Skermanella aerolata* sp. nov., isolated from air, and emended description of the genus *Skermanella*. Int J Syst Evol Microbiol. 2007;57:1539–1542. doi: 10.1099/ijs.0.64676-0.

83. Wu J, Brookes PC. The proportional mineralization of microbial biomass and organic matter caused by air drying and rewetting of a grassland soil. Soil Biol Biochem. 2005;37:507–515. doi: 10.1016/j.soilbio.2004.07.043.
84. Wu Y, Tan L, Liu W, Wang B, Wang J, Cai Y, Lin X. Profiling bacterial diversity in a limestone cave of the western Loess Plateau of China. Front Microbiol. 2015;6:244.
85. Wu Y, Zeng J, Zhu Q, Zhang Z, Lin X. pH is the primary determinant of the bacterial community structure in agricultural soils impacted by polycyclic aromatic hydrocarbon pollution. Sci Rep. 2017;7:40093. doi: 10.1038/srep40093.
86. Zhang L, Gao G, Tang X, Shao K, Bayartu S, Dai J. Bacterial community changes along a salinity gradient in a Chinese wetland. Can J Microbiol. 2013;59:611–619. doi: 10.1139/cjm-2013-0212.
87. Ziółkowska D, Shyichuk O, Cichurska M. Alkaline calcium carbonate from soda production. Composition and proposal of use. Przem Chem. 2013;92:239–243.

CHAPTER 14

MOLECULAR ECOLOGY OF HYPERSALINE MICROBIAL MATS: CURRENT INSIGHTS AND NEW DIRECTIONS

Hon Lun Wong [1,2], Aria Ahmed-Cox [1] and Brendan Paul Burns [1,2]

[1]School of Biotechnology and Biomolecular Sciences, The University of New South Wales, Sydney 2052, Australia

[2]Australian Centre for Astrobiology, University of New South Wales, Sydney 2052, Australia

ABSTRACT

Microbial mats are unique geobiological ecosystems that form as a result of complex communities of microorganisms interacting with each other and their physical environment. Both the microorganisms present and the network of metabolic interactions govern ecosystem function therein. These systems are often found in a range of extreme environments, and those found in elevated salinity have been particularly well studied. The purpose of this review is to briefly describe the molecular ecology of select model hypersaline mat systems (Guerrero Negro, Shark Bay, S'Avall, and Kiritimati Atoll), and any potentially modulating effects caused by salinity to community structure. In addition, we discuss several emerging issues in the field (linking function to newly discovered phyla and microbial dark matter), which illustrate the changing paradigm that is seen as technology has rapidly advanced in the study of these extreme and evolutionally significant ecosystems.

Keywords: hypersaline mats, microbial diversity, niche differentiation, microbial dark matter

INTRODUCTION

Microbial mats are a unique ecological niche representative of early life on Earth. This is due, in part, to the persistence of fossilised mat counterparts, stromatolites, which date back over 3 billion years [1]. Holistically, microbial mats are laminated organo-sedimentary biofilms which house specialized consortia of bacteria and archaea [2,3,4]. Rapid nutrient cycling across microgradients coupled with putative niche differentiation within mat layers enables diverse metabolic processes to occur in spatial proximity. These conditions thereby create an environment ideal for the establishment of unique community symbioses and cross-genera communication. In light of harsh abiotic factors, these relationships enhance consortia survival that would be less likely for individual species alone [2,5].

On a broad scale, these early ecosystems are posited to have played a major role in the oxygenation of the atmosphere, and thereby paved the way for evolution of oxygen-dependent life [6,7]. Beyond this anoxic origin, there is also the suggestion that early oceans were more saline [8], and thus early stromatolitic microbial ecosystems may have formed in hypersaline conditions. Microbial mats are therefore of great interest in the study of how microorganisms in the Precambrian period may have adapted and survived on an evolutionary scale, and for furthering our understanding of the role these primordial communities may have played in global biogeochemical cycling. In mapping the complex biotic and abiotic interactions within hypersaline microbial mats, ecological and functional models of the mat consortia can be explored, to investigate adaptations and evolution in such an extreme environment.

The understanding of microbial mat ecology has recently been deepened by the continued development of molecular methods. Pioneering studies focused on the morphology of microbial mats [9,10], or an examination of the abiotic environment including oxygen, sulfide, and light gradients, e.g., [11,12]. This initial research included a closer analysis of phylogeny targeted at the isolation of cultivatable, key players which are known to operate within multilayered consortia, namely cyanobacteria, sulphate-reducing bacteria and anoxygenic phototrophic bacteria [13,14,15,16]. As a result, functional studies typically reflected phylogenetic determinates, and were limited to oxygenic/anoxygenic photosynthesis by cyanobacteria/filamentous anoxygenic phototrophs (FAPs), aerobic/anaerobic degradation by aerobes/fermenters, sulphate reduction by sulphate-reducing bacteria, and sulphide oxidation by green/purple sulphur bacteria [17].

However, with the advent of high throughput sequencing technology, and the "omics revolution", knowledge of microbial mats has leapt into the molecular and genomic era, revealing niche differentiation events as well as novel microbes and metabolic pathways [18,19]. It suggests our understanding of these hypersaline ecosystems is continually evolving, as the methods to study them advance. This review is designed to briefly summarize current approaches (with an emphasis on molecular techniques) and discoveries in relation to the untapped potential of hypersaline microbial mats from several well-known settings, and proposes future study directions using a combination of different contemporary techniques that can also apply to a broad range of "extreme" environments.

MICROBIAL DIVERSITY IN REPRESENTATIVE HYPERSALINE MATS

The following is a short description of microbial mat communities—

with a focus on the 16S rRNA gene as a marker of diversity—in select hypersaline environments. The purpose is not to describe in detail all hypersaline microbial mats—as this is beyond the scope of this review—but to provide brief selected examples and context with the emerging issues that have arisen as technology has advanced in the field. We have focused on four well-studied environments: Guerrero Negro, Shark Bay, S'Avall, and Kirrimati Atoll. Figure 1 compares the bacterial composition (based on 16S rRNA gene data) across the four hypersaline mats, in which Proteobacteria is a major member across all four samples, in particular Shark Bay and S'Avall [2,20,21,22,23]. Guerrero Negro is dominated by Chloroflexi while Bacteroidetes predominates the microbial community in Kiritimati Atoll mats. Uncultured lineages of Bacteroidetes predominated the surface layers, suggesting potentially novel phototrophic pathways in Kiritimati Atoll [20]. S'Avall is the only hypersaline mat with no reported Chloroflexi, suggesting anoxygenic phototrophy may not be as prominent in these systems, or at least conducted by other community members [21]. The difference in phototrophic pathways is suggested to cause distinct microbial interactions which leads to different assembly patterns of disparate microbial communities [24]. Kiritimati Atoll mats had lower microbial diversity compared to those in Guerrero Negro, and it was suggested that this may be related to the higher salinity in Kiritimati Atoll [20].

In terms of archaeal distribution, Guerrero Negro mats are the only systems dominated by Crenarchaeota, while the other hypersaline mats are dominated by Euryarchaeota [25]. It is important to note that the publications on the hypersaline mats reviewed here do not contain information on newly reported archaeal phyla (discussed later), and thus comparison in the current review is only based on phyla reported in these publications. Further details on individual aspects of microbial diversity (e.g., lower abundance of bacterial nitrifiers in Shark Bay) and suggestions as to some of the underlying reasons certain groups predominate in a given system are given below. The difference between the microbial structures may be a result of different physiochemical conditions such as salinity, causing distinguishable microbial assembly that likely not only affects ecosystem function, but may also determine whether lithification is going to take place.

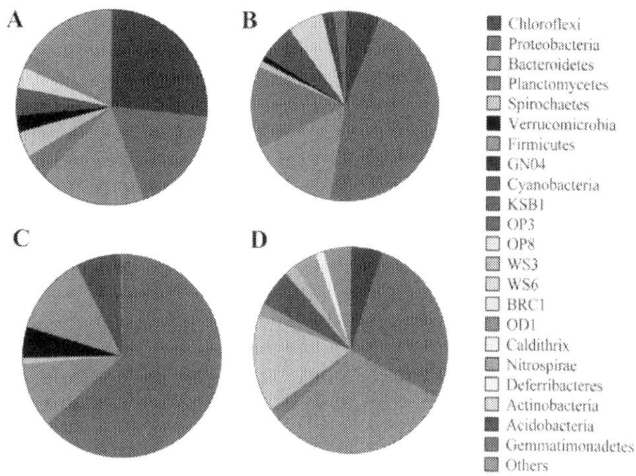

Figure 1: Bacterial diversity of hypersaline mats in (A) Guerrero Negro; (B) Shark Bay; (C) S'Avall; (D) Kiritimati Atoll. Proteobacteria are a major member in all four mats while Guerrero Negro is dominated by Chloroflexi, and Kiritimati Atoll dominated by Bacteroidetes. Both Shark Bay and S'Avall mats harbor abundant Proteobacteria. Data obtained from references [2,20,21,22].

Guerrero Negro. One of the most extensive and well-studied systems of hypersaline microbial mats is found in Guerrero Negro, Baja, Mexico. These mats are exposed to a range of salinities (47–125 PSU (Practical Salinity Units)), and have been characterized in detail in terms of microbial mat structure and biogeochemical cycling [7,26,27,28,29,30,31]. Halite crystallization and formation of gypsum-halite matrices were observed [32]. Early work proposed that microbial mats have a steep geochemical gradient at a millimeter scale, with oxygen decreasing rapidly with depth while sulfide increases with depth [33]. Ley and colleagues [22] conducted the first study to examine hypersaline microbial mats at successive mm intervals with depth. Total DNA was extracted and amplified by traditional 16S rRNA gene PCR, cloned, and sequenced. This study found that cyanobacteria dominated the surface layer, while Proteobacteria, Chloroflexi and Bacteroidetes were enriched in the deeper, anoxic parts of the mats [22,23]. Cyanobacteria were represented by *Microcoleus*, while other phyla (Chloroflexi, Bacteroidetes, Spirochates) have novel lineages at family to genus levels that were poorly described [23]. Of known archaeal sequences, archaea were represented by *Halorubrum* and *Methanohalophilus*, indicating potential methanogenesis in the mats [25]. Furthermore, a range of novel bacterial candidate phyla were found, designated GN01 through to GN15. Euryarchaeotes dominated the archaeal community at the surface, which decreased with depth while Crenarchaeotes dominated the anoxic parts of the mat [25]. In addition to the discovery of novel microbial diversity, Harris and colleagues reinforced that the mats are phylogenetically stratified with depth, according to light and geochemical gradients throughout the mats [22,23]. This approach is also useful to facilitate inferences of function of novel candidate phyla, e.g., novel candidates such as GN04 and Hyd24-12 are abundant in the deep, anoxic zone of the Guerrero Negro mats, suggesting anaerobic metabolisms [23].

Shark Bay. Hamelin Pool, located in Shark Bay, Australia, is a hypersaline pool (>60 PSU), which harbours microbial mats and stromatolites that are considered as potential modern analogues of ancient stromatolites dating back to the Archean period, e.g., [34,35]. Hamelin Pool has arguably the most extensive and diverse system of marine microbial mats [34]. Over the last 10 years, several microbial studies on these systems have been undertaken, with methods ranging from microbial isolation, 16S rRNA gene cloning, lipid analyses, to shotgun metagenome analyses, e.g., [2,3,35,36,37]. Early studies in Shark Bay utilising a clone library approach indicated these hypersaline microbial mats were dominated by Proteobacteria and Bacteroidetes [2]. Of particular interest Shark Bay appears to be novel compared with other marine microbial mat ecosystems that it contains a relatively high abundance of Archaea [3]. Euryarchaeota were the predominant members of these mats, primarily members of the haloarchaea that are known to possess high tolerance to hypersalinity [3,36]. Several novel archaea—namely *Halococcus hamelinensis*, *Haloferax elongans*, and *Haloferax mucosum*—were also isolated and characterised from these systems [38,39]. Several studies examining the stress response of *Halococcus hamelinensis* were carried out, including adaptation to high UV and salinity [40,41,42]. Interestingly, potassium uptake was not observed [40], a common adaptive response to salinity stress in archaea. Instead this archaeon demonstrated a high abundance of acidic amino acids in its proteome and a preference for glycine betaine accumulation as an osmoadaptive strategy [40,43]. Recent findings in our lab also suggests phylogenetic stratification of microbial communities at the millimeter scale in Shark Bay microbial mats, and it is likely salinity is one of the key factors shaping microbial community structure in these systems.

S'Avall, Mallorca Island. S'Avall solar saltern is a crystallizer pond in the Mediterranean with salt concentrations up to 35% *w/v* [21]. Halite is the main mineral found overlying the microbial mats. With exceptionally high sulphate concentration (> 150 mM), it is expected to harbor an abundance of sulphate reducing bacteria. 16S rRNA gene sequencing revealed that both the surface and the core of the mat are indeed dominated by Deltaproteobacteria, represented by *Desulfohalobiaceae*, an extremely halophilic sulphate-reducing bacterium [44]. Other bacterial members include Alphaproteobacteria, Cyanobacteria, Gammaproteobacteria, Bacteroidetes and Firmicutes [21]. Metagenome end-sequence analysis revealed that the mat has a high novelty of archaea with low degree of sequence similarity [45]. The archaeal community was dominated by Halobacteriales, Methanomicrobia, Thermoplasmata and Thermoprotei. However, 16S rRNA gene libraries only yielded a single phylotype in abundance, MSBL1, a novel candidate of methanogens [21]. This archeon is considered ecologically important as methanogenesis in hypersaline environments is typically outcompeted by sulphate reduction thermodynamically [46,47,48,49,50]. With higher than usual sulphate concentration and the dominance of sulphate reducing bacteria, it is still unclear how MSBL1 can effectively carry out methanogenesis at such high salinities—perhaps through the use of non-competitive substrates.

Kiritimati Atoll. Lake 21, Kiritimati Atoll located in the Central Pacific is an evaporitic, hypersaline lake (average salinity 170 g/L) that harbours unique mineralising mats [20,50,51]. In the lake, high concentrations of calcium and carbonate ions result in $CaCO_3$ mineral supersaturation [20]. Microbial mats were precipitated with calcite, gypsum and aragonite respectively [50]. The mats were characterized with nine layers separated by different colors, and was further classified into three zones according to different types of mineral precipitation, namely the calcification zone, gypsum zone and microbialite zone. Phylogenetic studies based on large-scale analysis of 16S rRNA gene sequences have revealed that the microbial community consists of primarily Cyanobacteria, Proteobacteria, Bacteroidetes, Chloroflexi and Nitrospira, while the surface mat layers were mainly composed of halophilic phototrophs (oxygenic and anoxygenic) as well as aerobic heterotrophs [20]. Bacteroidetes, represented by *Salinibacter* and *Saprospiraceae*, dominated the surface layers, which were suggested to be photoheterotrophic [20]. Nitrospira, a nitrite-oxidizing bacterium was found in high abundance at the bottom (*ca.* 10 cm) in the anoxic zone of the mats, indicative of potentially important nitrogen cycling at this depth. The archaeal community was dominated by Halobacteria and Thaumarchaeota, and also included numerous uncultured taxonomic lineages such as Marine Benthic Group D (MBGD) and Deep Sea Hydrothermal Vent Group 6 (DHVEG-6). However, the processes of lithification, whether caused by specific metabolic traits or specific physiochemical conditions, are still unknown. Interestingly, an isolate with closest match to *Halococcus hamelinensis*—an archaeon which was previously only found in Shark Bay [38]—was recently isolated from the Kiritimati Atoll mats [4]. The three distinguishable mineralization zones were thought to be the result of the interplay between different functional guilds [4], though this needs to be confirmed as to date community metagenomics analyses have yet to be performed in Kiritmati Atoll mats.

METAGENOMIC STUDIES IN REPRESENTATIVE HYPERSALINE MATS

While an important first step, examining microbial community structures (16S rRNA sequences) alone is not sufficient to elucidate habitat-specific functional pathways. Metagenomics, however, is able to assign functional profiles at the community-wide level. Metagenomics utilizes taxonomic and functional binning to cluster similar nucleotide patterns together, assigning taxonomic and metabolic functions to the sample population [52]. A range of techniques has been employed (*i.e.*, G + C content, short oligomer (kmer) frequency, tetramer frequency and read coverage) to assign sequence fragments of the metagenome into microbial populations [52,53,54]. The following summarizes some of the major findings from metagenomics studies performed on the hypersaline microbial mat systems reviewed here.

Guerrero Negro. Shotgun metagenomic sequencing has revealed insights into the functional genetic potential of the Guerrero Negro mats.

Photosynthesis-related genes were abundant in the top layers in these mats, while genes encoding ferredoxins, sulfatases and sugar degradation metabolisms (glycolysis, pentose and uronic acid degradation) were in abundance in the anoxic part of the mat [55]. This may indicate the heterotrophic metabolism of sulphated organic compounds and sugars are the main strategy for microorganisms to obtain energy in the anoxic zones in the Guerrero Negro microbial mats. This study found that millimeter-scale functional profiles were consistent with the physiochemical profile of the mat [55]. There was the suggestion by one study of acid-shifted isoelectric point amino acids detected in the Guerrero Negro mats, which were potentially caused by the enrichment of the amino acid aspartate [55]. It was proposed that this is a strategy utilized by the Guerrero Negro microbial community to adapt to their hypersaline environment, as it is known that this enrichment of amino acids allows proteins to function even in high salt concentrations [56]. However revaluation of this data suggests there may not be such a prevalence of acid shifted amino acids in these hypersaline mats as first thought [57], thus clarification is required as to whether this is indeed an adaptive strategy in the Guerrero Negro mats. Further experiments regarding putative novel functional pathways of the candidate members have also yet to be undertaken in this system.

Shark Bay. One recent study has delineated for the first time the functional metagenomic potential of the hypersaline mats in Shark Bay [3]. Analysis of Shark Bay mat metagenomes identified several putative osmoadaptive traits, with the majority of the sequences categorized as betaine biosynthesis and choline and betaine uptake [3]. It is likely that compatible solutes play major roles in osmoadaptive mechanisms in these communities, and this is further supported by studies on adaptation to salt in individual isolates from this environment described earlier [40,58]. It was also suggested that compatible solute turnover in the Shark Bay mats may be important both as an adaptive response as well as potential sources of energy and carbon under changing environmental conditions [3]. A range of metabolic pathways involved in sulphur cycling were identified, with the sulphur species delineated also found coinciding with cyanobacterial photosynthesis, suggesting metabolic cooperation at the surface [3,59]. Of further interest to ecosystem function, a lower abundance of common bacterial nitrifiers (*i.e., Nitrosomonas, Nitrobacter)* was observed in the

Shark Bay mats [3]. Hypersalinity has been shown in other systems to inhibit the growth of nitrifying bacteria [60], and it may be that salinity is an important factor regulating the nitrogen cycle in hypersaline settings such as the Shark Bay mats.

S'Avall, Mallorca Island. Metagenomic analyses were undertaken on the hypersaline microbial mats in the S'Avall solar saltern, and results indicated that carbohydrate degradation pathways (glycolysis, pyruvate metabolism and citrate cycle) were over-represented, followed by amino acid metabolism [45]. Result suggests that heterotrophic carbohydrate metabolism is the main strategy for microbial members to obtain energy, while the enrichment of amino acid metabolism is suggested to help maintain growth of the halophilic microbiota, since amino acids can also act as compatible solutes [45,56]. However, functional genes of MSBL1 (an organism described earlier as being in abundance in these systems) could not be accurately deduced due to the low degree of similarity of the annotated open reading frames [21].

Metagenomic studies from microbial mats in Shark Bay, Guerrero Negro and S'Avall have yielded similar findings of acidic amino acid enrichment, while glycine betaine uptake/biosynthesis was observed only in Shark Bay. This indicates hypersalinity is likely a selective pressure on the mat communities and amino acid enrichment may be a common osmoadaptive strategy. Heterotrophic sugar degradation pathways were found as the major metabolism in Guerrero Negro and S'Avall [21,55], while both Guerrero Negro and Shark Bay have considerable abundance of genes related to sulphur metabolism [3,55]. These putative differences in preferred metabolism and metabolic cooperation may be a reason for different microbial assemblies and thus distinct community structure.

EMERGING ISSUES

The major pitfall of metagenomic data from microbial communities such as described above lies in the existence of newly discovered "rare phyla", for which corresponding reference 16S rRNA gene sequences are not available in databases [61]. Thus only sequences of predominant (and therefore abundant) members in the communities will be captured and reconstructed, thus neglecting low abundance novel microbial groups [61]. Furthermore, binning of contigs into population-level is a major challenge with the absence of related reference genomes especially with novel lineages [53], referred as "microbial dark matter", which will be discussed below.

Thus despite the advances in technology that have facilitated a greater understanding of hypersaline mats at the molecular level, the field is moving rapidly and a number of issues are important to consider when analyzing these systems.

Correlating function with phylogeny. A logical approach to developing models of microbial mats systems is to correlate specific metabolic pathways to the microbial taxa, thereby assigning function(s) to a specific taxon or defined Operational Taxonomic Unit (OTU). This, in turn, leads to deciphering the likely functional roles of novel phylogenetic groups and any potential evolutionary significance within the studied ecosystem. Naturally, cultivation-dependent methods remain in high use due to their capacity to generate valuable microbial mass for isolation and genetic analysis [4,18]. However, as with historical studies there remains an uncultured candidate majority that modern studies have attempted to divulge via novel culturing techniques.

A recent study on the Kiritimati Atoll microbial mats successfully isolated 62 strains that cover a broad range of microorganisms, that have been used for genomic studies to assign known metabolic functions, thus correlating function with phylogeny [4]. However the dominant microbes identified previously in this system using 16S rDNA data including Chloroflexi, Nitrospira, Thermoplasmata, Thaumarchaeota and novel candidate phyla (Hyd24-12, OP9, WS3) could not be cultivated. Thus any putative metabolic pathways of the novel candidate phyla have yet to be determined. Furthermore, despite bioinformatics tools (*i.e.*, Picrust, Tax4Fun) which can predict functional roles based on phylogenetic data, function-related information on the recently proposed super-phyla (e.g., TACK, DPANN [18,62,63] cannot be provided due to a need for (and lack of) available reference data in the corresponding databases. As a result, a functional model of microbial diversity such as in the Kiritimati Atoll mats, amongst others, has been hindered by insufficient *de novo* genomics of candidate phyla.

Microbial dark matter (MDM). This discussion naturally leads into addressing the aforementioned bottle-neck in environmental microbiology: "microbial dark matter" (MDM). MDM was coined as a term to denote the uncultured microbial majority that represents over half of the lineages of bacteria and archaea [18,61,64,65,66]. MDM includes microorganisms with unknown ecological roles and metabolic capability, in which cultivation efforts have not succeeded. To date, only cultivation-independent methods have detected members of MDM, e.g., [18,62,63,66,67,68], though even this in itself has some weight in the illumination of the role of MDM in microbial ecology in general, as well as specifically in hypersaline mat systems.

Of particular interest in light of this discussion is the rapidly changing shift in our understanding of the phylogenetic diversity of bacteria and archaea [18,62,63,69]. One study recently clustered newly identified bacteria with small genome size into a group described as the candidate phyla radiation (CPR) [62]. This group has successfully reconstructed complete genomes of Parcubacteria (OD1), Microgenomates (OP11), Saccharibacteria (TM7), Berkelbacteria (ACD58) and the Kazan phyla, expanding genomic information in the bacteria domain [62]. Comparative metagenomic studies support the proposition of an archaeal phylum, the TACK super-phylum (including Thaumarchaeota, Aigearchaeota, Crenarchaeota, Korachaeota and a newly proposed phylum Bathyarchaeota) [63,70,71] as well as the DPANN archaeal super-phylum (consisting of Diapherotrites, Parvarchaeota (also Micrarchaeota), Aenigmarchaeota, Nanoarchaetoa and Nanohaloarchaeota) [12,63,65,69]. Furthermore, two new archaeal phyla, Woesearchaeota and Pacearchaeota, have also recently been defined [63]. This development had refined phylogenetic estimation (predicting a target microorganism's position on a phylogenetic tree) and expanded the archaeal and bacterial domains, leading to the need of further functional profiling of the newly proposed phyla.

Prior to these discoveries, identification of archaeal members in hypersaline microbial mats were restricted to three main phyla: Euryarchaeota, Crenarchaeota, and Thaumarchaeota respectively, e.g., [2,3,20,22,36]. As a result, the expanding genomic datasets continue to reveal further phylogenetic differentiation and novel clade formation at a phylum level. However the consistent issue of phylum level phylogenetic differentiation and metabolic/environmental interaction between co-localized populations in extreme environments remains a pertinent question to address, especially in the context of adaptation in extreme environments [72].

COMBINED APPROACHES FOR MOLECULAR ECOLOGY

Generally, it is often ideal to utilize multiple approaches to facilitate a more comprehensive view of complex microbial communities, thereby expanding our understanding of interactions and adaptations, such as those evident in hypersaline mat ecosystems. Examples of emerging approaches and their potential application in hypersaline mat systems are discussed below.

Niche differentiation. Niche differentiation is a strategy for microorganisms to orient themselves in the microbial community, either by interacting with other microorganisms and/or adaptation to environmental factors. The spatial arrangement and relationship between microorganisms may also reveal resource utilization patterns and metabolic specialization [73]. Furthermore, by recording the spatial distribution and niche segregation of phylogenetic diversity and comparing data with geochemical composition and functional biodiversity, microbial adaptation can be revealed as a major driver of ecosystem function [74]. For instance, an ecological model regarding the interrelationships between key functional groups was proposed for the Kiritimati Atoll mats based on niche partitioning [4]. The differentiation between different types of biomineralization at a millimeter scale (calcification, gypsum and microbialite zone) was suggested to be the result of the interplay between different functional microbial guilds and the corresponding nutrient and element cycling, including production and degradation of exopolymers, photosynthesis, sulfate reduction and fermentation. Metabolic processes occurring in unique niche spaces have been considered as a driver towards microbialite formation in hypersaline mats [4,20].

Bioinformatics tools including network correlation analysis of the co-occurrences of bacteria and archaea can also be useful for investigating specific niche differentiation in hypersaline mats. Documenting these potential interactions between taxa (co-occurrence patterns) across complex and diverse communities may help to ascertain the functional roles of environmental niches occupied by uncultured microorganisms [75,76,77], thus providing a potential method to indirectly investigate MDM.

Studies thus far regarding microbial niche differentiation have primarily been conducted on soil samples, and hypersaline samples over meter scales, where biogeochemical gradients change over large distances [59,74,78]. However, microbial mats are versatile ecosystems where complete nutrient and element cycles occur at a millimeter scale [12,22,23,79,80]. Consequently, a wide range of metabolic activities may impose steep chemical gradients and create niches with fine spatial resolution [55]. Microbial interactions, for instance, the exchange of certain nutrients and elements, can determine the presence of certain niches within the mat. To date, only a handful of published studies have examined potential niches at millimeter scales of hypersaline microbial mats [20,22,55].

Studies of niche differentiation in hypersaline mats have reinforced the existence of anoxic niches at the surface of hypersaline microbial mats, with a considerable amount of sulfate-reducing bacteria residing in the photic-oxic zone of the mats in Guerrero Negro and Kiritimati Atoll [20,22,23]. In addition, the localization of uncultured lineages of Bacteroidetes at the surface of Kiritimati Atoll and Guerrero Negro might suggest a phototrophic lifestyle of these phylotypes [20,23].

In addition to internal biochemical factors, physiochemical parameters, such as the chemical gradients of O_2, H_2S and NH_4^+, have been shown to regulate vertical stratification of a microbial community and this could influence niche formation in the hypersaline mat systems reviewed here [4,20,22,23]. Furthermore, a study of a hypersaline microbial mat in Salt Pond, San Salvador Island, Bahamas suggested that antagonistic interactions between bacteria from adjacent laminae also regulate niche formation and microbial stratification in addition to physiochemical properties [81].

Niche differentiation approaches have been shown to help delineate complex microbial communities, aiding the examination of microbial structures and proposal of ecological models in hypersaline microbial mats. These studies not only facilitated higher molecular resolution of the microbial community, but also provide a platform for further metagenomic and/or transcriptomic studies.

Single cell genomics (SCG). Single cell genomics (SCG) is the isolation of unculturable single cells from a microbial community, followed by whole genome amplification and subsequent sequencing of the genome [67,82]. Briefly, the isolated individual cells are subjected to whole genome amplification (WGA), by the most commonly used technique multiple displacement amplification (MDA). The resulting amplified DNA, known as single amplified genomes (SAG), can then be sequenced using any next-generation sequencing technology [61,82,83,84]. This alternative approach provides phylogenetic information as well as functional markers of the community, effectively linking function to a particular population [83,85]. SCG is a powerful tool to directly address the issue of MDM, in which novel genomes such as OP1, OP3, OP9, as well as the archaeal superphyla TACK and DPANN were all determined [18,61,66,86]. Furthermore, novel metabolic pathways can also be determined from the single cell genome, which helps to add further levels of detail to phylogenetic trees [81,82]. An example of this was the proposal of Nanohaloarchaea, a new archaeal class found in surface waters of hypersaline environments [87,88]. Iterative phylogenetic binning revealed composite genomes characterized by an ample amount of acidic amino acids, which is suggested to be the same "salt in" strategy observed in Guerrero Negro, Shark Bay and S'Avall [3,45,55,87] for microorganisms to adapt in hypersaline environments. Although to date there are no studies utilizing single cell genomics to examine hypersaline microbial mats, this approach could potentially be employed to investigate putative novel microbes.

However, despite being used as an effective tool, SCG provides only snapshots of the genomic content of uncultured single cells [65], and only targets one specific class or even genus within the novel phyla at a time. In addition to binning difficulties, it would therefore be impractical to attempt to obtain community-wide diversity via SCG. As a result, SCG and metagenomics can be considered complementary and have the potential to analyze samples in a broad *versus* narrow brush approach, the benefits of which are discussed in the following sections.

Multiple "omics". Metagenomics, in contrast to single cell genomics, enables the analysis of an entire microbial community, thereby providing holistic data of metabolic and functional markers. This approach focuses on community-wide dynamics and interactions, complementary to SCG's focus on data mining from a single cell [52,89]. In addition to previously discussed issues that binning and assembly of individual representative genomes from metagenomes is difficult, another hurdle is that a reconstructed genome from metagenomics databases is a composite genome and does not represent the single microorganism isolated from a specific environment [90]. This is a significant issue as the composite genome cannot reflect how individual microorganisms behave in a given environment or specific microenvironments, especially those that occur in hypersaline microbial mats due to fine spatial and functional resolution [91]. This is even more pronounced when annotating genes from phyla within microbial dark matter, leading to potential over-confidence in gene function prediction. Furthermore, despite the fact that metagenomics facilitates inferred predictions of gene function, DNA-based analyses cannot distinguish between expressed and non-expressed genes, hence this approach alone fails to reflect true metabolic activity within microbial mats [89,92,93].

However, when used in tandem with SCG, metagenomics is well complemented, as SCG genomic data of a single cell not only enables specific metabolic pathways to be constructed, but can also target the microbial dark matter that might be underrepresented in the sample [90]. Moreover, with SCG providing novel, rare phylogenetic and metabolic information, taxonomic binning for metagenomic databases can be improved (*i.e.*, TACK and DPANN phyla) [65]. Metagenomics can therefore make use of this novel data to enhance community-wide screening and more precisely predict potentially functional genes. Furthermore, SAGs only focus on the single cell genome and are independent of the genetic variations of the whole microbial community. As a result it can facilitate the detection of naturally occurring population-level diversity [83,94]. Coupling SCG and metagenomics can bridge this gap and associate population-level diversity with their specific functionalities, which is critical for investigating the process of niche partitioning and the construction of an accurate hypersaline microbial mat ecosystem model.

Beyond metagenomics is the realm of functional meta-analysis. Metatranscriptomics enables direct sequencing of gene transcripts from environmental samples, with quantitative measurements of community-wide microbial activities [65,94]. Metatranscriptomes can also identify which set of genes are being transcribed, and thereby illuminate complex regulatory or interactive gene systems. Complementary metagenomics and metatranscriptomics can therefore reveal how microbes respond and adapt to different physiochemical conditions at the population level [65,95]. This combined approach can simultaneously investigate taxonomic composition, functional content and gene expression patterns, revealing the overall functional and metabolic activities within the community. Although metranscriptomics have not been undertaken to date on the hypersaline mat systems that were the focus of this review, such analyses have been undertaken in other mat communities, e.g., [96,97], increasing our understanding of the complex interplay over a diel cycle of major phototrophic and heterotrophic pathways in these systems.

While metatranscriptomics elucidates genes that are actively transcribed and can provide valuable predictions of active metabolism, gene expression does not always correlate with protein content as it still indicates the potential for protein expression [98]. Thus metaproteomics can be employed to attempt to reveal the true functionality of mat communities and validate genome-based predictions. Although metaproteomic analyses have also not been undertaken on hypersaline mat communities to date, such analyses have been undertaken on hot spring microbial mat communities [98], revealing patterns of nutrient cycling over a diel cycle. Some proteins were also observed to have different transcriptional patterns when under different stress conditions (heat shock and starvation), thus metaproteomics is another powerful approach to examine ecosystem function in "extreme" environments.

Considering the tremendous diversity of uncultivated microorganisms, multiple omics approaches have the potential to identify large numbers of new, niche-specific gene and protein families among localised microbial communities in hypersaline microbial mats [92]. It can help decipher the interactions between hypersaline mat members and their environments, as well as strategies for adaptation and metabolic specialisation that may be employed [94]. A caveat is that all database-dependent analyses have their own inherent biases and limitations that need to be taken into account [64], and thus any conclusions on ecosystem function still need to be made with care and in context.

CONCLUSIONS AND FUTURE OUTLOOK

This review has sought to provide a brief overview of our current understanding of the molecular ecology of hypersaline microbial mats—with a specific focus on four disparate geographic systems—and highlighted the advances in technology that have facilitated a more holistic understanding of ecosystem function. While, for the sake of brevity, the focus in this review has been on hypersaline systems, many of the principles and approaches discussed that help delineate ecosystem function are equally applicable to microbial mats in other settings, such as those influenced by temperature and pH, e.g., [95,96,97,98,99]. In terms of future directions, major ecological roles/interactions can be further deciphered via complementary approaches [82,100], such as single cell genomics, metatranscriptomics, metaproteomics, and metabolomics (Figure 2). However it should be acknowledged that "omics" approaches also have their limitations—e.g., any sampling bias needs to be taken into account, depth of sequencing required to obtain sufficient coverage, ensuring the data is not just a "snap-shot" of the current microbial community, and considering any effects of gene transfer on community structure. Thus, to obtain the most accurate representation of overall community function in hypersaline mats, defined geochemical and physiological analyses (e.g., metabolic rates, EPS production, signaling, mineralization), as well as targeted approaches such as NanoSIMS-FISH, stable isotope probing, and qPCR of specific genes/pathways, should also be conducted to complement "omics" approaches. This will allow for a better understanding of spatial niche distribution of microbial populations within hypersaline mat environments, and can facilitate the construction of a more complete ecological model of these evolutionally significant systems.

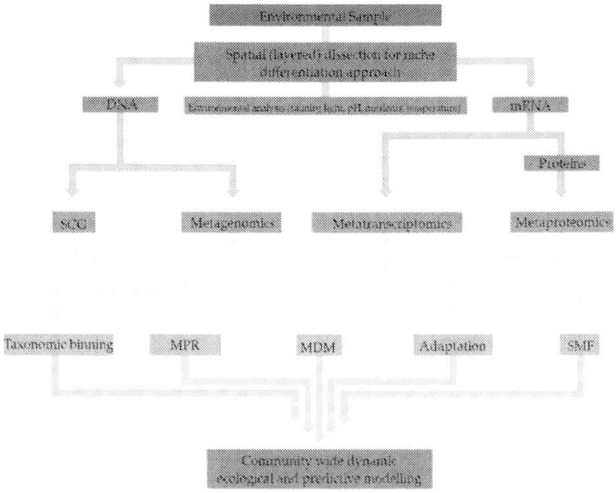

Figure 2: Schematic illustrating the evolving approaches and complementary pipelines in the study of hypersaline microbial mats and other "extreme" environments. MDM—Microbial dark matter; MPR—Metabolic pathway reconstruction; SCG—Single cell genomics; SMF—Specific metabolic function.

AUTHOR CONTRIBUTIONS

Hon Lun Wong wrote the initial review draft, and Aria Ahmed-Cox rewrote and edited major sections of the manuscript. Brendan Burns wrote specific sections and edited the final version of the submitted manuscript.

REFERENCES

1. Walter, M.R.; Buick, R.; Dunlop, J.S.R. Stromatolites 3400–3500 Myr old from the North Pole area, Western Australia. *Nature* 1980, *284*, 443–445.
2. Allen, M.A.; Goh, F.; Burns, B.P.; Neilan, B.A. Bacterial, archaeal and eukaryotic diversity of smooth and pustular microbial mat communities in the hypersaline lagoon of Shark Bay. *Geobiology* 2009, *7*, 82–96.
3. Ruvindy, R.; White, R.A., III; Neilan, B.A.; Burns, B.P. Unravelling core microbial metabolisms in the hypersaline microbial mats of

Shark Bay using high-throughput metagenomics. *ISME J.* 2015, *1*, 1–14.

4. Spring, S.; Brinkmann, N.; Murrja, M.; Spröer, C.; Reitner, J.; Klenk, H.P. High diversity of culturable prokaryotes in a lithifying hypersaline microbial mat. *Geomicrobiol. J.* 2015, *32*, 332–346.
5. Paerl, H.W.; Pinckney, J.L.; Steppe, T.F. Cyanobacterial-bacterial mat consortia: Examining the functional unit of microbial survival and growth in extreme environments. *Environ. Microbiol.* 2000, *2*, 11–26.
6. Hoehler, T.M.; Bebout, B.M.; des Marais, D.J. The role of microbial mats in the production of reduced gases on the early Earth. *Nature* 2001, *412*, 324–327.
7. DesMarais, D.J. Biogeochemistry of hypersaline microbial mats illustrates the dynamics of modern microbial ecosystems and the early evolution of the biosphere. *Biol. Bull.* 2003, *204*, 160–167.
8. Arp, G.; Reimer, A.; Reitner, J. Photosynthesis-Induced Biofilm Calcification and Calcium Concentrations in Phanerozoic Oceans. *Science* 2001, *292*, 1701–1704.
9. Logan, B.W. Cryptozoon and associate stromatolites from the recent, Shark Bay, Western Australia. *J. Geol.* 1961, *69*, 517–533.
10. Dravis, J.J. Hardened subtidal stromatolites, Bahamas. *Science* 1983, *219*, 385–386.
11. Visscher, P.T.; Prins, R.A.; van Gemerden, H. Rates of sulfate reduction and thiosulfate consumption in a marine microbial mat. *FEMS Microbiol. Ecol.* 1992, *86*, 283–294.
12. Visscher, P.T.; Stolz, J.F. Microbial mats as bioreactors: Populations, processes and products. *Palaeogeogr. Palaeocl.*2005, *219*, 87–100.
13. Bauld, J.; Chambers, L.A.; Skyring, G.W. Primary productivity, sulfate reduction and sulfur isotope fractionation in algal mats and sediments of Hamelin Pool, Shark Bay, WA. *Aust. J. Mar. Fresh. Res.* 1979, *30*, 753–764.
14. Moriarty, D.J.W. Bacterial biomass and productivity in sediments, stromatolites, and water of Hamelin Pool, Shark Bay, Western Australia. *Geomicrobiol. J.* 1983, *3*, 121–133.
15. Krumbein, W.E. Stromatolites—The challenge of a term in space

and time. *Precambr. Res.* 1983, *20*, 493.

16. Fründ, C.; Cohen, Y. Diurnal cycles of sulfate reduction under oxic conditions in microbial mats. *Appl. Environ. Microbiol.* 1992, *58*, 70–77.
17. Revsbech, N.P.; Madsen, B.; Jørgensen, B.B. Oxygen production and consumption in sediments determined at high spatial resolution by computer simulation of oxygen microelectrode data. *Limnol. Ocenogr.* 1986, *31*, 293–304.
18. Rinke, C.; Schwientek, P.; Sczyrba, A.; Ivanova, N.N.; Anderson, I.J.; Cheng, J.F.; Darling, A.; Malfatti, S.; Swan, B.K.; Gies, E.A.; *et al.* Insights into the Phylogeny and Coding Potential of Microbial Dark Matter. *Nature* 2013.
19. Berg, I.A.; Kockelkorn, D.; Buckel, W.; Fuchs, G. A 3-hydroxypropionate/4-hydroxybutyrate autotrophic carbon dioxide assimilation pathway in Archaea. *Nature* 2007, *318*, 1782–1786.
20. Schneider, D.; Arp, G.; Reimer, A.; Reitner, J.; Daniel, R. Phylogenetic analysis of a microbialite-forming microbial mat from a hypersaline lake of the Kiritimati Atoll, Central Pacific. *PLoS ONE* 2013, *8*, 1–14.
21. López-López, A.; Yarza, P.; Richter, M.; Suárez-Suárez, A.; Antón, J.; Niemann, H.; Rosselló-Móra, R. Extremely halophilic microbial communities in anaerobic sediments from a solar saltern. *Environ. Microb. Rep.* 2010, *2*, 258–271.
22. Ley, R.E.; Harris, J.K.; Wilcox, J.; Spear, J.R.; Miller, S.R.; Bebout, B.M.; Maresca, J.A.; Bryant, D.A.; Sogin, M.L.; Pace, N.R. Unexpected diversity and complexity of the Guerrero Negro hypersaline microbial mat. *Appl. Environ. Microbiol.* 2006, *72*, 3685–3695.
23. Harris, J.K.; Caporaso, J.G.; Walker, J.J.; Spear, J.R.; Gold, N.J.; Robertson, C.E.; Hugenholtz, P.; Goodrich, J.; McDonald, D.; Knights, D.; *et al.* Phylogenetic Stratigraphy in the Guerrero Negro Hypersaline Microbial Mat. *ISME J.* 2013, *7*, 50–60.
24. Cole, J.K.; Hutchison, J.R.; Renslow, R.S.; Kim, Y.M.; Chrisler, W.B.; Engelmann, H.E.; Dohnalkova, A.C.; Hu, D.; Metz, T.O.; Fredrickson, J.K.; *et al.* Phototrophic biofilm assembly in microbial-mat-derived unicyanobacterial consortia: Model

systems for the study of autotroph-heterotroph interactions. *Front. Microbiol.* 2014, *5*, 1–18.

25. Robertson, C.E.; Spear, J.R.; Harris, J.K.; Pace, N.R. Diversity and stratification of Archaea in a hypersaline microbial mat. *Appl. Environ. Microbiol.* 2009, *75*, 1801–1810.
26. Minz, D.; Fishbain, S.; Green, S.J.; Gerard, M.; Cohen, Y.; Rittmann, B.E.; Stahl, D.A. Unexpected population distribution in a microbial mat community: Sulfate-reducing bacteria localized to the highly oxic chemocline in contrast to a eukaryotic preference for anoxia. *Appl. Environ. Microbiol.* 1999, *65*, 4659–4665.
27. Minz, D.; Flax, J.L.; Green, S.J.; Gerard, M.; Cohen, Y.; Wagner, M.; Rittmann, B.E.; Stahl, D.A. Diversity of sulfate-reducing bacteria in oxic and anoxic regions of a microbial mat characterized by comparative analysis of dissimilatory sulfate reductase genes. *Appl. Environ. Microbiol.* 1999, *65*, 4666–4671.
28. Risatti, J.B.; Capman, W.C.; Stahl, D.A. Community structure of a microbial mat: The phylogenetic dimension. *Proc. Natl. Acad. Sci. USA* 1994, *9117*, 10173–10177.
29. Canfield, D.E.; des Marais, D.J. Biogeochemical cycles of carbon, sulphur, and free oxygen in a microbial mat. *Geochim. Cosmochim. Acta* 1993, *57*, 3971–3984.
30. Visscher, P.T.; Baumgartner, L.K.; Buckley, D.H.; Rogers, D.R.; Hogan, M.E.; Raleigh, C.D.; Turk, K.A.; des Marais, D.J. Dimethyl sulphide and methanethiol formation in microbial mats: Potential pathways for biogenic signatures. *Environ. Microbiol.* 2003, *5*, 296–308.
31. Reimer, J.J.; Huerta-Diaz, M.A. Phosphorus speciation and sedimentary fluxes in hypersaline sediments of the Guerrero Negro salt evaporation area, Baja California Sur, Mexico. *Estuar. Coasts* 2011, *34*, 514–528.
32. Spear, J.R.; Ley, R.E.; Berger, A.B.; Pace, N.R. Complexity in natural microbial ecosystems: The Guerrero Negro experience. *Biol. Bull.* 2003, *204*, 168–173.
33. Des Marais, D.J. The biogeochemistry of hypersaline microbial mats. *Adv. Microb. Ecol.* 1995, *14*, 251–274.
34. Jahnert, R.J.; Collins, L.B. Characteristics, Distribution and

Morphogenesis of Subtidal Microbial Systems in Shark Bay, Australia. *Mar. Geol.* 2012, *303*, 115–136.

35. Goh, F.; Allen, M.A.; Leuko, S.; Kawaguchi, T.; Decho, A.W.; Burns, B.P.; Neilan, B.A. Determining the specific microbial populations and their spatial distribution within the stromatolite ecosystem of Shark Bay. *ISME J.* 2009, *3*, 383–396.
36. Burns, B.P.; Goh, F.; Allen, M.; Neilan, B.A. Microbial diversity of extant stromatolites in the hypersaline marine environment of Shark Bay, Australia. *Environ. Microbiol.* 2004, *6*, 1096–1101.
37. Allen, M.A.; Neilan, B.A.; Burns, B.P.; Jahnke, L.L.; Summons, R.E. Lipid biomarkers in Hamelin Pool microbial mats and stromatolites. *Org. Geochem.* 2010, *41*, 1207–1218.
38. Goh, F.; Leuko, S.; Allen, M.A.; Bowman, J.P.; Kamekura, M.; Neilan, B.A.; Burns, B.P. *Halococcus hamelinensis* sp. nov., a novel halophilic archaeon isolated from stromatolites in Shark Bay, Australia. *Int. J. Syst. Evol. Microbiol.*2006, *56*, 1323–1329.
39. Allen, M.A.; Goh, F.; Leuko, S.; Echigo, A.; Mizuki, T.; Usami, R.; Kamekura, M.; Neilan, B.A.; Burns, B.P. *Haloferax elongans* sp. nov. and *Haloferax mucosum* sp. nov., isolated from microbial mats from Hamelin Pool, Shark Bay, Australia. *Int. J. Syst. Evol. Microbiol.* 2008, *58*, 798–802.
40. Goh,F.;Jeon,Y.J.;Barrow,K.;Neilan,B.A.;Burns,B.P.Osmoadaptive strategies of the archaeon *Halococcus hamelinensis* isolated from a hypersaline stromatolite environment. *Astrobiology* 2011, *11*, 529–536.
41. Leuko, S.; Neilan, B.A.; Burns, B.P.; Walter, M.R.; Rothschild, L.J. Molecular Assessment of UVC Radiation-Induced DNA Damage Repair in the Stromatolitic halophilic archaeon, *Halococcus hamelinensis*. *J. Photochem. Photobiol. B* 2011, *102*, 140–145.
42. Leuko, S.; Goh, F.; Ibáñez-Peral, R.; Burns, B.P.; Walter, M.R.; Neilan, B.A. Lysis efficiency of standard DNA extraction methods for *Halococcus* sp. in an organic rich environment. *Extremophiles* 2008, *12*, 301–308.
43. Gudhka, R.K.; Neilan, B.A.; Burns, B.P. Adaptaion, ecology, and evolution of the halophilic stromatolite archaeon *Halococcus hamelinensis* inferred through genome analyses. *Archaea* 2015.

44. Kjeldsen, K.U.; Loy, A.; Jakobsen, T.F.; Thomsen, T.R.; Wagner, M.; Ingvorsen, K. Diversity of sulfate-reducing bacteria from an extreme hypersaline sediment, Great Salt Lake (Utah). *FEMS Microbiol. Ecol.* 2007, *60*, 287–298.
45. López-López, A.; Richter, M.; Peña, A.; Tamames, J.; Rosselló-Móra, R. New insights into the archaeal diversity of a hypersaline microbial mat obtained by a metagenomics approach. *Syst. Appl. Microbiol.* 2013, *36*, 205–214.
46. Oremland, R.S.; Polcin, S. Methanogenesis and sulfate reduction: Competitive and non-competitive substrates in estuarine sediments. *Appl. Environ. Microbiol.* 1982, *44*, 1270–1276.
47. King, G.M.; Klug, M.J.; Lovley, D.R. Metabolism of acetate, methanol, and methylated amines in intertidal sediments of Lowes Cove, Maine. *Appl. Environ. Microbiol.* 1983, *45*, 1848–1853.
48. King, G.M. Metabolism of trimethylamine, choline, and glycine betaine by sulfate-reducing and methanogenic bacteria in marine-sediments. *Appl. Environ. Microbiol.* 1984, *48*, 719–725.
49. Kiene, R.P.; Oremland, R.S.; Catena, A.; Miller, L.G.; Capone, D.G. Metabolism of reduced methylated sulphur-compounds in anaerobic sediments and by a pure culture of an estuarine methanogen. *Appl. Environ. Microbiol.*1986, *52*, 1037–1045.
50. Buckley, D.H.; Baumgartner, L.K.; Visscher, P.T. Vertical distribution of methane metabolism in microbial mats of the Great Sippewissett Salt Marsh. *Environ. Microbiol.* 2008, *10*, 967–977.
51. Arp, G.; Helms, G.; Karlinska, K.; Schumann, G.; Reimer, A.; Reitner, J.; Trichet, J. Formation of microbialtes on the Atoll of Kiritimati, Republic of Kiribati, Central Pacific. *Geomicrobiol. J.* 2012, *29*, 29–65.
52. Dröge, J.; McHardy, A.C. Taxonomic binning of metagenome samples generated by next-generation sequencing technologies. *Brief. Bioinform.* 2012, *13*, 646–655.
53. Alneberg, J.; Bjarnason, B.S.; de Bruijn, I.; Schirmer, M.; Quick, J.; Kjaz, U.Z.; Lahti, L.; Loman, N.J.; Andersson, A.F.; Quince, C. Binning metagenomics contigs by coverage and composition. *Nat. Methods* 2014, *11*, 1144–1150.
54. Saeed, I.; Tang, S.L.; Halgamuge, S.K. Unsupervised discovery of

microbial population structure within metagenomes using nucleotide base composition. *Nucleic Acid Res.* 2012, *40*, e34.

55. Kunin, V.; Raes, J.; Harris, J.K.; Spear, J.R.; Walker, J.J.; Ivanova, N.; von Mering, C.; Bebout, B.M.; Pace, N.R.; Bork, P.; *et al.* Millimeter-scale genetic gradients and community-level molecular convergence in a hypersaline microbial mat. *Mol. Syst. Biol.* 2008, *4*, 198–203.
56. Soppa, J. From genomes to function: Haloarchaea as model organisms. *Microbiology* 2006, *152*, 585–590.
57. Bardavid, R.E.; Oren, A. Acid-shifted isoelectric point profiles of the proteins in a hypersaline microbial mat: An adaptation to life at high salt concentrations? *Extremophiles* 2012, *16*, 787–792.
58. Goh, F.; Barrow, K.D.; Neilan, B.A.; Burns, B.P. Identification and regulation of novel compatible solutes from hypersaline stromatolite-associated cyanobacteria. *Arch. Microbiol.* 2010, *192*, 1031–1038.
59. Pages, A.; Welsch, D.T.; Teasdale, P.R.; Grice, K.; Vacher, M.; Bennett, W.W.; Visscher, P.T. Diel fluctuations in solute distributions and biogeochemical cycling in a hypersaline microbial mat from Shark Bay, WA. *Mar. Chem.*2014, *167*, 102–112.
60. Jeffries, T.C.; Seymour, J.R.; Newton, K.; Smith, R.J.; Seuront, L.; Mitchell, J.G. Increases in the abundance of microbial genes encoding halotolerance and photosynthesis along a sediment salinity gradient. *Biogeosciences*2012, *9*, 815–825.
61. Hedlund, B.P.; Dodsworth, J.A.; Murugapiran, S.K.; Rinke, C.; Woyke, T. Impact of single-cell genomics and metagenomics on the emerging view of extremophile "microbial dark matter". *Extremophiles* 2014.
62. Brown, C.T.; Hug, L.A.; Thomas, B.C.; Sharon, I.; Castelle, C.J.; Singh, A.; Wilkins, M.J.; Wrighton, K.C.; Williams, K.H.; Banfield, J.F. Unusual biology across a group comprising more than 15% of domain Bacteria. *Nature* 2015, *523*, 208–211.
63. Castelle, C.J.; Wrighton, K.C.; Thomas, B.C.; Hug, L.A.; Brown, C.T.; Wilkins, M.J.; Frischkorn, K.R.; Tringe, S.G.; Singh, A.; Markillie, L.M.; *et al.* Genomic expansion of domain Archaea highlights roles for organsims from new phyla in anaerobic carbon

cycling. *Curr. Biol.* 2015, *25*, 690–701.

64. McDonald, D.; Price, M.N.; Goodrich, J.; Nawrocki, E.P.; DeSantis, T.Z.; Probst, A.; Andersen, G.L.; Knight, R.; Hugenholtz, P. An improved Greengenes taxonomy with explicit ranks for ecological and evolutionary analyses of bacteria and archaea. *ISME J.* 2012, *6*, 610–618.
65. Baker, B.J.; Dick, G.J. Omic Approaches in Microbial Ecology: Charting the Unknown. *Microbe* 2013, *3*, 353–360.
66. Dodsworth, J.A.; Blainey, P.C.; Murugapiran, S.K.; Swingley, W.D.; Ross, C.A.; Tringe, S.G.; Chain, P.S.G.; Scholz, M.B.; Lo, C.C.; Raymond, J.; *et al.* Single-cell and metagenomics analyses indicate a fermentative, saccharolytic lifestyle for members of the OP9 lineage. *Nat. Commun.* 2013, *4*, 1854.
67. Blainey, P.C. The future is now: Single-cell genomics of bacteria and archaea. *Microbiol. Rev.* 2013, *37*, 407–427.
68. Scholz, M.B.; Lo, C.C.; Chain, P.S.G. Next generation sequencing and bioinformatics bottlenecks: The current state of metagenomics data analysis. *Curr. Opin. Biotechnol.* 2012, *23*, 9–15.
69. Youssef, N.H.; Rinke, C.; Stepanauskas, R.; Farag, I.; Woyke, T.; Elshahed, M.S. Insights into the metabolism, lifestyle and putative evolutionary history of the novel archaeal phylum "Diapherotrites". *ISME J.* 2015, *9*, 447–460.
70. Meng, J.; Xu, J.; Qin, D.; He, Y.; Xiao, X.; Wang, F. Genetic and functional properties of uncultivated MCG archaea assessed by metagenome and gene expression analyses. *ISME J.* 2014, *8*, 650–659.
71. Guy, L.; Ettema, T.J. The archaeal "TACK" superphylum and the origin of eukaryotes. *Trends Micriobiol.* 2011, *19*, 580–587.
72. Nichols, D.; Cahoon, N.; Trakhtenberg, E.M.; Pham, L.; Mehta, A.; Belanger, A.; Kanigan, T.; Lewis, K.; Epstein, S.S. Use of ichip for high-throughput *in situ* cultivation of "uncultivable" microbial species. *Appl. Environ. Microbiol.* 2010, *76*, 2445–2450.
73. Grünke, S.; Felden, J.; Lichtschlag, A.; Girnth, A.C.; de Beer, D.; Wenzhöfer, F.; Boetius, A. Niche differentiation among mat-forming, sulfide-oxidizing bacteria at cold seeps of the Nile Deep Sea Fan (Eastern Mediterranean Sea). *Geobiology* 2011, *9*, 330–

348.

74. Vigneron, A.; Cruaud, P.; Roussel, E.G.; Pignet, P.; Caprais, J.C.; Callac, N.; Ciobanu, M.C.; Godfroy, A.; Cragg, B.A.; Parkes, J.R.; *et al.* Phylogenetic and functional diversity of microbial communities associated with subsurface sediments of the Sonora Margin, Guaymas Basin. *PLoS ONE* 2014, *9*, e104427.
75. Ruan, Q.; Dutta, D.; Schwalbach, M.S.; Steele, J.A.; Fuhrman, J.A.; Sun, F. Local similarity analysis reveals unique associations among marine bacterioplankton species and environmental factors. *Bioinformatics* 2006, *22*, 2532–2538.
76. Fuhrman, J.A.; Steele, J.A. Community structure of marine bacterioplankton: Patterns, networks, and relationships to function. *Aquat. Microb. Ecol.* 2008, *59*, 69–81.
77. Chaffron, S.; Rehrauer, H.; Pernthaler, J.; von Mering, C. A global network of coexisting microbes from environmental and whole—Genome sequence data. *Genome Res.* 2010, *20*, 947–959.
78. Andrei, A.Ş.; Robeson, M.S., II; Baricz, A.; Coman, C.; Muntean, V.; Ionescu, A.; Etiope, G.; Alexe, M.; Sicora, C.I.; Podar, M.; *et al.* Contrasting taxonomic stratification of microbial communities in two hypersaline meromictic lakes. *ISME J.* 2015, 1–15.
79. Ludemann, H.; Arth, I.; Liesack, W. Spatial changes in the bacterial community structure along a vertical oxygen gradient in flooded paddy soil cores. *Appl. Environ. Microbiol.* 2000, *66*, 754–762.
80. Stomp, M.; Huisman, J.; Stal, L.; Matthijs, H.C.P. Colorful niches of phototrophic microorganisms shaped by vibrations of the water molecule. *ISME J.* 2007, *1*, 271–282.
81. Long, R.A.; Eveillard, D.; Franco, S.L.M.; Reeves, E.; Pinckney, L.L. Antagonistic interactions between heterotrophic bacteria as a potential regulator of community structure of hypersaline microbial mats. *FEMS Microbiol. Ecol.* 2013, *83*, 74–81.
82. De Jager, V.; Siezen, R.J. Single-cell genomics: Unravelling the genomes of unculturable microorganisms. *Microb. Biotechnol.* 2011, *4*, 431–437.
83. Stepanauskas, R. Single cell genomics: An individual look at microbes. *Curr. Opin. Microbiol.* 2012, *15*, 613–620.

84. Dean, F.B.; Hosono, S.; Fang, L.H.; Wu, X.H.; Faruqi, A.F.; Bray-Ward, P.; Sun, Z.Y.; Zong, Q.L.; Du, Y.F.; Du, J.; *et al.* Comprehensive human genome amplification using multiple displacement amplification. *Proc. Natl. Acad. Sci. USA* 2002, *99*, 5261–5266.

85. Walker, A. Adding genomic "foliage" to the tree of life. *Nat. Rev. Microbiol.* 2014, *12*, 78.

86. Kolinko, S.; Jogler, C.; Katzmann, E.; Wanner, G.; Peplies, J.; Schüler, D. Single-cell analysis reveals a novel uncultivated magnetotactic bacterium within the candidate division OP3. *Environ. Microbiol.* 2012, *14*, 1709–1721.

87. Narasingarao, P.; Podell, S.; Ugalde, J.A.; Brochier-Armanet, C.; Emerson, J.B.; Brocks, J.J.; Heidelberg, K.B.; Banfield, J.F.; Allen, E.E. *De novo* metagenomics assembly reveals abundant novel major lineage of Archaea in hypersaline microbial communities. *ISME J.* 2012, *6*, 81–93.

88. Ghai, R.; Pašic, L.; Fernández, A.B.; Martin-Cuadrado, A.B.; Mizuno, C.M.; McMahon, K.D.; Papke, R.T.; Stepanauskas, R.; Rodriguez-Brito, B.; Rohwer, F.; *et al.* New abundant microbial groups in aquatic hypersaline environments. *Sci. Rep.* 2011, *1*, 135.

89. Simon, C.; Daniel, R. Metagenomic analyses: Past and future trends. *Appl. Environ. Microbiol.* 2011, *77*, 1153–1161.

90. Podar, M.; Abulencia, C.B.; Walcher, M.; Hutchison, D.; Zengler, K.; Garcia, J.A.; Holland, T.; Cotton, D.; Hauser, L.; Keller, M. Targeted access to the genomes of low-abundance organisms in complex microbial communities. *Appl. Environ. Microb.* 2007, *73*, 3205–3214.

91. Dupraz, C.; Reid, R.P.; Braissant, O.; Decho, A.W.; Norman, R.S.; Visscher, P.T. Processes of carbonate precipitation in modern microbial mats. *Earth Sci. Rev.* 2008, *96*, 141–162.

92. Sorek, R.; Cossart, P. Prokaryotic transcriptomics: A new view on regulation, physiology and pathogenicity. *Nat. Rev. Genet.* 2010, *11*, 9–16.

93. Bhaya, B.; Grossman, A.R.; Steunou, A.S.; Khuri, N.; Cohan, F.M.; Hamamura, N.; Melendrez, M.C.; Bateson, M.M.; Ward, D.M.; Heidelberg, J.F. Population level functional diversity in a microbial

community revealed by comparative genomic and metagenomics analyses. *ISME J.* 2007, *1*, 703–713.

94. Shi, Y.; Tyson, G.W.; Eppley, J.M.; DeLong, E.F. Integrated metatranscriptomic and metagenomics analyses of stratified microbial assemblages in the open ocean. *ISME J.* 2011, *5*, 999–1013.
95. Chen, L.X.; Hu, M.; Huang, L.N.; Hua, Z.S.; Kuang, J.L.; Li, S.J.; Shu, W.S. Comparative metagenomics and metatranscriptomic analyses of microbial communities in acid mine drainage. *ISME J.* 2015, *9*, 1579–1592.
96. Liu, Z.; Klatt, C.G.; Wood, J.M.; Rusch, D.B.; Ludwig, M.; Wittekindt, N.; Tomsho, L.P.; Schuster, S.C.; Ward, D.M.; Bryant, D.A. Metatranscriptomic analyses of chlorophototrophs of a hot-spring microbial mat. *ISME J.* 2011, *5*, 1279–1290.
97. Mobberley, J.M.; Khodadad, C.L.M.; Visccher, P.T.; Reid, R.P.; Hagan, P.; Foster, J.S. Inner workings of thrombolites: Spatial gradients of metabolic activity as revealed by metatranscriptome profiling. *Sci. Rep.* 2015, *5*, 12601.
98. Schaffert, C.S.; Ward, D.M.; Klatt, C.G.; Pauley, M.; Steinke, L.A. Identification and distribution of high-abundance proteins in the Octopus Spring microbial mat community. *Appl. Environ. Microbiol.* 2012, *78*, 8481–8484.
99. Jungblut, A.D.J.; Allen, M.A.; Burns, B.P.; Neilan, B.A. Lipid biomarker analysis of cyanobacterial dominated microbial mats in melt water ponds on the McMurdo Ice Shelf, Antarctica. *Org. Geochem.* 2009, *40*, 258–269.
100. Rubakhin, S.S.; Romanova, E.V.; Nemes, P.; Sweedler, J.V. Profiling metabolites and peptides in single cells. *Nat. Methods* 2011, *8*, 20–29.

REFERENCES

1. Azapagic, A., and S.Perdan. 2000. Indicators of sustainable development for industry. A general framework.
2. Process Safety and Environmental Protection 78(4):243-261.
3. An, J.Y., S.J. Sim, J.S. Lee, and B.W. Kim. 2003. Hydrocarbon production from secondarily treated piggery waste-water by the green alga Botryococcus braunii. Journal of Applied Phycology 15(2-3):185-191.
4. AlfaLaval. 2010. Multiple choices for fats and oils refining. Available online at http://local.alfalaval.com/en-us/key-industries/food/fatsandoils/Documents/PFT00355EN_Degumming_neutralization.pdf. Accessed December 21, 2011.
5. Alley, W.M. 2003. Desalination of Ground Water: Earth Science Perspectives. Reston: U.S. Geological Survey.
6. Anderson, A.L., K. O'Brien, and M. Hartwell. 2007. Comparisons of mosquito populations before and after construction of a wetland for water quality improvement in Pitt County, North Carolina, and data-reliant vectorborne disease management. Journal of Environmental Health 69(8):26-33.

7. A.A. Kiss, A.C. Dimian, G. Rothenberg, Solid acid catalysts for biodiesel production towards sustainable energy. Adv. Synth. Catal. **348**, 75–81 (2006)
8. A.A. Kiss, A.C. Dimian, G. Rothenberg, Biodiesel by reactive distillation powered by metal oxides. Energy Fuels **22**, 598–604 (2008)
9. A.A. Kiss, Separative reactors for integrated production of bioethanol and biodiesel. Comput. Chem. Eng. **34**(5), 812–820 (2010)
10. A.A. Kiss, A.C. Dimian, G. Rothenberg, 'Green catalysts' for enhanced biodiesel technology, catalysis of organic reactions. Chem. Ind. Ser. **115**, 405–414 (2006)
11. A.A. Kiss, G. Rothenberg, A.C. Dimian, F. Omota, The heterogeneous advantage: biodiesel by catalytic reactive distillation. Top. Catal. **40**, 141–150 (2006)
12. Assmann, A., A. Braun, S. John, A. Lei, and S. Southard. 2011. The Potential for Micro-Algae and Other "Micro-Crops" to Produce Sustainable Biofuels. Master of Science, University of Michigan, Ann Arbor.
13. Atkinson, K.M. 1972. Birds as transporters of algae. British Phycological Journal 7:319-321.
14. A.K. Singh, S.D. Fernando, Reaction kinetics of soybean oil transesterification using heterogeneous metal oxide catalysts. Chem. Eng. Technol. 30(12), 1–6 (2007)
15. Aresta, M.; Dibenedetto, A; Carone, M.; Colonna, T. and C. Fagale, "Production of Biodiesel from macroalgae by supercritical CO2 Extraction and Thermochemical Liquifaction," Environmental Chemistry Letters, Vol. 3, 2005, pp. 136-139. Briggs, M., "Widescale Biodiesel Production from Algae," Energy Bulletin, October 3, 2004. Brown, L. M. and K. G. Zeiler, " Aquatic Biomass and Carbon Dioxide Trapping," Energy Conversion Management, Vol. 34, No. 9-11, 1993, pp. 1005-1013. Bowman, M.; Hilligoss, D.; Rasmussen, S. and R. Thomas, "Biodiesel: A Renewable and Biodegradable Fuel," Hydrocarbon Processing, February, 2006, pp. 103-106. Aresta, M.; Dibenedetto, A; Carone, M.; Colonna, T. and C. Fagale, "Production of Biodiesel from macroalgae by supercritical CO2 Extraction and Thermochemical Liquifaction,"

Environmental Chemistry Letters, Vol. 3, 2005, pp. 136-139. Briggs, M., "Widescale Biodiesel Production from Algae," Energy Bulletin, October 3, 2004. Brown, L. M. and K. G. Zeiler, " Aquatic Biomass and Carbon Dioxide Trapping," Energy Conversion Management, Vol. 34, No. 9-11, 1993, pp. 1005-1013. Bowman, M.; Hilligoss, D.; Rasmussen, S. and R. Thomas, "Biodiesel: A Renewable and Biodegradable Fuel," Hydrocarbon Processing, February, 2006, pp. 103-106.

16. Adosraku RK, Choi GTY, Constantinou-Kokotos V, Anderson MM, Gibbons WA. 1994.
17. Atmospheric Environment 43:2084-2086. Hossain ABM, Shalleh A, Boyce AN, Naqiuddin M. 2008. Biodiesel fuel production from algae as renewable energy.
18. American Journal of Biochemistry and Biotechnology 4(3):250-254.
19. Armbrust EV. The life of diatoms in the World's oceans. Nature. 2009;459(7244):185–192.
20. Adosraku RK, Choi GTY, Constantinou-Kokotos V, Anderson MM, Gibbons WA. 1994. NMR lipid profiles of cells, tissues, and body fluids: proton NMR analysis of human erythrocyte lipids. Journal of lipid Research 35:1925 – 1931.
21. Atmospheric Environment 43:2084-2086. Hossain ABM, Shalleh A, Boyce AN, Naqiuddin M. 2008. Biodiesel fuel production from algae as renewable energy.
22. Applied Energy 88(11):3632-3635. Karmee SK, Mahesh P, Ravi R, Chadha A. 2004.
23. American Journal of Biochemistry and Biotechnology 4(3):250-254. Karatay SE, Donmez G. 2011.
24. Atmospheric CO2. Mauna Loa. http://co2now.org/images/stories/data/co2-mlo-monthly-noaa-esrl.pdf (accessed on 12th October 2013) NREL. 2012. NREL Creates New Pathways for Producing Biofuels and Acids from Cyanobacteria.
25. Amaro HM, Guedes AC, Malcata FX. Advances and perspectives in using microalgae to produce biodiesel. Appl Energy in press. Doi: 10.1016/ j.apenergy.2010.12.014.

26. Alabi AO, Tampier M, Bibeau E. Microalgae Technogies and Processes for Biofuels/Bioenergy Production in British Columbia. The BC Innovation Council; BC, Canada: 2009.
27. Abou-Shanab RAI, Hwang J-H, Cho Y, Min B, Jeon B-H. Characterization of microalgal species isolated from fresh water bodies as a potential source for biodiesel production. Appl Energy in press. doi:10.1016/ j.apenergy.2011.01.060.
28. Austria: OPEC. Patil V, Tran K, Giselrod HR. 2008. Towards Sustainable Production of Biofuels from Microalgae.
29. Apt KE, Kroth-Pancic PG, Grossman AR. Stable nuclear transformation of the diatom Phaeodactylum tricornutum. Mol Gen Genet. 1996;252(5):572–579.
30. B. M. S. Hossain, A. N. Boyce, A. Salleh, and S. Chandran, "Impacts of alcohol type, ratio and stirring time on the biodiesel production from waste canola oil," African Journal of Agricultural Research, vol. 5, no. 14, pp. 1851–1859, 2010. W. N. N. Wan Omar and N. A. Saidina Amin, "Optimization of heterogeneous biodiesel production from waste cooking palm oil via response surface methodology," Biomass and Bioenergy, vol. 35, no. 3, pp. 1329–1338, 2011. S. Gan, H. K. Ng, P. H. Chan, and F. L. Leong, "Heterogeneous free fatty acids esterification in waste cooking oil using ion-exchange resins," Fuel Processing Technology, vol. 102, pp. 67–72, 2012. E. Lotero, Y. Liu, D. E. Lopez, K. Suwannakarn, D. A. Bruce, and J. G. Goodwin, "Synthesis of biodiesel via acid catalysis," Industrial and Engineering Chemistry Research, vol. 44, no. 14, pp. 5353–5363, 2005.
31. Singh, B.M. Smyth, J.D. Murphy, A biofuel strategy for Ireland with an emphasis on production of biomethane and minimization of land-take. Renew. Sustain. Energy Rev. 14(1), 277–288 (2010)
32. A.F. Turhollow, E.O. Heady, Large-scale ethanol production from corn and grain sorghum and improving conversion technology. Energy Agric. 5(4), 309–316 (1986)
33. B. M. S. Hossain and A. N. Boyce, "Biodiesel production from waste sunflower cooking oil as an environmental recycling process and renewable energy," Bulgarian Journal of Agricultural Science, vol. 15, no. 4, pp. 312–317, 2009.

34. Demirbas, Comparison of transesterification methods for production of biodiesel from vegetable oils and fats. Energy Convers. Manag. 49, 125–130 (2008)
35. Kumari, P. Mahapatra, V. K. Garlapati, and R. Banerjee, "Enzymatic transesterification of Jatropha oil," Biotechnology for Biofuels, vol. 2, article 1, 2009. R. MacEiras, A. Cancela, M. Vega, and M. C. Márquez, "Enzymatic alholysis for biodiesel production from waste cooking oil," in Chemical Engineering Transactions, vol. 19, pp. 103–107.
36. M. S. Hossain, B. A. Nasrulhaq, A. Salleh, and S. Chandran, "Biodiesel production from waste soybean oil biomass as renewable energy and environmental recycled process," African Journal of Biotechnology, vol. 9, no. 27, pp. 4233–4240, 2010.
37. A.K. Agrawal, Biofuels (alcohols and biodiesel) applications as fuels for internal combustion engines. Prog. Energy Combust. Sci. 33, 233–271 (2007)
38. Okullo, A. K. Temu, P. Ogwok, and J. W. Ntalikwa, "The physico-chemical properties of jatropha and castor oils," International Journal of Renewable Energy Research, vol. 2, no. 1, pp. 47–52, 2012.
39. Demirbas, Progress and recent trends in biodiesel fuels. Energy Conserv. Manag. 50, 14–34 (2009)
40. Singh, D. Pant, N.E. Korres, A.S. Nizami, S. Prasad, J.D. Murphy, Key issues in life cycle assessment of ethanol production from lignocellulosic biomass: challenges and perspectives. Bioresour. Technol. 101(13), 5003–5012 (2010)
41. B.M. Choudary, M.L. Kantam, C.V. Reddy, S. Aranganathan, Pl. Santhia, F. Figueras, Mg–Al–O–t-Bu hydrotalcite: a new and efficient heterogeneous catalyst for transesterification. J. Mol. Catal. A Chem. 159, 411–416 (2000)
42. Rice, A. Fröhlich, and R. Leonard, "Bio-diesel production from camelina oil, waste cooking oil and tallow," The Science of Farming and Food, 1998.
43. K. Highina, I. M. Bugaje, and B. Umar, "Biodiesel production from Jatropha caucus oil in a batch reactor using zinc oxide as catalyst,"

Journal of Petroleum Technology and Alternative Fuels, vol. 2, no. 9, pp. 146–149, 2011.

44. Supple, R. Howard-Hildige, E. Gonzalez-Gomez, and J. J. Leahy, "The effect of steam treating waste cooking oil on the yield of methyl ester," Journal of the American Oil Chemists' Society, vol. 79, no. 2, pp. 175–178, 2002.
45. Delfort, I. Durand, G. Hillion, A. Jaecker-Voirol, X. Montagne, Glycerin for new biodiesel formulation. Oil Gas Sci. Technol. e Rev. IFP 63(4), 395–404 (2008)
46. Dale, Greening the chemical industry: research and development priorities for biobased industrial products. J. Chem. Technol. Biotechnol. 78, 1093–1103 (2003)
47. B.B. He, A.P. Singh, J.C. Thompson, A novel continuous-flow reactor using reactive distillation for biodiesel production. Trans. ASAE 49, 107–112 (2006)
48. Bøhn T, Primicerio R, Hessen DO, Traavik T. Reduced fitness of Daphnia magna fed a bt-transgenic maize variety. Arch Environment Contamin Toxicol. 2008;55(4):584–592.
49. Boynton JE, Gillham NW. Genetics and transformation of mitochondria in the green alga Chlamydomonas. Methods Enzymol. 1996;264:279–296.
50. Boynton JE, Gillham NW. Chloroplast transformation in Chlamydomonas. Methods Enzymol. 1993;217:510–536.
51. Blanco AM, Moreno J, Del Campo JA, Rivas J, Guerrero MG. Outdoor cultivation of lutein-rich cells of Muriellopsis sp. in open ponds. Appl Microbiol Biotechnol. 2007;73(6):1259–1266.
52. Borevitz JO, Maloof JN, Lutes J, *et al.* Quantitative trait loci controlling light and hormone response in two accessions of Arabidopsis thaliana. Genetics. 2002;160(2):683–696.
53. Bingham SE, Cox JC, Strem MD. Expression of foreign DNA in Chlamydomonas reinhardtii. FEMS Microbiol Lett. 1989;53(1–2):77–81.
54. Bloomberg. Retrieved from http://www.bloomberg.com/news/2013-04-18/chevrondefies-california-on-carbon-emissions.html Fenton, O., & Ó hUallacháin, D. (2012). Agricultural nutrient

surpluses as potential input sources to grow third generation biomass (microalgae): A review. Algal Research, 1(1), 49-56. Georgianna, D. R., Hannon, M. J., Marcuschi, M., Wu, S., Botsch, K., Lewis, A. J., . . . Mayfield, S. P. (2013). Production of recombinant enzymes in the marine alga dunaliella tertiolecta. Algal Research, 2(1), 2-9. Handler, R. M., Canter, C. E., Kalnes, T. N., Lupton, F. S., Kholiqov, O., Shonnard, D. R., & Blowers, P. (2012).

55. Biotechnology Advances 25:294- 306. Chisti Y. 2008. Biodiesel from microalgae beats bioethanol. Trends in Biotechnology 26(3):126-131. Daigo K, Nakano Y, Casareto Be, Suzuki Y, Shioi Y. 2008.
56. Benthic cyanobacteria of the genus Nodularia are non-toxic, with gas vacuoles, able to glide and genetically more diverse than planktonic Nodularia.
57. Berry JP, Gantar M, Perez MH, Berry G, Noriega FG. Cyanobacterial toxins as allelochemicals with potential applications as algaecides, herbicides and insecticides. Mar Drugs. 2008;6:117–146.
58. Bhadury P, Wright PC. Exploitation of marine algae: biogenic compounds for potential antifouling applications. Planta. 2004;219(4):561–578.
59. Borowitzka MA. Culturing microalgae in outdoor ponds. In: Andersen RA, editor. Algal Culturing Techniques. Vol. 218. Academic Press; NY, USA: 2005.
60. Benthic cyanobacteria of the genus Nodularia are non-toxic, with gas vacuoles, able to glide and genetically more diverse than planktonic Nodularia. International Journal of Systematic Microbiology 55:555-568.
61. Bernstein L, Bosch P, Canziani 0, Chen Z, Christ R, Davidson O, Hare W, Huq S, Karoly D, Kattsov V, Kundzewicz Z, Liu J, Lohmann U, Manning M, Matsuno T, Menne B, Metz B, Mirza M, Nicholls N, Nurse L, Pachauri R, Palutikof J, Parry M, Qin D, Ravindranath N, Reisinger A, Ren J, Riahi K, Rosenzweig C, Rusticucci M, Schneider S, Sokona Y, Solomon S, Stott P, Stouffer R, Sugiyama T, Swart R, Tirpak D, Vogel C, Yohe G. 2007.
62. Barba FJ, Grimi N, Vorobiev E: New Approaches for the Use of Non-conventional Cell Disruption Technologies to Extract Potential

Food Additives and Nutraceuticals from Microalgae. Food Eng Rev.2015; 7(1):45–62. 10.1007/s12393-014-9095-6

63. Borowitzka MA: High-value products from microalgae—their development and commercialisation. J Appl Phycol. 2013; 25(3):743–56. 10.1007/s10811-013-9983-9
64. Bhatnagar A, Chinnasamy S, Singha M, Das KC. Renewable biomass production by mixotrophic algae in the presence of various carbon sources and wastewaters. Appl Energy in press. doi:10.1016/j.apenergy.2010.12.064
65. Borowitzka MA. Algal biotechnology products and processes – matching science and economics. J Applied Phycology. 1992; 4:267–279.
66. Borowitzka MA. Commercial production of microalgae: ponds, tanks, tubes and fermenters. J Biotechnology. 1999; 70:313–321. Krichnavaruk S, Shotipruk A, Goto M, Pavasant P.
67. Byerlee D, De Janvry A, World Bank. Agriculture for Development. World Bank; DC, USA: 2007.
68. Berman-Frank I, Lundgren P, Falkowski P. Nitrogen fixation and photosynthetic oxygen evolution in cyanobacteria. Res Microbiol. 2003; 154(3):157–164.
69. Boyd PW, Law CS, Wong CS, *et al*. The decline and fate of an iron-induced subarctic phytoplankton bloom. Nature. 2004; 428(6982):549–553.
70. Brussaard CPD. Viral control of phytoplankton populations – a review. J Eurkaryotic Microbiol. 2004; 51(2):125–138.
71. Baumann, H., andA.M. Tillman. 2014. The Hitch Hikers Guide to LCA. Lund, Sweden: Studentliterature AB.
72. Benemann, J.R., R.P. Goebel, J.c Weissman, and D.C Augenstein.1982. Microalgae as a source of Liquid Fuel. FairField, CA: EnBio, Inc.
73. Bhatnagar, A., M. Bhapnagar, S.Chinnasamy, and K.C. Das. 2010. Chlorella minutissima- A promosing fuel alga for cultivation in municipal wastewater. Applied biochemistry and biotechnology 161(1-8):523.536.

74. Brune, D.E., T.J. Lundquist, and J.R. Benemann. 2009. Microalgal biomass for greenhouse gas reductions:

75. Bullard, C.W., and R.A.Herendeen. 1975. Energy cost of goods and services. Energy policy 3(4):268-278.

76. Bullard, C., P.Pennter, and D, Pilati. 1978. Net energy analysis: Handbook for combining process and input-output analysis Resources and Energy 1:267-313.

77. Boussiba, S. 2011. Questionnaire reply from AlgaTechnologies, Kibbutz Qetura. Received by the NRC Committee on the Sustainable Development of Algal Biofuels on July 6.

78. Beal, C.M., R.E. Hebner, M.E. Webber, R.S. Ruoff, and A.F. Seibert. 2011. The energy return on investment for algal biocrude: Results for a research production facility. Bioenergy Research: DOI: 10.1007/s12155-011-9128-4.

79. Benemann, J.R. 2008. Opportunities and challenges in algae biofuel production. Available online at http://www.fao.org/uploads/media/algae_positionpaper.pdf. Accessed October 21, 2011.

80. BioProcess Algae. 2011. Questionnaire reply from BioProcess Algae. Received by the NRC Committee on the Sustainable Development of Algal Biofuels on September 27.

81. Baral, A., B.R. Bakshi, and R.L. Smith. 2012. Assessing resource intensity and renewability of cellulosic ethanol technologies using Eco-LCA. Environmental Science and Technology 46 (4):2436-2444.

82. Batan, L., J. Quinn, T. Bradley, and B. Willson. 2011. Erratum: Net energy and greenhouse gas emissions evaluation of biodiesel derived from microalgae (Environmental Science and Technology (2010) 44 (7975-7980)). Environmental Science and Technology 45(3):1160.

83. Batan, L., J. Quinn, B. Willson, and T. Bradley. 2010. Net energy and greenhouse gas emission evaluation of biodiesel derived from microalgae. Environmental Science and Technology 44(20):7975-7980.

84. Benemann, J., P.M. Pedroni, J. J. Davison, H. Beckert, and P. Bergman. 2003a. Technology roadmap for biofixation of CO2 and

greenhouse gas abatement with microalgae. Paper read at the Second Annual Conference on Carbon Sequestration, May 5-8, Alexandria, VA.

85. Benemann, J.R., J.C. Van Olst, M.J. Massingill, J.C. Weissman, and D.E. Brune. 2003b. The controlled eutrophication process: Using microalgae for CO2 utilization and agricultural fertilizer recycling. Paper read at the 6th Greenhouse Gas Control Technologies Conference (GHGT6), October, Kyoto Japan.
86. BLM (Bureau of Land Management) and DOE (U.S. Department of Energy). 2010. Solar Energy Development Draft Programmatic Environmental Impact Statement (Draft Solar PEIS). Washington, D.C.: Bureau of Land Management and U.S. Department of Energy.
87. Boyd, C.E., and A. Gross. 2000. Water use and conservation for inland aquaculture ponds. Fisheries Management and Ecology 7(1-2):55-63.
88. Brentner, L.B., M.J. Eckelman, and J.B. Zimmerman. 2011. Combinatorial life cycle assessment to inform process design of industrial production of algal biodiesel. Environmental Science and Technology 45:7060-7067.
89. Babcock, B.A. 2009. Measuing unmeasurable land-use changes from biofuels. Iowa Agriculture Review Summer: 4-6.
90. Bachmann, R.W., B.L. Jones, D.D. Fox, M. Hoyer, L.A. Bull, and D.E. Canfield. 1996. Relations between trophic state indicators and fish in Florida (USA) lakes. Canadian Journal of Fisheries and Aquatic Sciences 53(4):842-855.
91. Beeman, P. 2007. Biofuel plants generate new air, water, soil problems for Iowa. Available onlineat http://www.desmoinesregister.com/article/20070603/BUSINESS01/706030325/Biofuel-plants-generate-new-air-water-soil-problems-for-Iowa. Accessed January 21, 2012.
92. Benemann, J.R. 2008. Opportunities and challenges in algae biofuel production. Available online at http://www.fao.org/uploads/media/algae_positionpaper.pdf. Accessed October 21, 2011.
93. BLM (Bureau of Land Management) and DOE (U.S. Department of Energy). 2010. Solar Energy Development Draft Programmatic

Environmental Impact Statement (Draft Solar PEIS). Washington, DC: Bureau of Land Management and U.S. Department of Energy.

94. Bollag, J.M., and S.T. Czlonkowski. 1973. Inhibition of methane formation in soil by various nitrogen-containing compounds. Soil Biology and Biochemistry 5(5):673-678.

95. Boyd, C.E. 1995. Bottom soils, sediment, and pond aquaculture. New York: Chapman and Hall.

96. Breitburg, D.L., D.W. Hondorp, L.A. Davias, and R.J. Diaz. 2009. Hypoxia, nitrogen, and fisheries: Integrating effects across local and global landscapes. Pp. 329-349 in Annual Review of Marine Science.

97. Briand, E., C. Yepremian, J.F. Humbert, and C. Quiblier. 2008. Competition between microcystin- and non-microcystin-producing Planktothrix agardhii (cyanobacteria) strains under different environmental conditions. Environmental Microbiology 10(12):3337-3348.

98. Buntjer, J. 2010. Heron Lake BioEnergy fined. Daily Globe, December 27. State and Regional News section.

99. Burkholder, J.M., K.M. Mason, and H.B. Glasgow. 1992. Water-column nitrate enrichment promotes decline of eelgrass Zostera-Marina—Evidence from seasonal mesocosm experiments. Marine Ecology-Progress Series 81(2):163-178.

100. Burns, J.C. 2008. ASAS centennial paper: Utilization of pasture and forages by ruminants: A historical perspective. Journal of Animal Science 86(12):3647-3663.

101. Benthic cyanobacteria of the genus Nodularia are non-toxic, with gas vacuoles, able to glide and genetically more diverse than planktonic Nodularia.

102. Biotechnology Advances 25:294- 306. Chisti Y. 2008. Biodiesel from microalgae beats bioethanol. Trends in Biotechnology 26(3):126-131. Daigo K, Nakano Y, Casareto Be, Suzuki Y, Shioi Y. 2008.

103. Biodiesel Production Technology. USA: NREL (SR-510-36244) http://www.nrel.gov/docs/fy04osti/36244.pdf (accessed on 10th December 2012) Gouveia L, 2011.

104. Bernstein L, Bosch P, Canziani 0, Chen Z, Christ R, Davidson O, Hare W, Huq S, Karoly D, Kattsov V, Kundzewicz Z, Liu J, Lohmann U, Manning M, Matsuno T, Menne B, Metz B, Mirza M, Nicholls N, Nurse L, Pachauri R, Palutikof J, Parry M, Qin D, Ravindranath N, Reisinger A, Ren J, Riahi K, Rosenzweig C, Rusticucci M, Schneider S, Sokona Y, Solomon S, Stott P, Stouffer R, Sugiyama T, Swart R, Tirpak D, Vogel C, Yohe G. 2007. Climate Change 2007: Synthesis Report – Summary for Policymakers.

105. Biotechnology Advances, 30(3), 673-690. Lane, J. (2013, April 9). A crash of algae, bioterrorism and the future of biofuels. Biofuels Digest. Retrieved from http://www.biofuelsdigest.com/bdigest/2013/04/09/a-crashof-algae-bioterrorism-and-the-future-of-biofuels/ Mahapatra, D. M., Chanakya, H. N., & Ramachandra, T. V. (2013).

106. Barros AI, Gonçalves AL, Simões M, *et al.* : Harvesting techniques applied to microalgae: A review.Renewable and Sustainable Energy Reviews. 2015;41:1489–500. 10.1016/j.rser.2014.09.037

107. Burkart MD: Green algae cultivation and engineering of its fatty acid synthase. In: 249th American Chemical Society National Meeting & Exposition.2015.

108. Barrick JE, Yu DS, Yoon SH, *et al.* : Genome evolution and adaptation in a long-term experiment withEscherichia coli. Nature. 2009;461(7268):1243–7. 10.1038/nature08480

109. Beyter D, Tang PZ, Becker S, *et al.* : Diversity, Productivity, and Stability of an Industrial Microbial Ecosystem. Appl Environ Microbiol. 2016;82(8):2494–505. 10.1128/AEM.03965-15

110. Biodiesel from microalgae.

111. Biotechnology Advances 25, 294–306 9. Huber, G.W., Iborra, S., Corma, A.(2006).

112. Coale KH, Johnson KS, Chavez FP, *et al.* Southern ocean iron enrichment experiment: carbon cycling in high- and low-si waters. Science. 2004; 304(5669):408–414.

113. Riordon, G. Love, N. Barron, P. Nigam, R. Marchant, L. McHale *et al.*, Production of ethanol from sucrose at 45 °C by alginate immobilized preparations of the thermotolerant yeast strain Kluyveromyces marxianus IMB 3. Bioresour. Technol. 55, 17–173

(1996)

114. C.L. Peterson, J.L. Cook, J.C. Thompson, J.S. Taberski, Continuous flow biodiesel production. Appl. Eng. Agric. 18(1), 5–11 (2002)

115. Coale KH, Johnson KS, Fitzwater SE, *et al.* A massive phytoplankton bloom induced by an ecosystem-scale iron fertilization experiment in the equatorial pacific ocean. Nature. 1996; 383(6600):495–501.

116. Chermack, T.J., S.A. Lynham and W.E.A. Ruona. A review of scenario planning literature. Available online at http://www.cse.buffalo.edu/-peter/refs/DataAssimilation/Multihypoyhesis/ReviewofSP.PDF. Accessed June 18, 2012.

117. Chinnasamy, S., A. Bhatnagar, R.W. Hunt, and K.C. Das.2010. Microalgae cultivation in wastewater dominated by carpet mill effluents for biofuel applications. Bioresource Technology 101(9):3097-3105.

118. Corbiere-Nicollier, T., I, Blanc, and S. Erkman. 2011. Towards a global criteria based framework for the sustainability assessment of bioethanol supply chains. Application to the swiss dilemma: Is local produced bioethanolmore sustainable than bioethanol imported from Brazil? Ecological indicators 11:1447-1458.

119. Craggs, R, I., S. Heubeck, T.J. Lundquist, and J.R. Benemann. 2011. Algal biofuels from wastewater treatment high rate algal ponds. Water Science and Technology 63(4):660-665.

120. CAI, X.M., X.A. Zhang, and D.B. Wang. 2011. Land availability for biofuel production. Environmental Science and Technology 45(1):334-339.

121. Campbell, P.K., T. Beer, and D. Batten. 2011. Life cycle assessment of biodiesel production from microalgae in ponds. Bioresource Technology 102(1):50-56.

122. Chiu, Y.-W., B. Walseth, and S. Suh. 2009. Water embodied in bioethanol in the United States. Environmental Science and Technology 43(8):2688-2692.

123. Clarens, A.F., H. Nassau, E.P. Resurreccion, M.A. White, and L.M. Colosi. 2011. Environmental impacts of algae-derived biodiesel and bioelectricity for transportation. Environmental Science and Technology 45:7554-7560.

124. Clarens, A.F., E.P. Resurreccion, M.A. White, and L.M. Colosi. 2010. Environmental life cycle comparison of algae to other bioenergy feedstocks. Environmental Science and Technology 44(5):1813-1819.

125. Cordell, D., J.O. Drangert, and S. White. 2009. The story of phosphorus: Global food security and food for thought. Global Environmental Change-Human and Policy Dimensions 19(2):292-305.

126. Craggs, R.J., S. Heubeck, T.J. Lundquist, and J.R. Benemann. 2011. Algal biofuels from wastewater treatment high rate algal ponds. Water Science and Technology 63(4):660-665.

127. Chisti, Y. 2007. Biodiesel from microalgae. Biotechnology Advances 25:294-306.

128. Chance, R.R., B. McCool, and J.D. Coleman. 2011. A Cyanobacteria-Based Photosynthetic Process for the Production of Ethanol. Presentation to the NRC Committee on Sustainable Development of Algal Biofuels on June 13.

129. Carpenter EJ, Janson S. Intracellular cyanobacterial symbionts in the marine diatom Climacodium frauenfeldianum (bacillariophyceae) J Phycol. 2000;36(3):540–544.

130. Chaisson Mj, Brinza D, Pevzner PA. De novo fragment assembly with short mate-paired reads: does the read length matter? Genome Res. 2009;19(2):336–346.

131. Casas-Mollano JA, Van Dijk K, Eisenhart J, Cerutti H. SET3P monomethylates histone H3 on lysine 9 and is required for the silencing of tandemly repeated transgenes in Chlamydomonas. Nucleic Acids Res. 2007;35(3):939–950.

132. Charles, C. & Wooders, P. (2012). Biofuels – At what cost? Mandating ethanol and biodiesel consumption in the United Kingdom. Geneva: International Institute for Sustainable Development (IISD). Chowdhury, R., Viamajala, S., & Gerlach, R. (2012). Reduction of environmental and energy footprint of microalgal biodiesel production through material and energy integration. Bioresource Technology, 108, 102-111.

133. Chisti Y. Biodiesel from microalgae beats bioethanol. Trends Biotechnol. 2008;26(3):126–131

134. Chisti Y. Fuels from microalgae. Biofuels 2010;1:233–5.

135. Climate Change 2007: Synthesis Report – Summary for Policymakers. Intergovernmental Panel on Climate Change. http://www.ipcc.ch/pdf/assessment-report/ar4/syr/ar4_syr.pdf (accessed on 12th October 2013). Campbell A, Cooke P, Cass K, Earl K. 2004. Plant and Animal Evolution. New Zealand, the University of Waikato. http://sci.waikato.ac.nz/evolution/aboutus.shtml (accessed on 2nd April 2013) Charpy L, Palinska Ka, Casareto B, Langlade Mj, Suzuki Y, Abed Rmm, Golubic S. 2010. Dinitrogen-Fixing Cyanobacteria in Microbial Mats of Two Shallow Coral Reef Ecosystems. Microb Ecol 59:174-186.

136. Conley DJ, Paerl HW, Howarth RW, *et al.* Ecology. Controlling eutrophication: nitrogen and phosphorus. Science. 2009;323(5917):1014–1015.

137. Chisti Y. Biodiesel from microalgae. Biotechnol Adv 2007;25:294–306.

138. Chisti Y. Biodiesel from microalgae beats bioethanol. Trends Biotechnol 2008;26:126–31.

139. Chen F, Li H, Wong R N, Jiang Y. 2005. Isolation and purification of the bioactive carotenoid zeaxanthin from the microalga Microcystis aeruginosa by high-speed counter-current chromatography.

140. Chu CY, Liao WR, Huang R, Lin LP. Haemagglutinating and antibiotic activities of freshwater microalgae. World J Microbiol Biotechnol. 2004;20(8):817–825.

141. Chen Y, Wang Y, Sun Y, Zhang L, Li W. Highly efficient expression of rabbit neutrophil peptide-1 gene in chlorella ellipsoidea cells. Curr Genet. 2001;39(5):365–370.

142. Clayton, M., "New Coal Plants Bury 'Kyoto,'" Christian Science Monitor, December 23, 2004. Christi, Y., "Biodiesel from Microalgae," Biotechnology Advances, Vol. 25, 2007, pp. 294-306. Crookes, R. J., "Comparative Bio-Fuel Performance in Internal Combustion Engines," Biomass and Bioenergy, Vol. 30, 2006, pp. 461-468.

143. Copping LG, Duke SO. Natural products that have been used commercially as crop protection agents. Pest Management Sci.

2007;63(6):524–554.

144. Chisti Y. Biodiesel from microalgae. Biotechnol Adv. 2007;25(3):294–306.
145. Chisti Y. Biodiesel from microalgae. Biotechnol Adv 2007;25:294–306.
146. Chisti Y. Biodiesel from microalgae beats bioethanol. Trends Biotechnol 2008;26:126–31.
147. Chisti Y. Biodiesel from microalgae. Biotechnol Adv. 2007;25(3):294–306.
148. Degray G, Rajasekaran K, Smith F, Sanford J, Daniell H. Expression of an antimicrobial peptide via the chloroplast genome to control phytopathogenic bacteria and fungi. Plant Physiol. 2001;127(3):852.
149. Dunahay TG, Jarvis EE, Roessler PG. Genetic transformation of the diatoms Cyclotella cryptica and Navicula saprophila. J Phycol. 1995;31(6):1004–1012.
150. De Riso V, Raniello R, Maumus F, Rogato A, Bowler C, Falciatore A. Gene silencing in the marine diatom Phaeodactylum tricornutum. Nucleic Acids Res. 2009;37(14):e96.
151. Department of Energy's Office of Fuels Development NREL/TP-580-24190, July 1978. U.S. Department of Energy, "Alternative Fuels & Advanced Vehicles Data Center", Last updated: Nov 12, 2007, tonto.eia.doe.gov/oog/info/gdu/gasdiesel.asp, Accessed: Nov 14, 2007. Xu, H.; Miao, X. and Q. Wu., "High Quality Biodiesel Production from a Microalga Chlorella protothecoides by Heterotrophic Growth in Fermenters," Journal of Biotechnology, Vol. 126, 2006, pp.499-507.
152. Y. C. Leung, X. Wu, and M. K. H. Leung, "A review on biodiesel production using catalyzed transesterification," Applied Energy, vol. 87, no. 4, pp. 1083–1095, 2010.
153. Darnoko, M. Cheryan, Continuous production of palm methyl esters. J. Am. Oil Chem. Soc. 77, 1269–1272 (2000)
154. Dragone G. Fernandes BD. Abreu AP. Vicente AA. Teixeira JA. Nutrient limitation as a strategy for increasing starch accumulation in microalgae. Appl Energy in press. doi:10.1016/j.apenergy.2011.03.012.

155. Dyni JR. Scientific Investigations Report 2005-5294. US Geological Survey; VA, USA: 2006. Geology and resources of some world oil-shale deposits.

156. Deschamps P, Moreira D. Signal conflicts in the phylogeny of the primary photosynthetic eukaryotes. Mol Biol Evol. 2009;26(12):2745–2753.

157. Douskova I, Doucha J, Livansky K, *et al*. Simultaneous flue gas bioremediation and reduction of microalgal biomass production costs. Appl Microbiol Biotechnol. 2009;82(1):179–185.

158. Murray, Ethanol's potential: looking beyond corn (Earth Policy Institute, Washington DC, 2005). http://www.earthpolicy.org/Updates/2005/Update49.htm

159. D. Brady, P. Nigam, R. Marchant, A.P. McHale, Ethanol production at 45 °C by immobilized Kluyveromyces marxianus IMB3 during growth on lactose-containing media. Bioprocess Eng. 16, 101–104 (1997)

160. D. Brady, P. Nigam, R. Marchant, D. Singh, A.P. McHale, The effect of Mn2þ on ethanol production from actose using Kluyveromyces marxianus IMB3 immobilized in magnetically responsive matrices. Bioprocess Eng. 17, 31–34 (1997)

161. D. Kusdiana, S. Saka, Kinetics of transesterification in rapeseed oil to biodiesel fuels as treated in supercritical methanol. Fuel 80, 693–698 (2001)

162. D. Kusdiana, S. Saka, Two-step preparation for catalyst-free biodiesel fuel production: hydrolysis and methyl esterification. Appl. Biochem. Biotechnol. 115, 781–792 (2004)

163. D. Royon, M. Daz, G. Ellenrieder, S. Locatelli, Enzymatic production of biodiesel from cotton seed oil using t-butanol as a solvent. Bioresour. Technol. 98, 648–653 (2007)

164. D. Brady, P. Nigam, R. Marchant, L. McHale, A.P. McHale, Ethanol production at 45 °C by Kluyveromyces marxianus IMB3 immobilized in magnetically responsive alginate matrices. Biotechnol. Lett. 18(10), 1213–1216 (1996)

165. D. Bajpai, V.K. Tyagi, Biodiesel: source, production, composition, properties and its benefits. J. Olio Sci. 55, 487–502 (2006)

166. Dubinsky Z, Rotem J. Relations between algal populations and the pH of their media. Oecologia. 1974;16:53–69.

167. Dalrymple, O. K., Halfhide, T., Udom, I., Gilles, B., Wolan, J., Zhang, Q., & Ergas, S. (2013). Wastewater use in algae production for generation of renewable resources: A review and preliminary results. Aquatic Biosystems, 9(1), 1-11. Dana, G. V., Kuiken, T., Rejeski, D., & Snow, A. A. (2012). Synthetic biology: Four steps to avoid a synthetic-biology disaster. Nature, 483(7387), 29. Elgin, B. & Waldman, P. (2013, April 18). Chevron Defies California On Carbon Emissions.

168. Debuchy R, Purton S, Rochaix JD. The argininosuccinate lyase gene of Chlamydomonas reinhardtii: an important tool for nuclear transformation and for correlating the genetic and molecular maps of the Arg7 locus. EMBO J. 1989;8(10):2803–2809.

169. Dinitrogen-Fixing Cyanobacteria in Microbial Mats of Two Shallow Coral Reef Ecosystems. Microb Ecol 59:174-186. Chen F, Li H, Wong R N, Jiang Y. 2005.

170. Davis, R., A. Aden, and P.T. Pienkos. 2011. Techno-economic analysis of autotrophic microalgae for fuel production. Applied Energy 88(10):3524-3531.

171. Demirbas, A., "Importance of Biodiesel as Transportation Fuel," Energy Policy, Vol. 35, 2007, pp. 4661-4670.

172. Dismukes GC, Carrieri D, Bennette N, Ananyev GM, Posewitz MC. Aquatic phototrophs: efficient alternatives to land-based crops for biofuels. Curr Opin Biotechnol. 2008;19(3):235–240.

173. Development in oil extraction from microalgae.

174. Demirbas MF. Biofuels from algae for sustainable development. Appl Energy, in press. doi:10.1016/j.apenergy.2011.01.059.

175. Dismukes GC, Carrieri D, Bennette N, Ananyev GM, Posewitz MC. Aquatic phototrophs: efficient alternatives to land-based crops for biofuels. Curr Opin Biotechnol. 2008;19(3):235–240.

176. Doucha J, Livansky K. Productivity, CO2/O2 exchange and hydraulics in outdoor open high density microalgal (Chlorella sp.) photobioreactors operated in a middle and southern European climate. J Applied Phycology. 2006;18(6):811–826.

177. Energy Information Administration. International Energy Outlook. Vol. 284. EIA; DC, USA: 2009.

178. European Journal of Lipid Science and Technology 113(5):539-547. Nakpong P, Wootthikanokkhan S. 2010.

179. Engineering Journal 16(5):158-166.

180. Ehimen EA, Sun ZF, Carrington CG, Birch EJ, Eaton-Rye J. Anaerobic digestion of microalgae residues resulting from the biodiesel production process. Appl Energy in press. doi:10.1016/j.apenergy.2010.10.020.

181. Energy return on investment for algal biofuel production coupled with wastewater treatment. Water Environment Research, 84(9), 692-710. Borowitzka, M. A., & Moheimani, N. R. (2013). Sustainable biofuels from algae. Mitigation and Adaptation Strategies for Global Change, 18(1), 13-25. Cai, T., Park, S. Y., & Li, Y. (2013). Nutrient recovery from wastewater streams by microalgae: Status and prospects. Renewable and Sustainable Energy Reviews, 19, 360- 369. Casey, T. (2013, February 8). EU Is Burning More Coal, Looks To "Burn" Seawater Instead. CleanTechnica. Algal biofuel production and mitigation potential in India. Mitigation and Adaptation Strategies for Global Change, 18(1), 113-136.

182. E.D. Larson, Biofuel production technologies: status, prospects and implications for trade and development. Report No. UNCTAD/DITC/TED/2007/10. United Nations Conference on Trade and Development, New York and Geneva, 2008

183. Lotero, Y.J. Liu, D.E. Lopez, K. Suwannakarn, D.A. Bruce, J.G. Goodwin, Synthesis of biodiesel via acid catalysis. Ind. Eng. Chem. Res. 44, 5353–5363 (2005)

184. Minami, S. Saka, Kinetics of hydrolysis and methyl esterification for biodiesel production in two-step supercritical methanol process. Fuel 85, 2479–2483 (2006)

185. EU, Directive 2009/28/EC of The European Parliament and of The Council of 23 April 2009 on the promotion of the use of energy from renewable sources and amending and subsequently repealing Directives 2001/77/EC and 2003/30/EC. Off. J. Eur. Union 52, 16–62 (2009)

186. Evaluation of environmental impacts from microalgae cultivation in open-air raceway ponds: Analysis of the prior literature and investigation of wide variance in predicted impacts. Algal Research, 1(1), 83-92. Henley, W. J., Litaker, R. W., Novoveská, L., Duke, C. S., Quemada, H. D., & Sayre, R. T. (2012).
187. Euglena sp. as a suitable source of lipids for potential use as biofuel and sustainable wastewater treatment. Journal of Applied Phycology, January 27, 1-11.
188. Energy Information Administrator, "Gasoline and Diesel Fuel Update", Last updated: Oct 1, 2007, www.eere.energy.gov/afdc/price_report.html, Accessed: Nov 14, 2007. Haag, A. L., "Algae Bloom Again," Nature, Vol. 447, May 31, 2007, pp. 520-521.
189. Energy Information Administration. International Energy Outlook. Vol. 284. EIA; DC, USA: 2009.
190. Fan, X., R. Burton, and G. Austic. 2010. Conversion of degummed soybean oil to biodiesel: Optimization of degumming methods and evaluation of fuel properties. International Journal of Green Energy 7(6):593-599.
191. Flynn KJ, Mitra A, Greenwell HC, *et al.* : Monster potential meets potential monster: pros and cons of deploying genetically modified microalgae for biofuels production. Interface Focus. 2013;3(1):20120037. 10.1098/rsfs.2012.0037
192. Fuel 89:1806-1811. Noaa Earth System Research Laboratory. 2013.
193. Fargione J, Hill J, Tilman D, Polasky S, Hawthorne P. Land clearing and the biofuel carbon debt. Science. 2008;319(5867):1235–1238.
194. Falkowski PG, Barber RT, Smetacek VV. Biogeochemical controls and feedbacks on ocean primary production. Science. 1998;281(5374):200–207.
195. Fierro S, Sanchez-Saavedra Mdel P, Copalcua C. Nitrate and phosphate removal by chitosan immobilized Scenedesmus. Bioresour Technol. 2008;99(5):1274–1279.
196. Ullah, A. Nosheen, I. Hussain, and A. Bano, "Base catalyzed transesterification of wild apricot kernel oil for biodiesel production," African Journal of Biotechnology, vol. 8, no. 14, pp. 3289–3293, 2009.

197. O. Licht, "The Global Renewable Fuels Alliance is a non-profit organization dedicated to promoting biofuel friendly policies internationally," Global Renewable Fuels Alliance, 2011.

198. Funderberg E. Why are nitrogen prices so high. Pasture Range. 2009;5

199. F.O. Licht, World ethanol & biofuels report (Agra Informa Ltd., Kent). http://www.agra-net.com/portal/puboptions.jsp?Option¼menu&pubId¼ag072

200. FAO, The state of food and agriculture. BIOFUELS: prospects, risks and opportunities(FAO, Rome, 2008)

201. F. Omota, A.C. Dimian, A. Bliek, Fatty acid esterification by reactive distillation. Part 1: equilibrium-based design. Chem. Eng. Sci. 58, 3159–3174 (2003)

202. FAO, in World agriculture: towards 2015/2030. An FAO perspective, ed. by J. Bruinsma. Food and Agriculture Organization (FAO) (Earthscan Publications Ltd, London, 2003).

CITATIONS

CHAPTER 6

Surajit De Mandal, Amrita Kumari Panda, Satpal Singh Bisht, and Nachimuthu Senthil Kumar, Microbial Ecology in the Era of Next Generation Sequencing, DOI: 10.4172/ 2469-9853.S1-001.

CHAPTER 7

Richter DC, Ott F, Auch AF, Schmid R, Huson DH (2008) MetaSim—A Sequencing Simulator for Genomics and Metagenomics. PLoS ONE 3(10): e3373. https://doi.org/10.1371/journal.pone.0003373.

CHAPTER 8

Stilianos Louca, Michael Doebeli, Calibration and analysis of genome-based models for microbial ecology, DOI: 10.7554/eLife.08208.

CHAPTER 9

Jorge Barriuso, Jose R Valverde and Rafael P Mellado, Estimation of bacterial diversity using next generation sequencing of 16S rDNA: a comparison of different workflows, doi: 10.1186/1471-2105-12-473.

CHAPTER 10

Catherine H. Ravenscroft, Raj Whitlock, and Jason D. Fridley, Rapid genetic divergence in response to 15 years of simulated climate change, doi: 10.1111/gcb.12966.

CHAPTER 11

Luis E. Castan˜eda and Olga Barbosa, Metagenomic analysis exploring taxonomic and functional diversity of soil microbial communities in Chilean vineyards and surrounding native forests, DOI 10.7717/peerj.3098.

CHAPTER 12

Marcin Gołębiewski, Edyta Deja-Sikora, Marcin Cichosz, Andrzej Tretyn, Borys Wróbel, 16S rDNA Pyrosequencing Analysis of Bacterial Community in Heavy Metals Polluted Soils, DOI 10.1007/s00248-013-0344-7.

CHAPTER 13

Agnieszka Kalwasińska, Tamás Felföldi, Attila Szabó, Edyta Deja-Sikora, Przemysław Kosobucki, and Maciej Walczak, Microbial communities associated with the anthropogenic, highly alkaline environment of a saline soda lime, Poland, doi: 10.1007/s10482-017-0866-y.

CHAPTER 14

Hon Lun Wong, Aria Ahmed-Cox, and Brendan Paul Burns, Molecular Ecology of Hypersaline Microbial Mats: Current Insights and New Directions, doi:10.3390/microorganisms4010006.

INDEX

A

Abundance-based Coverage Estimator (ACE)
Amplified fragment length polymorphism (AFLP)

B

BEAST (Bayesian Evolutionary Analysis by Sampling Trees)
Buxton Climate Change Impacts Laboratory (BCCIL)

C

Candidate phyla radiation (CPR)
Canonical correspondence analysis (CCA)
Colony forming units (CFU)
Competitor/Stress-tolerator/Ruderal (CSR)

D

Deep Sea Hydrothermal Vent Group 6 (DHVEG-6).
Denaturing Gradient Gel Electrophoresis (DGGE)
Dynamic flux balance analysis (DFBA)

E

Earth Microbiome project (EMP)
Electrical conductivity (EC)
Emulsion PCR (emPCR)
Expressed sequence tags (ESTs)

F

Filamentous anoxygenic phototro-

phs (FAPs)
Fluorescence-activated cell sorting (FACS)

H

Human Microbiome Project (HMP)

I

Inter-transcribed spacer (ITS)

K

KEGG (Kyoto Encyclopedia of Genes and Genomes)

L

Large subunit (LSU)
Linkage disequilibrium (LD)
Local Similarity Analysis (LSA)

M

Marine Benthic Group D (MBGD)
Maximum Likelihood (ML)
Maximum Parsimony (MP)
MCM (Microbial Community Modeler)
MEGA (Molecular Evolutionary Genetics Analysis Software)
Metropolis-Hastings Markov-chain Monte Carlo (MH-MCMC)
Microbial community (MC)
Microbial Diversity (MD)
Molecular Ecological Network Analysis (MENA)
Multiple displacement amplification (MDA)
Multiple Displacement Amplification (MDA)
Multiple sequence alignment (MSA)
Multiple Sequence Alignment (MSA)
Multiplex identifier (MID)

N

Next generation sequencing (NGS)
Next Generation Sequencing (NGS)
Next Generation Sequencing technologies (NGS)

O

Operational Taxonomic Unit (OTU)
Operational taxonomic units (OTUs)
Operational Taxonomic Units (OTUs)
Optical density (OD)

P

Phospholipid fatty acids (PLFA)
Polymerase Chain Reaction (PCR)
Predictive metagenomics profiling (PMP)

R

RDP (Ribosomal database project)
Ribosomal RNA (rRNA)

S

Single amplified genomes (SAG)
Single cell genomics (SCG)
Single molecule real time (SMRT)